THE TECHNOLOGY TRAP

THE TECHNOLOGY TRAP

CAPITAL, LABOR, AND POWER
IN THE AGE OF AUTOMATION

CARL BENEDIKT FREY

PRINCETON UNIVERSITY PRESS

PRINCETON & OXFORD

Published by Princeton University Press
41 William Street, Princeton, New Jersey 08540
6 Oxford Street, Woodstock, Oxfordshire OX20 1TR

press.princeton.edu

Library of Congress Control Number 2018966069
ISBN 978-0-691-17279-8

British Library Cataloging-in-Publication Data is available

Jacket art: Diego Rivera, *Detroit Industry Murals* (1932–33), detail from north wall, fresco © 2018 Banco de México Diego Rivera Frida Kahlo Museums Trust, Mexico, D.F. / Artists Rights Society (ARS), New York / Alamy

This book has been composed in Arno

Printed on acid-free paper. ∞

Printed in the United States of America

10 9 8 7 6 5 4 3 2 1

For Sophie, with love

CONTENTS

PREFACE

Future historians may wonder why we failed to learn from the past. Historically, when large swaths of the population have found their livelihoods threatened by machines, technological progress has brought fierce opposition. We are now living through another episode of labor-replacing progress, and resistance is seemingly looming. According to a 2017 Pew Research Center survey, 85 percent of Americans now favor policies to restrict the rise of robots.[1] And Andrew Yang has just announced his bid for the White House in 2020 on a campaign to protect jobs from automation.[2] The underlying concern is not hard to understand. Aided by advances in artificial intelligence (AI), robotics, machine vision, sensor technology, and so on, computers have become capable of performing a wide range of tasks that could be done only by humans a few years ago. Top-down programming is no longer required for automation to happen. In the age of AI, computers can learn themselves. What used to be distant moon shoots in computing are now reality.

In September 2013, my Oxford friend and colleague Michael Osborne and I published a research paper estimating the potential impacts of advances in AI on jobs. We found 47 percent of American jobs to be at high risk of automation as a consequence.[3] A few months later, I was invited to speak at a conference in Geneva. I was in good company, with one former prime minister, one chancellor, and a couple of labor ministers. After my talk, a well-known economist in the audience—let's call him Bill—approached me and dismissively remarked, "Is this not just like the Industrial Revolution in England? ... Didn't machines displace jobs then as well?" Bill was right, of course, but it was only on my way back to the airport that I realized how right he actually was in suggesting that things are no different this time around. Some jobs will disappear, but people will find new things to do, as they always have, and therefore there is nothing to worry about. Unfortunately, that is only half the story.

The long-term economic benefits of the Industrial Revolution, to which Bill alluded, are uncontested. Before 1750, per capita income in the world doubled every 6,000 years; since then, it has doubled every 50 years.[4] But the industrialization process itself was a different matter. While economic historians are still debating whether the pains inflicted on the workforce by the Industrial Revolution were worth it, for later generations they surely were. Yet many contemporary laborers, who saw their livelihoods vanish as their skills became obsolete, would just as surely have been better off had the industrial world never arrived. As the mechanized factory displaced the domestic system, traditional middle-income jobs dried up, the labor share of income fell, profits surged, and income disparities skyrocketed. Sound familiar? Indeed, so far our age of automation largely mirrors the early days of industrialization in economic terms. It took over half a century until average people saw the benefits of the Industrial Revolution trickle down. And unsurprisingly, as many citizens experienced a reversal of fortunes, the consequence was cascading opposition to machines. The Luddites, as they were called, raged against mechanization, and they did everything they could to resist it. If this is "just" another Industrial Revolution, alarm bells should be ringing.

The idea underpinning this book is straightforward: attitudes toward technological progress are shaped by how people's incomes are affected by it. Economists think about progress in terms of enabling and replacing technologies.[5] The telescope, whose invention allowed astronomers to gaze at the moons of Jupiter, did not displace laborers in large numbers—instead, it enabled us to perform new and previously unimaginable tasks. This contrasts with the arrival of the power loom, which replaced hand-loom weavers performing existing tasks and therefore prompted opposition as weavers found their incomes threatened. Thus, it stands to reason that when technologies take the form of capital that replaces workers, they are more likely to be resisted. The spread of every technology is a decision, and if some people stand to lose their jobs as a consequence, adoption will not be frictionless. Progress is not inevitable and for some it is not even desirable. Though it is often taken as a given, there is no fundamental reason why technological ingenuity should always be allowed to thrive. The historical record, as we shall see, shows that technology's acceptance depends on whether those affected by it stand to gain from it. Episodes of job-replacing technological change have regularly brought social unrest and, at times, a backlash against technology itself. In this regard, the age of automation, which took off with the computer revolution in the 1980s, resembles the Industrial Revolution, when the mechanized factory replaced middle-income artisans in large numbers. Now, as then, middle-income jobs have been taken over by machines, forcing many people into lower-paying jobs or causing them to drop out of the workforce altogether.

To capture attitudes toward technology over the centuries, this book brings together much of the technical economics literature with historical accounts of technological change and popular commentary. Though it concerns the future, it is not a prediction of it. Prophets may be able to foretell the future; economists cannot. The objective here is to provide perspective, and perspective we get from history. As Winston Churchill once quipped, "The longer you can look back, the farther you can look forward."[6] Thus, before looking forward, we shall begin by looking back. The Industrial Revolution was a defining moment, but few people grasped its enormous consequences at the time. We are now in the midst of another technological revolution, but fortunately this time around, we can learn from previous episodes. Bill dismissed our study as Luddite. And indeed, parallels are often drawn between the Industrial Revolution and now to suggest that the Luddites were wrong in trying to halt the spread of the mechanized factory. Artisan craftsmen, whose feelings were stronger than their judgment, rebelled against the very machinery that came to deliver unprecedented wealth for the commoner, or so the story goes. This story is an accurate description of the long run, but in the long run we are all dead. Three generations of working Englishmen were made worse off as technological creativity was allowed to thrive. And those who lost out did not live to see the day of the great enrichment. The Luddites were right, but later generations can still be grateful that they did not have it their way. History is made in the short run because the decisions we make today shape the long run. Had the Luddites been successful in bringing progress to a halt, the Industrial Revolution would probably have happened somewhere else. And if not, economic life would most likely still look similar to the way it did in 1700.

This brings us to the second theme of this book: whether replacing technologies will be blocked depends on who stands to gain from them and the societal distribution of political power. During the Industrial Revolution, the Luddites and other groups did what they could to stop the spread of labor-replacing technologies, but they were unsuccessful, as they lacked political clout. In fact, as we shall see, one of the reasons why the Industrial Revolution first happened in Britain is that for the first time, political power was with those who stood to gain from mechanization. The hegemony of landed wealth was challenged by the mobile fortunes of merchants, who came to form a new industrial class with growing political influence.[7] The mechanized factory was deemed critical to Britain's competitive position in trade and thus to merchants' fortunes, which its government would do nothing to jeopardize. But for most of history, the politics of progress were such that the ruling classes had little to gain and much to lose from the introduction of labor-replacing technology. They rightly feared that angry workers might rebel against the government. In the seventeenth century, for example, the craft guilds had become a force of growing political power in Europe, and they vehemently resisted technologies that threatened their livelihoods. And fearing social unrest, European governments typically sided with the guilds. Consequently, economic incentives to invest in labor-saving technology were few. And because mechanization would put the incomes of parts of the population at risk, prompting social unrest and possibly a challenge to the political status quo, the ruling classes did their best to restrict it.

One reason economic growth was stagnant for millennia is that the world was caught in a technology trap, in which labor-replacing

technology was consistently and vigorously resisted for fear of its destabilizing force. Could countries in the industrial West experience a return of the technology trap in the twenty-first century? While it may
seem unlikely, it certainly looks more likely than it did when I began
writing this book four years ago. Proposals to tax robots in order to slow
down the pace of automation now feature in the public debate on both
sides of the Atlantic. And unlike the situation in the days of the Industrial Revolution, workers in the developed world today have more political power than the Luddites did. In America, where Andrew Yang
is already tapping into growing anxiety about automation, an overwhelming majority now favor policies to restrict it. The disruptive
force of technology, Yang fears, could cause another wave of Luddite
uprisings: "All you need is self-driving cars to destabilize society. . . .
[W]e're going to have a million truck drivers out of work who are
94 percent male, with an average level of education of high school or
one year of college. That one innovation will be enough to create riots
in the street. And we're about to do the same thing to retail workers, call
center workers, fast-food workers, insurance companies, accounting
firms."[8]

The point is not fatalism or pessimism. And it is surely not that we
would be better off slowing down the pace of progress or restricting
automation. The Industrial Revolution was the beginning of an unprecedented transformation that benefited everyone over the long run. AI
systems have the potential to do the same, but the future of AI depends
on how we manage the short run. If we seek to understand the challenges ahead rather than glossing over them in the belief that in the long
run everyone will come out ahead, we will be in a much better position
to shape the outcome. Yang may be extremely unlikely to be elected
president, but as the observant Rana Foroohar writes, automation is
likely to become a major topic in the 2020 election.[9] As there has been
a populist backlash against globalization, we should be concerned that
populists might easily and effectively tap into growing anxiety about
automation as well, unless we address it. Fortunately, for all the parallels
between the age of automation and the days of the Industrial Revolution, which I shall push shamelessly throughout this book, there are also

many differences. But our fascination with our own day and age, and our preoccupation with both the promise and peril of new technology, often leads us to think that we are experiencing something entirely new. Seen through the lens of the long record of human history, however, this seems unlikely to be true.

INTRODUCTION

Progress would be wonderful—if only it would stop.

—ROBERT MUSIL

When looms weave by themselves, man's slavery will end.

—ARISTOTLE

Had it not been for the deeds of six hundred lamplighters, the streets of New York City at night in 1900 would have been lit by nothing but the moon. Equipped with torches and ladders, they were the force ensuring that pedestrians could see more than a burning cigar a block off when they left their homes. But on the night of April 24, 1907, most of the twenty-five thousand gas lights in the streets of Manhattan were never lit. The lamplighters, who would normally start carrying the torch of civilization around 6:50 P.M., left the lights out and went on strike. No violence was reported. But as it grew darker, New Yorkers poured in complaints to the gas companies and the local police. Policemen were sent in to light up the neighborhoods, yet without ladders this proved a difficult task. Many officers were too obese to climb the lampposts. And they got little help from the public. In Harlem, crowds of boys invented a new sport: whenever an officer was successful in firing up a lamp, they would climb the post, turn out the light, and run. On Park Avenue one youngster was arrested after having put a light out after an officer got it burning. Few lamps burned for long. Even by 9:00 P.M., the only bright public spots were a few transverse roads in Central Park, which had been equipped with electric streetlights.[1]

Citizens who took up work as lamplighters that year were unlucky. Oil and gas lamps had always required personal attention, but with the mysterious force of electricity, the touch of the lamplighter was no longer a skill that had any value. Electric streetlights brought light and nostalgia. Many citizens still felt that a young man must turn lights on at dusk and off at dawn. In New York City, lamplighters had become a neighborhood institution alongside the police and the postman. Their profession had existed since the first streetlights were inaugurated in London in 1414, but it was about to become a distant memory. As the *New York Times* noted in 1924, "The lamplighting business in the great metropolis has been victim of too much progress."[2] To be sure, the first electric streetlights in New York City had already been installed in the late nineteenth century, but they had hardly made lamplighters redundant. Each lamp was equipped with its own switch, which had to be turned on manually. Early electrification just made the job easier, as lamplighters no longer had to carry long torches to ignite the lamps. Still, the men who used to light the gas lamps were not the beneficiaries of progress. The mastery of light had once allowed a working man to support his family. Now, turning on the lights had become a task so simple that it could be done by young boys on their way home from school. And as so often in history, simplification was merely a step toward automation. As electric streetlights were increasingly regulated from substations, the jobs of lamplighters were cut in large numbers. By 1927, electricity had a monopoly on illumination in New York City, and the last two gas lamplighters left their craft, ending the story of their profession and that of the Lamplighters Union.[3]

Thomas Edison's invention of the light bulb surely made the world better and brighter. In his laboratory in Menlo Park, oil lamps and candles still polluted the air on the day of his breakthrough. As William Nordhaus, winner of the Nobel Prize in Economics in 2018, has shown, the price of light fell dramatically thereafter, as electricity spread to Chicago's Academy of Music, London's House of Commons, Milan's La

Scala, and the trading floor of the New York Stock Exchange.[4] For the purpose of streetlighting, even the New York lamplighters, some of whom were forced into early retirement, willingly admitted that the new system was more expeditious. One lamplighter could at best attend to some fifty lamps per night. Now, several thousand lamps could be switched on by one substation employee in seconds. Yet nothing could be more natural than resisting a threat to one's livelihood. For most citizens, their skills are their capital, and it is from that human capital that they derive their subsistence. Thus, despite all the virtues of the new system, it is not surprising that electric light wasn't welcomed by everyone everywhere. When the municipality of Verviers in Belgium announced the switch to electricity, for example, lamplighters took to the streets in fear of losing their jobs. To banish the tyranny of darkness, the local government enrolled another team of lamplighters, but they were soon attacked by the strikers—who threatened to keep breaking lamps till doomsday. Intervention by local police ended with angry lamplighters raiding police headquarters. The Belgian government had to call in the army to resolve the situation.[5]

Some surely paid the price for progress. But over the course of the twentieth century, the vast majority of citizens in the West have accepted technology as the engine of their fortunes. They have recognized that it improved working conditions by eliminating the most hazardous and servile jobs. They realized that their wages depended on the use of mechanical power. And they benefited from the continuous flow of new goods and services that became available to them. Revolutionary technologies like automobiles, refrigerators, radios, and telephones—to name just a few—were all unavailable to European monarchs in the Renaissance, but by 1950 they were common features of Western life. In 1900, the average housewife could still only dream of living like the upper classes, who had servants to do the most tedious household tasks for them. In the following decades every home suddenly got equal access to the electric servant. Washing machines, electric irons, and a host of other electric appliances took over hours of drudgery in the home. In short, the capitalist achievement, as the great economist Joseph Schumpeter observed, did not consist of providing "more silk stockings for queens but in bringing them within the reach of factory girls in return for steadily decreasing amounts of effort."[6]

It is easy to oversimplify history. However, if there is one predominant factor underlying economic and social change over the past two centuries, it is surely the advancement of technology. Without technological change, "capital accumulation would amount to piling wooden plows on top of wooden plows," to borrow Evsey Domar's phrase.[7] Economists estimate that over 80 percent of the income differences between rich and poor countries can be explained by differential rates of technology adoption.[8] And relying on income alone hugely understates the transformation that has taken place. It is quite extraordinary to think that in the world my great-grandmother was born into, people could not travel faster than horses or trains could carry them. The only escape from darkness during night was the candle and the oil lamp. Jobs were physically demanding. Few women did paid work. The home was the woman's workplace, where meals were prepared on an open hearth, and trees had to be chopped down for fuel to cook with and keep the house warm. And buckets of water had to be carried indoors from a stream or well. Unsurprisingly, people felt much enthusiasm for progress, not to say euphoria. A 1915 article published in *Literary Digest* confidently predicted that with electrification, it "will become next to impossible to contract disease germs or get hurt in the city, and country folk will go to town to rest and get well."[9] Edison himself was convinced that electricity would help us overcome the greatest hurdle to further progress: our need to sleep. Technology was the new religion of the people. There was the sense that there was no problem that technology could not solve.

In hindsight, and in the light of the gains brought by technology, it is astounding to think that economists of the early nineteenth century like Thomas Malthus and David Ricardo did not believe that technology could improve the human lot. The technological virtuosity of the nineteenth and early twentieth centuries took some time to trickle down to the economics profession. But in the 1950s, Robert Solow, who would go on to win the Nobel Prize in Economics in 1987, found that virtually all economic advance over the twentieth century had been thanks to technology. And others documented that those gains had been widely shared. Simon Kuznets found that America had become more equal and advanced his theory of capitalist development in which inequality automatically decreases along the industrialization path. Nicholas Kaldor observed that labor had consistently reaped about two-thirds of the gains of growth. And Solow developed a theoretical framework in which progress delivered equal benefits for every social group around that time. Seen through the lens of today, such optimism might seem absurd. But the economist of the 1950s had much to be optimistic about.

What do the jobs of a few lamplighters matter if society as a whole can become both richer and more equal simply by letting technological creativity thrive? Many displaced lamplighters probably even found less hazardous and better-paying jobs. And even if some lost out to technology, it seems right that society willingly accepted progress for the many at the expense of the few. But would we feel that way if the victims of progress had been more plentiful? What if the majority of replaced workers were forced to move into jobs that paid less well? After all, the "special century" was not just special in that it excelled in economic growth.[10] Just as important was the fact that almost everyone gained from progress. While there were clearly labor-replacing technologies, most were of the enabling sort. Overall, technology served to make workers more productive and their skills more valuable, allowing them to earn better wages. And even those who lost their jobs to the force of mechanization had a greater abundance of less physically demanding and better-paying jobs to choose from as a consequence. In the age of artificial intelligence (AI), as this book will argue, such optimism about technology can no longer be taken for granted. Nor has it been the historical norm. Economists of the golden age were right to be optimistic about the time in which they lived. Their mistake was in thinking that what they witnessed would continue indefinitely. There is no iron law that postulates that technology must benefit the many at the expense of the few. And quite naturally, when large swaths of the populace are left behind by technological change, they are likely to resist it.

The price of progress has varied greatly throughout history. Simplifications of human advancement like figure 1, which are often used to illustrate the great leap forward, miss all the action. The point is not that the figure is incorrect. It rightly shows that per capita growth in gross domestic product (GDP) was stagnant for millennia and took off in an extraordinary fashion around 1800. Thus, tracking progress purely in terms of average incomes leads one to conclusions like this one: "Modern humans first emerged about 100,000 years ago. For the next 99,800 years or so, nothing happened. . . . Then—just a couple of hundred years ago—people started getting richer. And richer and richer still. Per capita income, at least in the West, began to grow at the unprecedented rate of about three quarters of a percent per year. A couple of decades later, the same thing was happening around the world. Then it got even better."[11]

This standard narrative is unfortunate. Because of it, we often forget that during the extraordinary upward trend in growth that began in eighteenth-century England, millions of people were adjusting to change. And some had more cheerful stories than others. There were even those who would have been better off had mechanization not been allowed to progress. Figure 1 leads us to think that everyone living today must be better off than the previous generation, just as the generation born in 1800 must have seen staggering improvements in their living standards relative to those of their grandparents. Figure 1 also suggests that we were not very inventive before the eighteenth century. Otherwise, why would growth have been so slow? Yet a closer examination of preindustrial times reveals some pathbreaking inventions and ideas. And if we zoom into different episodes of progress, as this book will, we find that people fared very differently in the winds of change.

The "takeoff" depicted in figure 1 began with the arrival of the mechanized factory. Italy could take some credit for its inception. Drawings of the silk-throwing machines that led to the first factories came from Piedmont through an episode of industrial espionage for which Thomas Lombe received a knighthood from the British government. But England was first to exploit machinery on a mass scale. Indeed, while the Industrial Revolution had its origins in silk production, its true

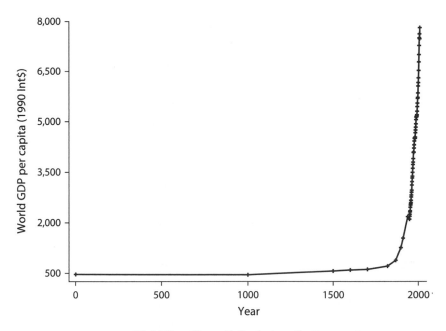

FIGURE 1: World Gross Domestic Product per Capita, 1–2008
Source: J. Bolt, R. Inklaar, H. de Jong, and J. L. Van Zanden, 2018, "Rebasing 'Maddison': New Income Comparisons and the Shape of Long-Run Economic Development," Maddison Project Working Paper 10, Maddison Project Database, version 2018.

beginnings were in the cotton industry. As the historian Eric Hobsbawm famously remarked, "Whoever says Industrial Revolution says cotton."[12] After the mechanization of cotton production, change begat change as a self-reinforcing cascade of progress created the modern world. As technology progressed in the early days of industrialization, however, living standards for many regressed. Our vocabulary bears witness to the changes that signify the century after 1750. Words like "factory," "railroad," "steam engine," and "industry" first emerged then. But so did "working class," "communism," "strike," "Luddite," and "pauperism." What began with the arrival of the first factories ended not only with the construction of the railroads, but also with the publication of the *Communist Manifesto*. Just as the Industrial Revolution was responsible for many revolutionary technologies, so it was responsible for many political revolutionaries along the way.[13]

The point is not to downplay the significance of the British Industrial Revolution. It is rightly regarded as the main event in human history because it eventually allowed humanity to escape the life that Thomas Hobbes described as "nasty, brutish, and short."[14] Eventually was nonetheless a long time. The "Great Escape," as the economist Angus Deaton has called it, didn't immediately turn the cottage of the commoner into a Garden of Eden.[15] During the early days of industrialization, the lives of many commoners got nastier, more brutish, and shorter. Material standards and living conditions for the masses in Britain failed to improve before 1840. The poet William Blake's phrase "dark, satanic mills" captures the long working hours in the factories and the hazardous conditions that embodied the industrialization process.[16] In major industrial cities like Manchester and Glasgow, life expectancy at birth was some staggering ten years shorter than the national average. The wages that workers took home in industrial cities hardly compensated for the dirty and unhealthy conditions in which people lived and worked. Although output expanded, the gains from growth didn't find their way into the pockets of ordinary people. Real wages were stagnant or even falling for some. The only thing workers saw expanding was the number of hours spent in the "dark, satanic mills." The gains of progress overwhelmingly went to industrialists, who saw their rate of profit double. Consequently, the average amount of food consumed in Britain during the Industrial Revolution did not increase until the 1840s. The share of households with a surplus for nonessentials declined among low-wage agricultural laborers and factory workers over the first half of the nineteenth century. And poor nutrition meant that people grew shorter by the generation. These were the glorious decades in which modern growth began.[17]

The cause of the living standards crisis in Britain was the downfall of the domestic system of production, which was gradually displaced by the mechanized factory. Artisan craftsmen were highly skilled and earned decent wages. But with the rise of the factory, one artisan after another saw his income vanish. And while new jobs were created in the factories, spinning machines were specifically designed for children, who could do the job for a fraction of the cost of adults and thus became a growing share of the workforce. They were the robots of the Industrial

Revolution. Besides working for very little, they did not have any bargaining power and were easy to control.[18]

As the old artisanal skills were made obsolete by advances in mechanization, adult male workers lost out: the share of children workers rapidly expanded, reaching about half of the workforce employed in textiles during the 1830s. The social costs inflicted upon the workforce—including vanishing incomes, deteriorating health and nutrition, forced occupational and geographical migration, and in some cases unemployment—were not negligible. Not to mention the suffering of children. In an interview, Robert Blincoe, a former child laborer, stated that he would rather have his children deported to Australia than let them experience life working in the factories.[19] But from a purely economic point of view, adult artisans were without question the prime victims of industrialization. And there were many of them. As one leading scholar of the Industrial Revolution, David Landes, writes, "If mechanization opened new vistas of comfort and prosperity for all men, it also destroyed the livelihood of some and left others to vegetate in the backwaters of the stream of progress. . . . The victims of the Industrial Revolution numbered in the hundreds of thousands or even millions."[20]

Historians have puzzled over why ordinary English people would voluntarily agree to take part in an industrialization process that reduced their living standards. The simple answer is that they didn't. British governments at times clashed with workmen raging against the machine. But their efforts were unsuccessful, as British governments took an increasingly stern view of anything that might diminish England's competitive position in trade. All the Luddites achieved during the risings of 1811–16 is prompting the government to deploy an even larger army against them: the twelve thousand troops sent to resolve the machinery riots amounted to more people than the army Wellington took into the Peninsular War against Napoleon in 1808.

As we shall see, before the late nineteenth century, resistance to technologies that threatened workers' skills was the rule rather than the exception. While much commentary tends to focus on the Luddite riots, they were just part of a long wave of riots that swept across Europe and China. And the history of opposition to labor-replacing technologies goes back much further. Vespasian, the emperor of Rome in 69–79, refused to adopt machinery for transporting columns to the Capitoline Hill due to employment concerns. And in 1589, Elizabeth I famously refused to grant William Lee a patent for his stocking-frame knitting machine, fearing unemployment as a result of the technological advance. The gig mill, which saved considerable amounts of labor, had been prohibited in Britain in 1551. And elsewhere in Europe opposition was just as fierce. Many European cities banned automatic looms in the seventeenth century. Why? Where they were adopted (for example, in the city of Leiden), riots followed. The ruling classes feared that angry workers like those in Leiden would start to rebel against the government. And this concern was by no means just European. One reason why China was so late to industrialize, economic historians have argued, is that resistance to technologies that threatened workers' skills persisted up until the closing decades of the nineteenth century, when imported sewing machines were destroyed by native workers. In fact, the British government was the first to side with the pioneers of industry rather than rebelling workers, providing one explanation for why Britain was the first country to industrialize.[21]

Back in 2012, Bill Gates took note of what has been called the paradox of our age: "Innovation is faster than ever before . . . yet Americans are more pessimistic about the future."[22] Indeed, according to the Pew Research Center, just over a third of Americans still believe that their children will be better off financially than they were.[23] If the past few decades are any guide to the future, some people surely have much to be pessimistic about. Only half of Americans born in 1980 are economically better off than their parents, compared to 90 percent of those born in 1940.[24] Despite this fact, slogans like "the greatest country on earth" continued to be the norm in presidential election campaigns. It was only in 2016 that the Republican presidential candidate won with the slogan "Make America Great Again." At last a candidate spoke the truth—or so it must have felt in those parts of the country where opportunities had long since faded.

As the Industrial Revolution illustrates, the Gates paradox is not really a paradox. Like in the early days of industrialization, workers today are no longer reaping the gains of progress. Worse, many have been left behind in the backwaters of progress. In the same way that opportunity dried up for middle-income artisans as a consequence of the industrialization process, the age of automation has meant diminishing opportunities for the American middle class. Like the victims of the early factories, many Americans have adjusted to the computerization of work by unwillingly shifting into lower-paying jobs or have failed to adjust and dropped out of the workforce completely. And similar to the victims of the factories, the losers to automation have primarily been men in the prime of life. Up until the 1980s, manufacturing jobs allowed ordinary working men to attain a middle-class lifestyle without going to college. As employment opportunities in manufacturing receded, a path of upward mobility was closed to many citizens.[25]

What's more, the adverse consequences of automation have so far primarily been a local phenomenon. Focusing too closely on national statistics disregards the fact that if you put one hand in the freezer and the other on the stove, you should feel quite comfortable on average. The same was true of the Industrial Revolution. While the local cloth industry in Northamptonshire was left in ruins, factories were almost unheard of in 1800 in the pastoral areas of southern England, where Jane Austen resided. This time around, the social and economic fabric has been torn apart in old manufacturing cities, where automation has deprived middle-aged men of opportunity. Communities that have seen manufacturing jobs vanish, due either to automation or globalization, have endured persistent increases in joblessness. They have also seen public services deteriorate, greater increases in property crime and violent crime, and worse health outcomes. They have seen mortality rates increase due to suicide and alcohol-related liver disease. They have seen marriage rates collapse, leaving more children in single-parent households, with dismal future prospects. Rates of social mobility are significantly lower in places where middle-class jobs have evaporated.[26] And where jobs have disappeared, people have become more likely to vote for populist candidates. Indeed, studies have shown that both in America and in Europe, the appeal of populism has been greater where jobs have become more exposed to automation.[27] Just like the days of the Industrial Revolution, the losers to technology are demanding change.

We should have seen it coming. In 1965, when the first electronic computers entered offices, Eric Hoffer warned in the *New York Times* that "a skilled population deprived of its sense and usefulness would be the ideal setup for an American Hitler."[28] Perhaps somewhat ironically, Hitler and his government were well aware of the disruptive force of labor-replacing technology. His appointment as chancellor of Germany on January 30, 1933, heralded the return of preindustrial policies, which sought to restrict the use of machinery. In Danzig, where the Nazi Party won over 50 percent of the votes that year, such efforts became a major priority. To deal with the issue of technological unemployment, the Senate decreed that machinery would not be installed in factories without special permission of the government. Failure to comply would lead to heavy penalties or even being forced to shut down by the government.[29] In August 1933, Alfred von Hodenberg, leader of the Nazi Labor Front, made clear that machines would not be allowed to threaten workers' jobs in the future. "Never again," he reassured the public, "must the worker be replaced by a machine."[30]

Technology at Work

Our path to riches is best understood in terms of the adoption of a steady flow of labor-saving technologies over the centuries. As the economist Paul Krugman once quipped, "depressions, runaway inflation, or civil war can make a country poor, but only productivity can make it rich."[31] Productivity growth happens when technology allows us to produce more with less. If the adoption of machines makes labor productivity grow by 2.5 percent per year, output per person will double every twenty-eight years. The notion that the product of an hour of work can double in just about half of a working lifetime is surely sufficient justification for the disruptive force of technology, which has shrunk that timescale visibly. But while productivity is a prerequisite for growing incomes for the commoner, it is not a guarantee of such growth. And, if machines replace workers in existing functions, some people may be left worse off as technology progresses. Despite this fact, textbook economics treats technological progress as a Pareto improvement: in other words, the assumption is that when machines take workers' jobs, new and better-paying jobs become available for everyone at the same time. As evidenced by the historical record, such models are utterly irrelevant for understanding episodes when technological progress is labor replacing. These technologies have brought higher material standards but also worker dislocation.

The extent to which labor-saving technologies will cause dislocation depends on whether they are enabling or replacing. Replacing technologies render jobs and skills redundant. Enabling technologies, in contrast, make people more productive in existing tasks or create entirely new jobs for them. Thus, the term "labor saving" has two closely associated but not identical meanings, and the difference between the two has important implications for labor.[32] As the economist Harry Jerome noted in 1934, if the 1929 tonnage of iron and steel were produced with the technology available in 1890, a million and a quarter workers would have been needed instead of four hundred thousand. Does this mean that eight hundred thousand men had lost their jobs by 1929? Surely not. At the onset of the Great Depression, employment in steel had grown.[33] Better technology reduced the number of workers required to produce a given amount of steel, but the steadily growing demand for steel meant that the number of jobs in the industry grew, too. Clearly, the nature of steel production changed as the industry mechanized, but there was probably little job displacement. Unlike replacing technologies, which take over the tasks previously done by labor, augmenting technologies increase the units of a worker's output without any displacement occurring, unless demand for a given product or service becomes saturated.[34] There are many examples of enabling technologies. Computer-aided design software has made architects, engineers, and other skilled professionals more productive by helping rather than replacing them.

Statistical computer programs like Stata and Matlab have made statisticians and social scientists better analysts without reducing the demand for them. And office machines like the typewriter created clerical jobs that did not previously exist.

To see how outcomes differ for labor when a technology is labor replacing, consider the arrival of the elevator. Without elevators, there would be no skyscrapers and no elevator operators. When the first elevators arrived, more elevators meant more jobs for people with a good sense of timing, capable of stopping the elevator when it was aligned with the floor. Things changed when a replacing technology emerged: the automatic elevator, which got rid of the human operator. All of the sudden, the job of elevator operator disappeared, even though we now use elevators more than ever. The demand for elevators has evidently not become saturated, just as we still demand many manufactured goods. But in a world where the jobs of machine operators have been taken over by robots, having more automobiles leave the factories does not inevitably mean more jobs for machine operators. Thus, it stands to reason that the effects of replacing technologies on jobs and wages will be very different from those of enabling technologies. Yet until recently, economists did not make such distinctions. Since the pioneering work of Jan Tinbergen—the first winner of the Nobel Prize in Economics—economists have tended to conceptualize technological progress in a purely augmenting way. According to the augmenting view of progress, new technologies will help some workers more than others but will never replace labor, meaning that workers cannot see their wages fall as technology progresses. This was a reasonable approximation of economic reality for much of the twentieth century. Indeed, most economic theory reflects the patterns of the particular times economists observe around them. The work of Tinbergen, which was published in 1974, before the age of computerization, was no exception. For much of the twentieth century, wages rose at all levels. What makes economic analysis hard is that there are few models that apply to every time and place.

The fact that wages have been falling for large groups in the American labor market for more than three decades has prompted economists to think differently about technological change. Pathbreaking work by the economists Daron Acemoglu and Pascual Restrepo provides a helpful formal model for understanding periods of falling wages, as well as times when wages are growing for everyone, by conceptualizing technological progress as either enabling or labor replacing. This book looks at the historical record through the lens of their theoretical framework.[35] The notion of machines being capable of taking over human work is important, because it means that technology can reduce wages and employment unless it is counterbalanced by other economic forces. Even though growing productivity still raises total income—offsetting the displacement effect in part, as more spending in the economy creates other jobs elsewhere—it does not fully counterbalance the negative effects of technological displacement. In Acemoglu and Restrepo's framework, the creation of new tasks is essential to raise the demand for labor, workers' wages, and the share of national income going to labor rather than owners of capital. How workers fare, in other words, in large part depends on the race between task replacement and new task creation, and how easily workers can transition into emerging jobs.

Historically, as we shall see, the extent to which technology is labor replacing or enabling has varied greatly, leading to very different outcomes for average people. When new technologies replace workers in existing tasks, those workers' skills become obsolete. Even when technologies are replacing for some but augmenting for others, workers might suffer hardships. In recent years, the creation of new jobs for robotics engineers has provided little relief to those who lost their jobs to industrial robots on the assembly lines. The arrival of the power loom, in similar fashion, replaced the jobs of hand-loom weavers, while creating new jobs for power-loom weavers. But while hand-loom weavers' incomes diminished almost immediately, it took decades for the wages of power-loom weavers to rise, as they had to acquire new skills and a new labor market had to develop for those skills.[36] Because replacing technological progress often comes with what Schumpeter called a "perennial gale of creative destruction," there are always winners and losers.[37] The overwhelming focus of popular commentary on unanswerable questions like whether there will be enough jobs in 2050 is unfortunate.

In fact, it misses the point. Even if new jobs emerge as old ones are lost to automation, that might be little reassurance for the person who loses his or her job. Modernist writers didn't fail to take note of the automation dilemma. In *Ulysses*, for example, James Joyce's hero Leopold Bloom points out that "a pointsman's back straightened itself upright suddenly against a tramway standard by Mr. Bloom's window. Couldn't they invent something automatic so that the wheel itself much handier? Well but that fellow would lose his job then? Well but then another fellow would get a job making the new invention?"[38]

A new job was created for someone to make the new invention. But the someone was "another fellow": making the invention required a different breed of worker. Both the Industrial Revolution and the computer revolution primarily created jobs for another fellow, whose skills could not have been more different from those of the displaced worker. The first episode of industrialization is best described by the wit of the economic historian Gavin Wright, who reckoned that "in the limit we could devise an economy in which technology is designed by geniuses and operated by idiots."[39] Early factory machines, it is true, were simple enough to be operated by young boys. And as a result, middle-income artisan craftsmen were replaced by children working for a fraction of their wages in the factories. The difference this time around is obviously that children are no longer needed to operate the machines. Computer-controlled machines can run on their own. Yet computerization has also given rise to new tasks, requiring an entirely different set of skills like those of audiovisual specialists, software engineers, database administrators, and so on. Thus, we seem to have devised an economy designed by geniuses to be operated by other geniuses. Some jobs have become automated, but computers have also led to greater demand for workers with highly developed cognitive skills. Indeed, a common misconception is that automation is an extension of mechanization. Automation has replaced precisely the semiskilled machine-tending jobs that mechanization created, which once supported a large and stable middle class. Broadly speaking, those fortunate enough to have gone to college have thrived in the age of computers. But as middle-income jobs have dried up, many semiskilled workers have struggled to find decent job. During the Industrial Revolution as well as the more recent revolution in computing, middle-aged men in middle-income jobs were the victims of progress, because their skills were unsuitable for the new jobs that emerged.

When technological change is labor replacing, how workers fare depends on their other job options. In Henrik Ibsen's play *The Pillars of Society*, written in 1877, parallels are drawn between the economic consequences of the Industrial Revolution and those of Johannes Gutenberg's printing press. One of the characters, Konsul Bernick, assumes that the fates of artisan craftsmen in the nineteenth century were similar to those of copyists when the printing press arrived, suggesting that "when printing was discovered, many copyists had to starve." The shipyard foreman Aune bluntly replies, "Would you have admired the art so much, Consul, if you had been a copyist?"[40] Though Ibsen's question was meant as a rhetorical one, copyists rarely opposed printing technology. As we shall see in chapter 1, unlike weavers—who suffered hardships from the mechanization of industry—copyists and scribes were more likely to benefit from Gutenberg's invention. Many of them did not make a living producing manuscripts. To them, the movable printing press didn't mean any loss of income. And those who copied books for a living either specialized in shorter texts that were uneconomical to produce with printing technology or became binders and designers of books. Thus, while weavers and other craftsmen, who faced worsening job options, smashed textile machines all over Europe in the eighteenth and nineteenth centuries, copyists rarely resisted the printing press in the late 1400s. Of course, the art of printing was not adopted with the same enthusiasm everywhere. Fearing that a literate population would undermine his leadership, Sultan Bayezid II issued an edict banning printing in Arabic in the Ottoman Empire in 1485, with dismal longlasting consequences for literacy and economic growth in the region.[41] But in the light of the hostility to replacing technologies that was so widespread in Europe before the twentieth century, episodes of labor unrest accompanying the adoption of the printing press were few.

The case of the printing press illustrates a broader point: when people have good alternative job options, they are less likely to rebel against machines. Job displacement is never painless, but if people have reason to believe that they will eventually come out ahead, they are more likely to accept the endless churn in the labor market. As we shall see, the explosive growth of middle-class jobs in the mass-production industries of the twentieth century was one key reason mechanization was allowed to progress uninterrupted: an abundance of manufacturing jobs was the best unemployment insurance people could get. In this period, a wave of enabling technologies and soaring productivity growth allowed working-class people to climb the economic ladder. Automobiles and electricity spawned new gigantic industries, and with more capital tied up in machinery, firms began to raise wages to keep workers from leaving for better jobs elsewhere. People at the top and the bottom of the income distribution saw their standard of living improve enormously, and, consequently, middle-class people accepted the reshufflings in the labor market with the expectation that they would benefit too.

Another reason people may not oppose technologies that threaten their jobs is obviously that almost everyone will benefit in their capacity as consumers. Even those who worked on Ford's and General Motors's assembly lines have to some extent benefited from the cheapening of automobiles as robots have taken their jobs. Yet machines only cheapen goods and services after they have been introduced, so that if a technology is labor replacing, consumer benefits will arise only after displacement has already occurred. More important, the individual costs from displacement, in terms of distress and lost income, will be much greater than any consumer benefits unless those workers have decent outside job options. The cheapening of textiles, for example, did not provide sufficient relief to the Luddites, who rioted against the introduction of machinery despite the consumer benefits brought by mechanization. The point is surely not that replacing technologies will be bad for people over the long run. The very opposite is true. But that alone does not provide much relief for those who see their jobs disappear, unless they can expect to find new work of equal pay.

Most economists will acknowledge that technological progress can cause some adjustment problems in the short run. What is rarely noted is that the short run can be a lifetime. And ultimately, the long run depends on policy choices made in the short run. The mere existence of better machines is not sufficient for long run growth. As Daron Acemoglu and the political scientist James Robinson point out in *Why Nations Fail*, economic and technological development will move forward only "if not blocked by the economic losers who anticipate that their economic privileges will be lost and by the political losers who fear that their political power will be eroded."[42] Workers alone might struggle to block new technologies effectively. But the ruling elites slowed labor-replacing progress for millennia.[43] Political incumbents, for the most part, had little interest in the destabilizing process of creative destruction, as groups of economic losers could challenge the political status quo. As the eminent economic historian Joel Mokyr has argued in separate accounts:

> Any change in technology leads almost inevitably to an improvement in the welfare of some and a deterioration in that of others. To be sure, it is possible to think of changes in production technology that are Pareto superior, but in practice such occurrences are extremely rare. Unless all individuals accept the verdict of the market outcome, the decision whether to adopt an innovation is likely to be resisted by losers through non-market mechanism and political activism.[44]
>
> Britain's edge during the Industrial Revolution did not lie in the absence of resistance against technological change, but in its government's consistently and vigorously siding with the "party" for innovation. . . . Resistance to technological progress in France appears to have been more successful than in Britain, and perhaps this difference offers another explanation why Britain's Industrial Revolution was first.[45]

As I will argue in a similar vein, the early decision of British governments to consistently squash any resistance to mechanization helps explain why Britain was the first to industrialize. This decision, as we shall see, was much the result of a shift in political power. As the discovery of the New World gave rise to international trade and commerce, the power of landed wealth was challenged by a new class of "chimney aristocrats," who stood to gain from mechanization.[46] And more broadly, cascading competition among nation-states made it harder to align technological conservatism with the political status quo. The outside threat of political replacement became greater than the threat of rebelling workmen from below. Even when workers managed to solve the so-called collective action problem and take to the streets in protest, their case was hopeless. They did not stand a chance against the British army. Many Luddites ended up being imprisoned and then sent to Australia.

The Reform Acts of 1832 and 1867 were surely important events, but they did not turn Britain into a liberal democracy. Property rights were regarded as most important, and civil rights and political rights were still lagging behind. Few people had access to education, and property ownership remained a requirement for voting—meaning that most ordinary people were politically disenfranchised. Had Britain been a liberal democracy, the case of the Luddites would surely have been much less hopeless. As Wassily Leontief, winner of the Nobel Prize in Economics, once joked, "If horses could have joined the Democratic party and voted, what happened on farms might have been different."[47] Horses might have used their political rights to bring the spread of the tractor to a halt. In similar fashion, if the Luddites had had their way, the Industrial Revolution would not have happened in Britain. Of course, there is no way of knowing exactly what would have happened; all we know is that many citizens tried to bring progress to halt by every means they had.

The Plan of the Book

In the age of AI, as we shall see, technological progress has become increasingly labor replacing. Thus, to understand the future of progress, we must understand its political economy. The notion that technology can leave groups in the labor market worse off for the rest of their working lives is quite sufficient to justify their resistance to automation. And for governments seeking to avoid social unrest, it is also quite sufficient to justify restricting some technologies. For these reasons, the long run cannot be disconnected from the short run. Our long-run trajectories can be interrupted and changed by short run events, with dismal consequences for our long-term prosperity.

We all know that human history has proceeded very differently in different parts of the world. Economists and economic historians have devoted considerable attention to the question of why some places have grown rich while others have remained poor. This book is not quite as ambitious. It examines why people have fared differently in the places of the world where the frontiers of technology have been allowed to progress throughout the centuries. The relationship between new technology and the wealth of humans has never been tidy and linear. History never quite repeats itself. But as Mark Twain noted, sometimes it surely rhymes. As I write, middle-income jobs are disappearing, and real wages are stagnating, just like in the classic period of industrialization. Of course, the computer technologies of the twenty-first century could not be more different from the machines that made modern industry. But many of their economic and social effects now look exceedingly similar. The Industrial Revolution has made us infinitely wealthier and better off over the long run. AI, similarly, has the potential to make us much wealthier, but just as in the Industrial Revolution, there is concern that it is leaving large swaths of citizens behind, possibly causing a backlash against technology itself. Many observers of current affairs have pointed out that the recent populist renaissance cannot be explained without reference to the losers of globalization. But technology has been just as important in driving down the wages of the middle class. And we have seen nothing yet. As AI becomes more pervasive, so will automation and its effects.

Economic historians have long debated why the technology boom of the 1760s in Britain took so long to produce higher standards of living, and economists are now engaged in a strikingly similar debate about why staggering advances in automation so far have failed to show results in the pockets of average people. This book is an attempt to connect two large bodies of scholarly research to put the Gates paradox in historical perspective. It tracks the expanding frontiers of technology from the invention of agriculture to the rise of AI, tracing the fates of humans as technology has progressed. I should warn the reader that this is not a balanced account. A book of this scope must be selective and carefully prioritize what it discusses. The history of technology is the subject of an extensive literature that I cannot do justice to here. Rather, by reviewing some of the most important technological advances, I shall try to convince the reader that the price of progress paid by the workforce has varied greatly in history, depending on the nature of technological change, and has increased in the twenty-first century—which explains many of the discontents people now face.

The reader should also be aware that because the Industrial Revolution happened in Britain and technological leadership has remained firmly in Western hands since then (though it remains to be seen for how long that will be the case), this book is a Western-biased account. The West caught up with more advanced Islamic and Oriental civilizations only in the fifteenth century. But to paint the contrast of the West before and after the Industrial Revolution, I shall primarily focus on the Western experience. I should also say that most of the history in this book concerns Britain and later America. The simple reason is that the Industrial Revolution first happened in Britain. America took over world technological leadership during the so-called Second Industrial Revolution, and I shall primarily focus on the U.S. experience thereafter. As the economic historian Alexander Gershenkron noted, catch-up growth, which rests on adopting existing technologies invented elsewhere, is fundamentally different from growth that rests on expanding the frontiers of technology into the unknown, and this book focuses on the latter. Some readers may also find it disappointing that many major technological breakthroughs are not even mentioned. To take just one example, the rise of modern medicine has arguably been the greatest boon to humanity but is shamelessly left out here. Technological developments in recent years, including advances in AI, mobile robotics, machine vision, 3-D printing, and the Internet of things, are all labor saving. The purpose of this book is to shed light on present times and challenges facing the workforce today, and for this reason labor-saving technology will receive the bulk of the attention.

It must also be emphasized that though the focus in the later chapters is much on the American experience, technology is not a soloist but part of an ensemble. It interacts with institutions and other forces in society and the economy, which explains why the rise of economic inequality has been less dramatic in other industrial nations over the past three decades. Yet stagnant wages, disappearing middle-income jobs, and a falling labor share of income are common features of Western countries, and they are all related to trends in technology. While there can be no doubt that numerous forces have shaped the income distribution, my focus here is on the long run rather than cyclical matters, and it is about the 99 percent rather than the top 1 percent. And over the grand sweep of history, average people's incomes have come to depend on technology more than any other factor.

The main challenge this book faces, however, is probably to convince the reader that we can learn from the past. Economists and economic historians alike tend to treat this idea with skepticism. As one anonymous reviewer of this manuscript put it:

> Economists are the obvious "History deniers." They are reluctant to accept that economists could learn anything from the past even as analysed by economic historians. The humbling experience of failing to predict (indeed perhaps unwittingly helping to create) the 2008 financial crisis produced an uncharacteristic expression of interest in economic history as economists sought insight into events that otherwise seemed unpredictable and disturbing. But the interest (and the humility) was temporary and superficial. However, economic historians too are reluctant to claim present-day insight from their studies of the past; it claims too much for their humble discipline. So, both of the disciplines that Frey addresses will be uneasy with his central proposition. Behind this issue is the bigger one of the communication difficulties between these two disciplines. Both have similar technical toolboxes but economics has honed its contents to a fine edge and is hostile to other approaches whereas history's contents are sometimes not on the technological frontier and have to be used in the context of a narrative. Any author wanting to make the point that we can learn from history to these two different audiences faces serious challenges.

In the remainder of this book I shall nonetheless try to convince the reader that history is more than one damn fact after another. There are broad patterns that we can learn from. When technological progress is labor replacing, history tells us, hostility and social upheaval is more likely to follow. When progress is of the enabling sort, in contrast, and the gains from growth are more widely shared, there tends to be greater acceptance of new technologies. The chapters that follow divide economic history into four episodes. Part 1, titled "The Great Stagnation," consists of three chapters that concern preindustrial technologies and their effects on people's standard of living. Chapter 1 gives a succinct summary of advances in technology from the invention of agriculture some 10,000 years ago up until the dawn of the Industrial Revolution. It shows that many significant technologies emerged before the eighteenth century, but they failed to improve material conditions for ordinary people. Chapter 2 demonstrates that though living standards had improved before the Industrial Revolution, growth was predominantly based on trade. The Schumpeterian growth of our modern age, based on labor-saving technology, creative destruction in employment, and the acquisition of new skills, was not the engine of economic progress. Chapter 3 seeks to explain why this was the case. As we shall see, innovation also flourished at times before the Industrial Revolution, but it rarely served to replace workers—and when it did, it was vehemently opposed or even blocked. A powerful explanation for why the technologies of the Industrial Revolution did not arrive earlier is the widespread opposition to machines that threatened citizens' livelihoods. The landed classes, whose members controlled the levers of political power, had little to gain and much to lose from replacing technologies, as workers might rebel against the government in fear of losing their jobs.

The second part, called "The Great Divergence," provides a whirlwind tour of the Industrial Revolution in Britain. It shows that preindustrial monarchs were right to fear the disruptive force of machinery. As the mechanized factory displaced the domestic system, working people raged against the machine. Chapter 4 zooms in on the technologies that made the Industrial Revolution, showing that nearly all of them served to replace workers. Chapter 5 shows that the result was the hollowing out of middle-income artisan jobs, causing a great divergence within Britain—which explains why industrialization brought so much conflict. But the ruling classes now had more to gain from allowing mechanization to progress, and effectively enforced the first machine age on the populace. Workers' resistance ended only when people began to see their wages rise in the closing decades of the Industrial Revolution.

Part 3, titled "The Great Leveling," shifts the focus to the American experience. With the Second Industrial Revolution, America took over technological leadership from Britain and the world. The purpose of this part is to examine why the twentieth century did not see the same hostility to mechanization, even as the frontiers of technology were advanced at an accelerating pace. Chapter 6 sketches the technological changes that accompanied the Second Industrial Revolution. It examines the enormous shifts that took place in the labor market as factories electrified, households mechanized, and people left the countryside for mass production industries in the city. We all know that these transitions weren't painless. Chapter 7 shows how machinery anxiety returned temporarily, as parts of the workforce struggled to adjust as some occupations vanished. But even though concerns about new technologies taking people's jobs were widespread at times, few people seriously believed that restricting the use of machines was a good idea. Why? America perhaps had the most violent labor history of the industrial world, but after the 1870s, workers rarely, if ever, targeted machines when violence erupted. Chapter 8 is devoted to the question of why labor didn't oppose machines in the way it did in the nineteenth century. I harbor no illusions that I have succeeded in providing a full answer to that question, but technology is certainly part of the story. A flow of enabling technologies pulled people into new and better-paying jobs in the smokestack cities of the Second Industrial Revolution. As labor began to see technology as working in its self-interest, the rational response became to seek to minimize the adjustment costs imposed on the workforce rather than retarding technological progress. Labor de facto accepted a laissez-faire regime with regard to mechanization but insisted on establishing a welfare and educational system to help people adjust while making individual costs for those who lost their jobs more narrowly constrained. This became the social contract of the twentieth century.

Part 4, called "The Great Reversal," concerns the era of computers. Chapter 9 shows that the age of automation was not a continuation of twentieth-century mechanization. On the contrary, it was a complete reversal of it. The first three-quarters of the twentieth century has rightly been regarded as producing "the greatest levelling of all time."[48] It was a period of egalitarian capitalism when workers' wages rose at all ranks, to the point where Karl Marx's proletariat could join the middle class. In the 1970s, the American middle class had become a diverse blend of blue- and white-collar citizens. Many of these workers tended machines of some kind, in offices and factories. As we shall see, robots and other computer-controlled machines cut out precisely the middle-income factory and office jobs that mechanization created. Chapter 10 shifts the focus from the aggregate to the communities that have seen jobs disappear. Despite the promise of digital technology to flatten the world, it has done the opposite. Since the dawn of the computer revolution, new jobs have overwhelmingly been clustered in cities with skilled populations, while automation has replaced jobs in old manufacturing powerhouses, amplifying the polarization of the American social fabric along geographic lines. And as America has become increasingly polarized along economic lines, it has also become more politically polarized.

Chapter 11 turns to the question of why citizens who have seen their wages fall have not demanded more compensation, as the median voter theorem would predict. If the middle class declines and inequality rises, we would expect workers to vote for more redistributive policies. One reason that they haven't, I shall argue, is that they have lost political influence. Growing socioeconomic segregation has made people who have suffered hardships increasingly detached from the rest of American society. Meanwhile, the would-be working class, whose members would have flocked into the factories during the postwar boom years, has become increasingly detached from both labor unions and mainstream political parties. The growing populist appeal, it seems, in large part reflects diminishing opportunities for the losers to globalization and automation and the lack of a political response to address their concerns. Globalization has already become a populist target. Looking forward, however, more and more workers are becoming shielded from the force of globalization, as a growing percentage of the workforce is employed in nontradable sectors of the economy. But they are not shielded

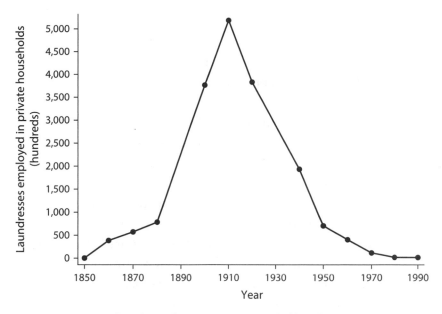

FIGURE 2: Number of Laundresses in Private Households in the U.S., 1850–1990
Source: M. Sobek, 2006, "Detailed Occupations—All Persons: 1850–1990 [Part 2]. Table B1396-1439," in Historical Statistics of the United States, Earliest Times to the Present: Millennial Edition, *ed. S. B. Carter et al. (New York: Cambridge University Press).*

from automation. If current economic trends persist for several years more or even decades, as they did during the Industrial Revolution, there is nothing that shields automation from becoming a target, as globalization already has.

Part 5 is titled "The Future," although it does not attempt to predict what will happen. As discussed above, much depends on the race between replacing and enabling technologies, but obviously the next three decades must not mirror the past three. The idea here is not that we can simply extrapolate from current trends, which is what economists usually do. Nor is my ambition to predict future technological breakthroughs. The best I can do is examine the prototypical technologies coming out of the labs today that have not yet found widespread use. Take, for example, the employment prospects of laundresses, which peaked around 1910—the year when Alva J. Fisher took out a patent for the first electric washing machine, called Thor (figure 2). If economists had extrapolated from the recent past in 1910, they would have inferred that there would be jobs in abundance for people doing laundry in the coming decades. By looking to trends in technology, in contrast (which is what chapter 12 will do), one might have concluded that the electric washing machine would replace laundresses in this task.

After reviewing many recent technological developments, including in machine learning, machine vision, sensors, various subfields of AI, and mobile robotics, my conclusion is that while these technologies will spawn new tasks for labor, they are predominantly replacing technologies and will continue to worsen the employment prospects for the already shattered middle class. Thus, assuming that the positive attitudes toward technological progress of the twentieth century will continue to hold, regardless of how working people fare from automation, is an exceedingly strong assumption. As we shall see, people are already turning more pessimistic about the future and even about automation. A majority of Americans would vote for policies to restrict it, and populists may well tap in to growing automation anxiety. How events play out will likely depend on policy choices. To that end, chapter 13 concludes by sketching some strategies and pathways to help people adjust.

PART I

THE GREAT STAGNATION

No craftsman shall think up or devise any new invention, or make use of such a thing, but rather each man shall, out of citizenly and brotherly love, follow his nearest and his neighbour, and practise his craft without harming another's.

—KING SIGISMUND I OF POLAND

Inequality was pervasive in the agrarian economies that dominated the world in 1800. The riches of a few dwarfed the pinched allocations of the masses. Jane Austen may have written about refined conversations over tea served in china cups. But for the majority of the English as late as 1813 conditions were no better than for their naked ancestors of the African savannah. The Darcys were few, the poor plentiful.

—GREGORY CLARK, *A FAREWELL TO ALMS*

The wealth of humans is best understood as the cumulative effect of technologies that allow us to produce more with fewer people. Yet before the Industrial Revolution, living standards were less dependent on the spread of technologies that substituted mechanical power for human muscle. This is not to suggest that technological progress commenced in the eighteenth century. Disparities in the adoption of technology across societies reveal substantial technological progress during the preindustrial era. No single event illustrates this better than when the Dutch explorer Abel Tasman discovered Tasmania in 1642, ending the longest isolation of a population recorded in history. Elsewhere in the world, the diffusion of technologies had significantly altered human development. In Tasmania, by contrast, people still lacked agriculture, metal, pottery, fire-making equipment, and even mounted stone tools.[1]

Based on what we know from recorded history, the lack of technological creativity was not the key obstacle to economic growth. The windmill, horse technology, the printing press, telescope, barometer, and mechanical clock—to name just a few—were all invented before the eighteenth century. The reason that we tend to attribute the beginning of meaningful technological change to the Industrial Revolution is that then, for the first time, there was progress that eventually translated into significantly higher average incomes. Hence, even though the historical record is one of uneven technological progress, 99 percent of human history can with some exaggeration be regarded as a great stagnation in economic terms. The first part of this book seeks to explain why. By reviewing the key technological advances in the West—where the Industrial Revolution first took place—and their respective applications in production, the objective is to shed some light on why technological progress in the preindustrial world failed to achieve anything like the comfort and prosperity that followed the technological achievements of the eighteenth century. The explanations in the literature are of course plentiful. One popular theory is that before the Industrial Revolution the world was caught in a Malthusian trap, where greater prosperity simply translated into larger populations, leading to no real gains in per capita income. The Malthusian view is not irrelevant, but living standards in Britain had already improved between 1500 and 1800, albeit slowly. The real conundrum is that most of the wave of gadgets that we associate with the Industrial Revolution could have been developed and put into widespread use long before the eighteenth century, yet they were not. Besides the steam engine, the eighteenth century didn't witness any breakthroughs that would have "puzzled Archimedes."[2]

The preindustrial history of technology illustrates an important point: resistance to worker-replacing technologies has been the norm rather than the exception. Innovation flourished before the eighteenth century, but it rarely took the form of capital that replaced labor—and when it did, fierce opposition typically followed. This should not be taken to imply technological backwardness. However, it does help explain why the job-replacing technologies of the Industrial Revolution did not arrive earlier.

1

A BRIEF HISTORY
OF PREINDUSTRIAL PROGRESS

Though preindustrial societies were clearly less productive than later ones, technological creativity has existed in some form throughout human history. The majority of our most basic technologies, such as fire-making equipment, tools for hunting and fishing, the domestication of animals, agriculture, irrigation, pottery and glazing, the wheel, and spinning and weaving techniques were invented before any historical record was made. The most transformative of these inventions was agriculture, because it allowed the first civilizations to emerge. As Bertrand Russell explains: "Civilized man is distinguished from the savage by *prudence*, or, to use the slightly wider term, *forethought*. He is willing to endure present pains for the sake of future pleasures. . . . This habit began to be important with the rise of agriculture."[1]

Before the Neolithic revolution, which began about 10,000 years ago, hunter-gatherers were preoccupied with finding food. Hunting required no planning but did involve sharing what was caught, as there was no technology for storing meat or other foods that had been foraged for—which meant that instant consumption was the only option. Therefore, there were no property rights in the modern sense and no need for them. Like chimpanzees, hunter-gatherers would inhabit and often fight over a territory, but because no one was able to accumulate any meaningful surplus, there were no assets to establish ownership over. The development of agriculture—the growing of crops and the cultivation of animals—changed that and allowed food to be stored in granaries and in the form of livestock for the first time. This in turn enabled people

33

to accumulate significant food surpluses, which led to the development of the concept of ownership and new forms of social organization for the protection of property rights.

Like hunter-gatherers, early Neolithic communities consisted of family members, but instead of foraging for plants and hunting wild animals, everyone worked in farming. Technological efforts were therefore mainly directed toward agricultural needs, and the tools and skills needed for farming were very different from those of hunter-gatherers. Farmers needed axes to clear the land of trees, digging sticks and stone-bladed hoes to cultivate the soil, and sickles with sharp edges for harvesting. By definition, the tools of the Neolithic era were made out of stone. Although these tools were simple, the megaliths and stone monuments that remain from that time reveal that people were capable of building impressive structures even before the first major civilizations emerged. But because people devoted most of their time to cultivating the land to produce the food they needed, construction took many years. A large food surplus was required to feed full-time construction workers before large-scale hydraulic engineering projects and the construction of cities became feasible. In due course, improvements in agricultural productivity allowed for the production of more food and the expansion of cities—in which the jobs of artisans, smelters, smiths, and others became full-time occupations—where increasingly skilled workers specialized in the development of better technologies that allowed agricultural productivity to increase further.[2] This meant that larger populations could be supported, permitting additional specialization that gave rise to more technologically sophisticated civilizations.

The first great civilizations to appear were the Minoan civilization, which was destroyed by a volcano eruption on Crete, and the civilizations of Mesopotamia and Egypt. In these societies the bulk of the population were still farmers, who grew an abundance of beans, wheat, lentils, barley, onions, and so on. They also raised cows, pigs, sheep, donkeys, and goats. And most importantly, they produced a food surplus that enabled people to engage in activities other than farming. Some people were construction workers, artisans, merchants, or warriors. Others worked as servants to the ruling classes, whose members were the political, religious, and military leaders. With a growing share of the population outside the agricultural sector, inventive contrivances were no longer confined to agriculture. For the modern world, the most important enabling technology inherited from ancient civilizations is writing, which still allows us to store and transmit information across time and space. Among other significant inventions was the potter's wheel, which first appeared in Mesopotamia during the fifth millennium. Although wheeled carts and wagons, drawn by oxen, became increasingly common in Mesopotamia around 3000 B.C., their wheels were made of heavy planks, which could not be used in rocky terrain and often sank into soft soil. The wheel's impact on productivity at this time was therefore negligible. Long after the invention of the wheel, caravans of donkeys were still used for the transportation of goods.[3]

In terms of labor-saving technology, the most important achievements of ancient civilizations were probably in the discovery and exploitation of metals. Copper was first to be exploited, and there were a number of innovations in the techniques used to harden it, by adding tin to make bronze (thus initiating the Bronze Age, which lasted from around 4000 to 1500 B.C.) or zinc to make brass. Gold and other soft metals were discovered and became the basis for currencies. And iron eventually emerged (thus starting the Iron Age, from around 1500 to 500 B.C.) and was soon widely adopted, as ancient smiths found it to be a much stronger and harder metal. These developments, in turn, led to a range of other technological advances. Tools previously made of wood or stone could now be made with more durable but also malleable metal. And entirely new tools could be made—such as saws, scythes, picks, and shovels—that would have been unthinkable without advances in metallurgy.[4] Though there were no machines to relieve workers of their burdens, even the simplest tools could save a considerable amount of labor: "A man with a spade will do as much work as twenty men who only have their nails to scratch the ground with."[5] All the same, while these tools clearly helped people, advances in metallurgy also led to some undesired disruptions. Warriors with steel weapons were able to conquer civilizations whose weapons were made of stone and wood. The old civilizations of Eurasia had lasted for millennia, in part because their elites had little to gain and plenty to lose from new technologies that could threaten their leadership. Their status came to be challenged only after the invention of iron and the domestication of the horse. The nomadic warriors that disrupted Mesopotamia were the first to use iron weapons. Thus, at the height of the Roman Empire, Pliny the Elder describes iron as

> the most precious and at the same time the worst metal for mankind. By its help we cleave the earth, establish tree-nurseries, fell trees, remove the useless parts from vines and force them to rejuvenate annually, build houses, hew stone and so forth. But this metal serves also for war, murder and robbery; and not only at close quarters, man to man, but also by projection and flight; for it can be hurled either by ballistic machines, or by the strength of human arms or even in the form of arrows. And this I hold to be the most blameworthy product of the human mind.[6]

Much like the fears surrounding the disruptive effects of artificial intelligence (AI) today, with scholars like Stephen Hawking and Nick Bostrom suggesting that it could spell the end of human civilization, people in preindustrial times worried that technology could destroy their much smaller and more isolated world. This was not just a worry of Pliny the Elder but the intuition that shaped attitudes toward technological progress among elites throughout classical antiquity (around 500 B.C. to A.D. 500). For political leaders concerned with conserving their power, technology was not always welcome.

Oppressed by Tradition

While most early scholars have argued that classical civilizations did not achieve much technological progress, such accounts are now seen as understating the breakthroughs of classical times.[7] This perception was largely due to the fact that new technologies were rarely developed with economic objectives. As the classical scholar Moses Finley has argued, our view of technological progress in antiquity often involves imposing our own value systems on civilizations that had little interest in industrial pursuits.[8] Because the chief function of technology since the Industrial Revolution has been to improve industrial processes, products, and services, we tend to think of technological progress in such terms. In contrast, technological advances in classical times typically served the public sector, rather than private interests. Instead of promoting technological development to increase productivity, leaders focused on advancing public works that helped them gain popularity and safeguarded their political power.[9] As documented by the historian Kyle Harper, "A proud inventory from the fourth century claimed that Rome had 28 libraries, 19 aqueducts, 2 circuses, 37 gates, 423 neighborhoods, 46,602 apartment blocks, 1,790 great houses, 290 granaries, 856 baths, 1,352 cisterns, 254 bakeries, 46 brothels, and 144 public latrines. It was, by any measure, an extraordinary place."[10]

In particular, classical civilizations are famous for their advances in civil and hydraulic engineering and architecture.[11] "The Rome of 100 A.D. had better paved streets, sewage disposal, water supply, and fire protection than the capitals of civilized Europe in 1800."[12] Water conduits to supply fresh water first emerged in early classical Greece and later spread to Rome.[13] Beginning with Appius Claudius in 312 B.C., the water system in Rome was gradually expanded, and by around A.D. 100, when the water superintendent Frontinus was writing, Roman homes were being supplied with running water. A central heating system was developed to serve public bathhouses, and the demand for bath buildings led to technological advances in heating methods, such as the hypocaust for heating floors.[14] An enabling technology for many of the grand structures of Rome was the discovery of cement masonry, which has been called the only great invention of the Romans.[15] That is certainly an exaggeration, but it is true that the Romans barely made any contributions to industrial development. This was not because they lacked the technological creativity or the technical skills. Roman rulers simply had no interest in industry. To paraphrase the historian Herbert Heaton, Roman leaders regarded war, politics, finance, and agriculture as the only activities to which they might put their hands.[16] Even the advances made in mechanics—including the development of cranes, pumps, and water-lifting devices—were largely a set of ancillary inventions to support construction and hydraulic engineering efforts. As far as we can tell, these devices did not have any meaningful impact on private-sector productivity. Spillovers to the private sectors of the economy were rare, although hydraulic engineering found some applications in irrigation and drainage. Labor-replacing inventions in agriculture were few: there is evidence of some harvesting machines, but they were last mentioned in the fifth century A.D., and their disappearance suggests that they failed to find widespread use.[17] In textile production, there were no noteworthy advances in mechanization. Spinning and weaving remained highly labor-intensive activities. Spinning was done using a spindle and whorl, which meant that it took some ten spinners in continuous employment to keep one loom supplied with yarn. Even the waterwheel—the most famous invention of the Roman Empire—probably had no significant impact on aggregate productivity. The waterwheel described by the Roman engineer Vitruvius during the first century B.C. was primarily used for flour milling by the fifth century A.D., and even in flour milling its use was seemingly limited.[18]

It is quite telling that most classical writers did not bother much with machinery. Vitruvius, who wrote extensively on technical matters, devoted only one of the ten books of his *De Architectura* to mechanical devices, and about half of that book was on military machines. The relative importance of military machinery speaks to the fact that technology in classical civilizations served as a tool to conserve and extend political power, rather than serving economic interests: even Roman roads and bridges were built mainly for military purposes.[19] Later perceptions of *De Architectura* also well summarize the important achievements of the time. While it came to have profound impacts on leading writers and architects of the Renaissance (including the likes of Filippo Brunelleschi, Leon Battista Alberti, and Niccolò de' Niccoli) its impacts on later developments in machinery were insignificant. The famous drawing of the Vitruvian man by Leonardo da Vinci—one of the great inventors of the Renaissance—was based on the concepts of proportion put forward by Vitruvius, as the name suggests. But da Vinci found the inspiration for his ideas of machines elsewhere.

The main mechanical achievements of classical civilizations were to understand some of the principles and features of machines. By applying mathematics to discover the law of lever, and the principles of hydrostatics, Archimedes (287–212 B.C.) laid the foundations for some of Galileo's later work, which would become essential to the development of more complex machines.[20] Moreover, *Mechanika* (commonly attributed to Aristotle, but presumably written by someone else) includes extensive discussions of the lever, wheel, wedge, and pulley, but the applications discussed suggest limited interest in their practical use. And other elements that can be found in classical literature—such as the gear, cam, and screw—were mostly applied to war machines.

In other words, classical civilizations witnessed a number of technological advances that had virtually no meaningful economic impact, the reason being that for inventions to improve material standards of living, they need to serve economic purposes and must be applied in production. To assert that this was not a period of technological creativity would therefore be severely misguided. In fact, classical times were an era of tremendous technological sophistication. Brilliant inventors, such as Hero of Alexandria, developed the first vending machine, the first steam turbine, and a wind wheel that operated an organ.[21] While these inventions were mere toys, they show the sparks of technological genius of classical times. In particular, the discovery of the Antikythera mechanism, an astronomical computing machine used to predict astronomical positions and eclipses, on a wreck near Crete in 1900, reveals the astounding technological creativity of Hellenism. The mechanism, which was built in the first century B.C., led Derek Price, who reconstructed it, to urge historians to "completely rethink our attitudes toward ancient Greek technology. Men who could have built this could have built almost any mechanical device they wanted to."[22]

The key question therefore is why so little of this technological creativity was translated into economic progress. Part of the answer probably lies in the fact that slavery provided disincentives for the introduction of worker-replacing technology. Although the historian Bertrand Gille has been critical of this thesis—arguing that science and technology flourished in the ancient world—the abundance of slaves might still explain why few technological insights were applied to production.[23] In addition, the persistence of slavery meant that a large share of the population in classical civilizations was not free to pursue industrial activities. A related explanation put forward by John Bernal, a scientist and historian, suggests that the reason why the classical period failed to produce the machines of the Industrial Revolution was lack of economic incentive. The wealthy, he argues, could afford handmade items, and slaves could not afford to buy anything that wasn't a necessity.[24]

Furthermore, technological advances were blocked at times. For example, Pliny the Elder tells a story from the reign of the Emperor Tiberius, when a man had invented unbreakable glass. Instead of rewarding the inventor for his creativity, Tiberius had the man executed, fearing the possibility of angry workmen rebelling. More direct evidence of the government's seeking to control technological progress is provided by Suetonius, who describes how Emperor Vespasian, who ruled in 69–79 A.D., reacted to the introduction of worker-replacing technology. When Vespasian was approached by a man who had invented a device for transporting columns to the Capitoline Hill, Vespasian refused to use the technology, declaring: "How will it be possible for me to feed the populace?"[25] Because columns were large and heavy, transporting them from the mines to Rome required thousands of workers. Even though this was a huge expense to the government, the concern that depriving Romans of work might be politically destabilizing made conserving jobs by maintaining the technological status quo the more politically appealing option. Transporting columns provided workers with livelihoods, kept them busy, and thereby minimized chances of social unrest.[26]

What is beyond dispute is that there was little cultural and political interest in driving industrial development. As the economic historian Abbott Usher has argued, classical civilizations were "oppressed by tradition" and therefore showed little interest in new technology.[27] While classical civilizations were clearly technologically creative, there were few incentives for them to invent anything for industrial purposes in general and labor-replacing technologies in particular. However, the absence of such innovation does not imply economic backwardness. Growth derived from things for which the Greeks and Romans are famous, including organization, trade, order, and law. Such institutions can take an economy a long way, and they surely did. As the economist Peter Temin has documented, the Roman Empire had a market economy. Pax Romana stimulated Mediterranean trade, and living conditions were certainly better than in most places before the Industrial Revolution.[28] But this was primarily growth based on trade. And when the political foundations upon which this growth was built were undermined, as was the case after the collapse of the Roman Empire, living standards rapidly deteriorated.[29]

Light in the Dark Ages

Ironically, technological progress increasingly came to serve an economic purpose during the Middle Ages, when government control of it diminished and technological efforts shifted from the public to the private sector. To many historians the fall of the Roman Empire marks the end of the ancient world and the onset of the Middle Ages, the early part of which is sometimes still referred to as the Dark Ages. During the early Middle Ages (A.D. 500–1100), the economic and cultural environment in Europe was more primitive than it had been in classical civilizations: literacy declined, law enforcement diminished, violence became more frequent, commerce deteriorated, and the roads and aqueducts of Rome fell into disrepair. The collapse of the Roman Empire was accompanied by the rise of the feudal order, with the crown at the top, the nobility beneath, and the peasantry at the bottom. Relative to the Roman Empire, the crown was weak, as the feudal order meant that political power was split among highly decentralized lords, who maintained their own armies. The lords allocated their land to peasants, often referred to as serfs, who had to perform extensive unpaid labor but, unlike slaves, were allowed to retain some of the product of their labor. Like slaves, however, serfs were subject to many restrictions, including being unable to leave the estate without the permission of the lord and being unable to litigate in courts presided over by nobles. Under this system, incentives to work hard and innovate were probably very low. Yet this period "managed to break through a number of technological barriers that had held the Romans back."[30] To be sure, medieval Europe did not have anything like the extravagant structures of the Roman Empire, but there was no need for expensive roads and bridges, as there were no great armies to maintain and use.[31] Instead, medieval technological efforts increasingly targeted economic problems, although these were mostly modest by modern standards. Unlike "the amusing toys of Alexandria's engineers or the war engines of Archimedes," medieval technology reduced daily toil.[32]

In particular, the increased willingness to imitate and adopt technologies from foreign lands was an early sign of a more technologically progressive society. Europe in early medieval times was by no means at the forefront of technology, but it was gradually catching up.[33] In the Middle Ages, improvements in agricultural technology were particularly important. Because most workers were still engaged in farming, agricultural inventions had the greatest impact on aggregate productivity, although the prevalence of serfdom held back technological development. The transformation of agriculture was a gradual process that would continue for many centuries, but eventually it shaped the world of work in Europe.

The drivers of this transformation were the introduction of the heavy plow and the establishment of the three-field system.[34] The heavy plow was an enabling technology: with it, huge tracts of land that could not be cultivated in Roman times could now be used for farming. But besides expanding the pool of farmland, the heavy plow also boosted productivity. As the medieval historian Lynn White writes, it constituted an "agricultural engine which substituted animal power for human energy and time."[35] But like most inventions, it came with new challenges—in this case, because several oxen were required to pull it.[36] The growing dependency on animate power in farming meant that peasants needed to find better and cheaper ways of feeding their animals. Part of the solution was found in the new three-field system that gradually spread across Europe, allowing the animals to graze on the land while fertilizing the soil. The productivity gains from this system were substantial. Compared to the two-field rotation system, it is estimated to have increased productivity by up to 50 percent, though it did so mainly by saving capital.[37] Furthermore, it greatly stepped up the production of certain crops, such as oats, that were particularly suitable for feeding horses, thereby increasing the quantity and quality of surplus food needed for horse technology. By the end of the Middle Ages, there appears to be a tight correlation between the adoption of the three-field system and the use of horses in agriculture.[38]

Following a series of ancillary inventions, horse technology greatly improved throughout the Middle Ages. The invention of the nailed horseshoe, for example, enabled the more widespread use of horses for commercial transportation, and the shoe's protection of the hooves from moist soil allowed for greater adoption of horsepower in agriculture. Another important improvement was the invention of the stirrup. Although its purpose was largely military, making it possible for a knight to fight on horseback, it equally benefited civilian riders in terms of stability and comfort. But in terms of economic impact, it was probably the arrival of the modern horse collar that made the largest contribution—though its true significance was not recognized until the beginning of the twentieth century, when it was documented by Richard Lefebvre des Noëttes, a retired French cavalry officer. Comparing the use of horses in antiquity and the Middle Ages, he found that the throat-girth harness used by the Greeks and the Romans, with two straps around the belly and neck of the horse, meant that the horse lost about 80 percent of its efficiency.[39]

The importance of these technological advances cannot be overstated, as 70 percent of all energy in Britain still came from animals in the eleventh century, with the remainder coming from water mills. But even though horses were increasingly employed in agriculture, the effects of horse technology on productivity are not entirely clear, as oxen were often used as well. What seems clear is that the switch to horse technology, when it occurred, was associated with substantial gains in productivity.[40] Modern experiments show that although the horse and ox perform similarly in terms of pull, the horse moves much faster, allowing it to produce 50 percent more foot-pounds per second, while being able to work up to two hours more per day. The impact on productivity in transportation is likely to have been just as significant, as horse technology helped facilitate Smithian growth by giving a boost to land transport and trade. With the new harness and nailed shoes for the horse, the cost of grain is estimated to have increased only 30 percent for each hundred miles of overland carriage in the thirteenth century, which is more than three times better than in Roman times.[41]

Much progress was also made using power from wind and water to replace animate power. During the Middle Ages—especially between the seventh and the tenth centuries—larger and better waterwheels spread throughout Europe and found applications in a growing number of industries. *The Domesday Book* of 1086, completed by order of William the Conqueror, lists 5,624 water mills for some three thousand British communities, or roughly two mills for every one hundred households.[42] These were used to drive fulling mills, breweries, sawmills, bellows, hemp treatment mills, cutlery grinders, and so on. Though *The Domesday Book* does not allow us to estimate average horsepower for these mills, their long and widespread use underlines their economic importance: water mills would remain a prime source of energy in Britain even throughout the Industrial Revolution.[43] Their arrival thus meant lasting progress relative to its predecessor civilizations. The late Middle Ages has indeed been described as a "medieval industrial revolution based on water and wind."[44]

While wind power had previously been used in sailing, windmills had been unknown to classical civilizations and were not invented until the time of the Norman Conquest (1066): the first windmills with credible documentation date back to 1185. The economic significance of these windmills is suggested by the associated disputes that emerged. A wealthy cleric named Burchard complained directly to Pope Celestine III that one knight had refused to pay tithes (one tenth of a person's annual earnings, taken as a tax to support the church and clergy) on the income from his windmill. Even though windmill owners argued that they were dealing with new circumstances that were not covered by the existing regulations, it took the pope only until 1195 to impose tithes on them.[45]

Overall, medieval Europe was clearly capable of achieving higher levels of productivity than predecessor civilizations, both in manufacturing and agriculture. Yet some of its most revolutionary technologies—like the mechanical clock and the printing press—had little impact on economic activity at the time. The weight-driven clock, which had emerged by the end of the thirteenth century, became of economic importance only after 1500. In the Middle Ages, domestic clocks were rare. They were nice toys for the rich and helpful instruments to scientists. Usher writes: "By 1500, few towns were without some tower clock, but domestic clocks, though widely diffused among the wealthy, were not common in Europe as a whole until a later period. Later writers imply that clockmaking was so highly developed in Nuremberg in the fifteenth century that domestic clocks came into more general use in central and southern Germany than elsewhere in Europe. These German clocks of the fifteenth century were among the first made to indicate minutes and seconds, and some use was made of them by astronomers."[46]

Public clocks, however, are a different story. The tower clocks of late medieval towns were built mainly for status and reputation rather than for economic reasons. They were financed by wealthy noblemen who wanted to showcase the progressiveness of their towns. But they had unintended economic consequences. The economic historians Lars Boerner and Battista Severgnini have shown that early adopters—cities that had a tower clock before 1450—grew faster between 1500 and 1700 than those that didn't.[47] Over the long run, the clocks' contribution to economic growth was significant, but their impact was delayed:

Building a clock in a town was motivated by prestige and not by economic needs—towns did not forecast any of the benefits clocks would bring in the long run, or what can be seen ex post as an economically efficient application. Consequently, the economic use of clocks was a slow process of adoption. Whereas the use of clocks for coordination activities, such as market times or administrative town meetings, can already be observed during the 14th and 15th centuries, the use of clocks to monitor and coordinate labour processes evolved only slowly, in particular during the 16th century. Finally, a cultural adoption reflected in the daily cultural and philosophical thinking of the time can be observed from the middle of the 16th century, for instance with the Protestant movement (in particular with John Calvin's propagation of the concept of "scarce time"). The 17th century also brought forth scientists and philosophers such as Robert Boyle and Thomas Hobbes, who used the clock as a metaphor for the functioning of the world and to explain how institutions such as the state should work. Looking at this slow process it is not surprising that it took some time before the complementary organisational, procedural, and cultural behavioural innovations transformed into economic growth.[48]

Many historians have pointed to the significance of accurate time measurement to economic progress. The French historian Jacques Le Goff has called the birth of the public clock a turning point in Western society.[49] And the historian Lewis Mumford has gone so far as to suggest that not the steam engine but the mechanical clock was the machine that made the industrial age.[50] While this might seem exaggerated, there can be no doubt that the clock changed Western life in general, and the pace of work in particular. New cultural attitudes about punctuality already began to emerge in the late Middle Ages. Of course, the practice of dividing the day into measurable time units existed before the clock, but the length of the hour was not fixed. It depended on the length of the day, meaning that it varied significantly between summer and winter. Thus, people still tended to follow the position of the sun for time guidance. And while medieval people had sun or water clocks, these did not play any meaningful role in business. Markets opened at sunrise and closed down at noon when the sun was at its zenith. It was only after the spread of public clocks that market times were set by the stroke of the hour. Public clocks thus greatly contributed to public life and work by providing a new concept of time that was easy for everyone to understand. This, in turn, helped facilitate trade and commerce. Interactions and transactions between consumers, retailers, and wholesalers, became less sporadic. Important town meetings began to follow the pace of the clock, allowing people to better plan their time and allocate resources in a more efficient manner.[51]

In industry, clocks grew in importance much later with the birth of the factory system in the eighteenth century (see chapter 4). Though the role of clock makers in designing the textile machines that powered the early Industrial Revolution has probably been overstated, there can be no doubt that the mechanical clock was a key enabling technology for the factory system, with its fixed working hours. The coordination of factory work rested on regularity, routine, and accurate time measurement. And many later advances in steam engines and other machinery required the precision lathes and measuring tools that were developed during the Renaissance to produce scientific and navigational instruments. The close connection of clock and watch making with the instrument-making sector facilitated much of the progress that took place around 1800. Karl Marx and Max Weber were right in thinking that clocks had an enormous impact on the evolution of capitalism.[52]

The first metal movable-type printing press, invented by Johannes Gutenberg in 1453, was another landmark achievement of late medieval times, whose main contributions to productivity came much later. Instead of creating one immensely complicated stamp for each page to be printed, Gutenberg made metal stamps for individual letters and symbols, which were set in the desired sequence. The virtue of Gutenberg's invention is evident from changes in the price of books, which soon fell by two-thirds—making them accessible to a growing share of the populace.[53] Yet what the historian of technology Donald Cardwell has called "the first revolution in information technology" cannot be attributed to Gutenberg alone.[54] The printing press was made economically feasible by a number of enabling technologies, including paper (which was introduced from China), cheap printing ink, the press (most likely adopted from ancient winepresses), and the Roman alphabet (which had become universal in Europe and was particularly suitable for printing, with its twenty-six letters). It is nonetheless undisputable that Gutenberg's invention was one of the most important in human history. Toward the end of the century, there were over 380 printing presses in Europe, producing a tsunami of books. More books were published in the fifty years following Gutenberg's invention than in the millennia before.[55]

Economic historians like Gregory Clark have concluded that the effects of the printing press on economic growth at the macroeconomic level were "unmeasurably small."[56] But even though the industry of printing did not appear in aggregate statistics, we know from recent work by the economist Jeremiah Dittmar that the printing press became a motor of urban growth in the sixteenth century.[57] In cities where the printing press was adopted, the spread of business textbooks allowed people to better transmit commercial know-how, including how to make currency conversions, determine interest payments, and calculate profit shares—which in turn helped foster the spread of valuable commercial skills. In the words of Gaspar Nicolas, author of the first Portuguese arithmetic textbook, published in 1519: "I am printing this arithmetic because it is a thing so necessary in Portugal for transactions with the merchants of India, Persia, Ethiopia, and other places."[58]

We all know that the printing press also facilitated the spread of science. But as we shall see, science didn't become a pillar of technological progress until the nineteenth century. In the 1500s, as Dittmar writes, "the role of print media in the diffusion of industrial innovations was probably more limited."[59] It was primarily a facilitator of trade. Where commerce was flourishing, the virtues of printing were the greatest. That the movable printing press was a force of Smithian growth is underlined by the fact that cities with access to waterborne transport were best positioned to profit from it. Indeed, the work of Dittmar shows that the printing press delivered special benefits to port cities, as they gained disproportionately from innovations in commercial practice. More broadly, early adopters of the printing press saw face-to-face interactions become more important, as printing for the first time brought mechanics, scholars, merchants, and craftsmen together in a commercial setting. Bookshops became meeting places for intellectuals. And cities that adopted the new printing technology also attracted paper mills, illuminators, and translators. Like the computer revolution, to which we shall return in chapter 10, the first revolution in information technology did not spell the death of distance. Just like computing, printing made the tyranny of geography all the more apparent, prompting people to cluster together and increasing urbanization. Thus, like the computer revolution, the revolution in printing, if anything, made the world less flat.

Though the printing industry itself was too small to drive aggregate growth, there can be no doubt that printing experienced a Schumpeterian transformation, as scribes, who made copies of manuscripts before the invention of printing, found their skills being rendered redundant. So why was the printing press adopted so enthusiastically in the West when people who face displacement usually oppose new technologies? In 1397, for example, when tailors protested, the city of Cologne banned the use of machines that automatically pressed pinheads. And in 1412, in response to resistance by silk spinners' guild to the adoption of a silk-twisting mill, the city declared that "many persons who earn their bread in the guild in this town would fall into poverty, for which reason the town council agreed that neither this mill nor in general any similar mill shall be made or erected, either now or in future."[60]

Why, then, didn't the scribes oppose the printing press in the same way? One reason might be that printing with movable type was a largely unregulated infant industry. As the historian Stephan Füssel has pointed out, in the early days of the industry, people in most cities were free to invent without restrictions imposed by the guilds or government regulations.[61] As we shall see, in places and industries where the guilds got stronger, they often tried to restrict replacing technologies, and printing was no exception: in the sixteenth century, the scribes' guild of Paris triggered a revolt against labor-replacing printing technology.

To be sure, everyone was not content with the advent of Gutenberg's invention in the fifteenth century. We know of some episodes of labor unrest accompanying its adoption, like the protests of professional writers in Genoa in 1472, the opposition from the card makers of Augsburg in 1473, and uprisings among the stationers of Lyons in 1477. But on the whole, the rapid diffusion of the printing press suggests that resistance was weaker than one might have expected. In an article titled "Why Were There No Riots of the Scribes?," Uwe Neddermeyer argues that the reason is simple: for the most part, the scribes benefited from the arrival of the printing press. The majority of handwritten manuscripts had been produced by people writing books for themselves without commercial motives. Few scribes and religious communities made their living by reproducing books. Thus, for most people affected, the printing press did not mean any loss of income. And for those who did see their incomes disappear, there were mostly good alternative options: "Many professional scribes continued to be in a position to earn their living from writing documents, inventories, letters, minutes, etc.— i.e. texts that were uneconomical to be reproduced by printing."[62] Perhaps more importantly, the printing press, which created an ever-growing demand for books, also created new jobs from which many scribes themselves benefited.

Contemporaries did not fail to take notice. In *Expositiones in Summulas Petri Hispani*, published around 1490 in Lyons, the editor Johann Treschel writes that as the new art of printing ends the careers of scribes, "they have to do the binding of books now."[63] Indeed, during the closing decades of the fifteenth century, many monasteries whose scriptoria had long churned out new books shifted their focus to cover design and bindings. Some even set up their own printing presses. Hence, some scribes even celebrated the new art of printing, which relieved them of tedious writing and allowed them to specialize in the design and binding of books. As Neddermeyer writes, if "asked whether they approved of the new craft, most scribes in the era of Gutenberg would have replied with a definite *Yes*."[64] As will be discussed in chapter 8, one reason resistance to labor-replacing technologies was so feeble in the twentieth century was that workers for the most part had good alternative job options, much thanks to the steady expansion of manufacturing operations. But clearly, that was not always the case.

All the same, on balance, the technical advances of the Middle Ages probably did more to foster trade than they did to save labor. In particular, advances in shipbuilding and navigation—including the three-master ship, the development of the movable rudder to replace the steering oar, and the invention of the mariner's compass—constituted enabling technologies for the age of discovery and the surge in international trade associated with it. What's more, so-called caravel construction culminated in the Portuguese caravel ship in the fifteenth century, the type of ship that was used by Vasco da Gama, Christopher Columbus, and Ferdinand Magellan, to discover new trade routes. By that time, Europe had gone some way toward catching up with previously more advanced Islamic and Oriental civilizations. And while Europe was still an imitator of foreign technologies, despite some sparks of technological brilliance, it was soon to turn from imitator to innovator.[65]

Inspiration without Perspiration

Between 1500 and 1700, the technological gap between the West and the rest widened. Europe was no longer a technological backwater. It was expanding the frontiers of technology long before the Industrial Revolution. The bridge between the Middle Ages and the industrial era was built by the Renaissance, which started in medieval Italy and gradually spread across Europe. Although it began as a cultural movement, it was equally a force of profound technological change. Still, as we shall see, hardly any of its key inventions served to replace workers. And when those inventions did, they were fiercely opposed.

The technological advances of the Renaissance owed much to one of the late medieval inventions: Gutenberg's printing press. For the first time, a vast technical literature emerged, containing detailed descriptions of dams, pumps, conduits, and tunnels and making technical knowledge more communicable and cumulative. This literature clearly shows that the practical relevance of machines was well understood among some of the Renaissance's leading figures. Leonardo da Vinci— who among other things was responsible for many inventions—refers to mechanics as "the paradise of the mathematical sciences, because it is in mechanics that the latter find their realization."[66] But the gap between best practice and the machines that were adopted and put into widespread use was large, and few inventions recorded in the stream of technical writings therefore had any significant impact on economic growth. For example, in *De Re Metallica*, Georg Bauer wrote extensively on various mining machines, and Vittorio Zonca describes an astoundingly sophisticated silk-throwing machine—which almost a century later inspired John Lombe to travel to Italy to discover the precious secret. However, like most machines described in the technical literature, they did not become standard equipment in Renaissance Europe. Similarly, while working for the British Royal Navy, the Dutch engineer Cornelis Drebbel built the first navigable submarine and demonstrated it to King James I in 1624, more than two centuries before the technology would be put into use. But although it was tested several times in the Thames, the vessel didn't generate sufficient enthusiasm for the idea to be further developed.[67]

The suggestion of Thomas Edison that invention is 1 percent inspiration and 99 percent perspiration was evidently not true of Renaissance Europe. It was rather the other way around. Few ideas and drawings were ever translated into prototypes. Indeed, the Renaissance is best described as an age of novel technical ideas and plenty of imagination, but little realization. As Joel Mokyr points out, "If inventions were dated according to the first time they occurred to anyone, rather than the first time they were actually constructed, this period may indeed be regarded just as creative as the Industrial Revolution. But the paddlewheel boats, calculating machines, parachutes, fountain pens, steam-operated wheels, power looms, and ball bearings envisaged in this age—interesting as they are to the historian of ideas—had no economic impact because they could not be made practical."[68]

The best that can be said about Renaissance technology in economic terms is that it paved the way for one of humanity's most important technological breakthroughs to date: the steam engine. The science of the steam engine started with Galileo and his secretary Evangelista Torricelli, who developed the first barometer. In 1648, Torricelli discovered that the atmosphere has weight. A number of subsequent experiments by Otto von Guericke in 1655 showed that the weight of air can be used to do work: von Guericke found that if air is pumped out of a cylinder, this pushes the piston down into it, allowing it to lift a load of weights. Denis Papin discovered that filling a cylinder with steam and then condensing it could achieve the same effect, and he built the first, albeit very simple, steam engine in 1675. This series of discoveries eventually culminated in Thomas Newcomen's steam engine, whose design built on the insight that the atmosphere has weight. No single discovery of the Renaissance would be more important to later industrial development, but it was by no means the only scientific achievement with applications in industry.[69]

From the viewpoint of the history of machines, Galileo's theory of mechanics was another landmark achievement. During antiquity, Archimedes had made some progress in describing the principles of the lever, but he had not considered more complex machines in motion. Galileo's theory of mechanics, in contrast, showed that all machines—systems of pulleys, gears, and so on—have the common function of applying force as efficiently as possible. Before Galileo, each machine had a unique description, as the general laws governing all machines had not yet been recognized. The significance of this shift is pointed out by Franz Reuleaux, the father of kinematics, who suggests that "in earlier times men considered every machine as a separate whole, consisting of parts peculiar to it; they missed entirely, or saw but seldom, the separate groups of parts we call mechanisms. A mill was a mill, a stamp a stamp and nothing else, and thus we find the older books describing each machine separately, from the beginning to the end."[70] What's more, before the theory of mechanics, machines could be evaluated only qualitatively; after, they could be evaluated quantitatively. What makes Galileo's theory of mechanics particularly interesting from an economic point of view is that it aims at efficiency. The function of a machine is to deploy and use the powers that nature makes available—such as water, wind, and animal power—to do a certain amount of work in the most efficient way.[71] But at the time, this intuition was rarely put into practice. The frequent confusion between mechanics and magic suggests that the principle of using the powers of nature to perform mundane tasks was typically not well understood. Machines were widely regarded as devices for cheating nature and the machine maker as a person with the power of the magician. The legend of the mechanic magician would last for a long time—for example, it was perpetuated in the character of the inventor Spallanzani in Jacques Offenbach's opera *The Tales of Hoffman*.[72]

In terms of productivity-enhancing technological improvements, the Renaissance was largely a continuation of the Middle Ages in that most technologies seemingly saved more capital than labor. Some progress was made in mining, including the introduction of underground rail transport and a variety of pumping devices.[73] Mining was probably the industry that benefited the most directly from scientists and science. Galileo as well as Isaac Newton were concerned with many mining engineering problems, ranging from air circulation to the raising of coal, but their insights did nothing to reduce the number of workers required in the mines. All the same, agriculture still constituted the largest sector of the economy, and improvements in farming techniques had the largest impact on aggregate productivity. The most important agricultural invention was the new husbandry—including the introduction of stalls for feeding cattle, new crops, and the elimination of fallowing—which allowed farmers to maintain more cattle and produce animal products in larger quantities. But few inventions served to reduce the number of workers in farming. The new iron plows, for example, reduced the number of animals required for plowing, and thus probably saved more capital than labor. Other agricultural inventions, such as the modern seed drill—whose invention is commonly attributed to Jethro Tull around 1700—similarly saved more capital by improving the use of farmland in terms of ensuring a more even distribution of seeds.[74]

When worker-replacing technologies emerged, as they did in the textile industry, they were typically subject to opposition and were frequently blocked by political authorities. The gig mill, for example, which is estimated to have allowed one man and two boys to do the work of eighteen men and six boys, was prohibited in Britain by a statute of 1551, although almost a century later, King Charles I issued another proclamation against them, which suggests that some of the mills were still in use and that penalties for employing them were being avoided.[75] The landmark labor-replacing invention of the time—the stocking-frame knitting machine, invented by the clergyman William Lee in 1589—faced considerable opposition, too. Queen Elizabeth I refused to grant Lee a patent, claiming: "Thou aimest high, Master Lee. Consider thou what the invention could do to my poor subjects. It would assuredly bring to them ruin by depriving them of employment, thus making them beggars."[76] The queen's decision reflected the hosiers' guild's opposition to the new technology: the hosiers feared that their skills would be rendered redundant. The guild's opposition to Lee's invention was so intense that he had to leave the country.

There is no shortage of examples of resistance to worker-replacing technologies. Beyond the textile industry, the Privy Council commanded the abandonment of a needle-making machine in 1623 and ordered the destruction of any needles made with it. Similarly, nine years later, Charles I banned the casting of buckets, suggesting that it might ruin the livelihoods of the craftsmen that were still making buckets the traditional way.[77] Elsewhere in Europe opposition was just as fierce. Many cities across Europe issued edicts against automatic looms during the seventeenth century, and the city of Leiden experienced riots in 1620 because of their use.[78] In Germany, automatic looms were prohibited entirely between 1685 and 1726. And as is well known, in 1705, Papin's steam digester was smashed by angry Fulda boatmen:

> At that time, river traffic on the Fulda and Weser was the monopoly of a guild of boatmen. Papin must have sensed that there might be trouble. His friend and mentor, the famous German physicist Gottfried Leibniz, wrote to the Elector of Kassel, the head of state, petitioning that Papin should be allowed to "pass unmolested" through Kassel. Yet Leibniz's petition was rebuffed and he received the curt answer that "the Electoral Councillors have found serious obstacles in the way of granting the above petition, and, without giving their reasons, have directed me to inform you of their decision, and that in consequence the request is not granted by his Electoral Highness." Undeterred, Papin decided to make the journey anyway. When his steamer arrived at Münden, the boatmen's guild first tried to get a local judge to impound the ship, but was unsuccessful. The boatmen then set upon Papin's boat and smashed it and the steam engine to pieces. Papin died a pauper and was buried in an unmarked grave.[79]

Craft guilds, like that of the boatmen of Fulda, controlled apprenticeship and production across cities and townships in preindustrial

Europe. In London in the mid-sixteenth century, for example, roughly 75 percent of workers belonged to a guild.[80] According to Sheilagh Ogilvie, an economic historian, "During the eight centuries before European industrialization, guilds were central institutions setting the rules of the game for economic activity."[81] And they blocked the introduction of replacing technologies, sometimes legally and sometimes violently, to safeguard their skills and self-interest. Indeed, while economic historians disagree about the attitudes of the guilds toward new technologies, there is an emerging consensus that their attitudes depended on how the technologies affected their skills. Guilds did not seek to slow down the march of technology in general, but they did forcefully resist it when it threatened their members' jobs.[82] They quietly accepted the new technologies they benefited from but bitterly fought against those that might affect them adversely, though there were instances when opposition failed. For example, the economic historian Stephen Epstein has argued that technologies that merely saved capital or made workers' skills more valuable were not frowned upon, while replacing technologies were more likely to be resisted.[83] But in practice, Epstein points out, the reaction of individual guilds was often the outcome of political rather than market forces: "There was a fundamental difference in outlook between the poorer craftsmen, who had low capital investments and drew their main source of livelihood from their skills, and who therefore (frequently in alliance with the journeymen) opposed capital-intensive and labor-saving innovations, and the wealthier artisans who looked on such changes more favourably."[84]

Pathbreaking work by Ogilvie that traces craft guilds and their activities over the centuries also shows that there were circumstances when the political economy of technological change was such that a new technology was adopted, even if it meant that some craftsmen lost out. At times, if a more powerful branch of a guild stood to benefit from the technology, it was adopted at the expense of the weaker faction. Sometimes craft guilds were overruled by powerful merchants. And there were cases when the political authorities granted a privilege to an inventor for economic gain—either because they would receive a direct payment for the benefits conferred or because they expected a share of the

profits. But for the most part, guilds vehemently and successfully resisted technologies that they perceived threatened their skills and rents.
Ogilvie explains:

> Guilds blocked horse-driven machines where they took work from
> guild masters, as in Cologne where horse-powered twisting-wheels
> were banned in 1498 because they threatened masters of the linen-
> twisters' guild. The multi-shuttle ribbon frame was successfully
> banned by most guilds in early modern Europe, but spread in the
> Northern Netherlands after 1604 thanks to vigorous support by fac-
> tions inside Dutch ribbon-weavers' guilds. It also spread in London
> after 1616 thanks to its adoption by a minority of politically con-
> nected liverymen inside the Weavers' Company before the hostile
> guild yeomen could mobilize resistance.[85]

Opposition to innovation is the most salient feature of how guilds
interacted with disruptive new processes and products. Pre-modern
people often complained that guilds blocked innovations. Guilds
themselves openly conducted lobbying campaigns to prevent guild
members and outsiders from producing things in new ways. Municipal, princely, seigneurial, and imperial governments were constantly
considering guild petitions against innovations, and often passed
legislation to deal with the issue.[86]

In a detailed study of English patents and legal cases in Elizabethan and
Jacobean times, the legal scholar Chris Dent found that "the legal decisions of the period confirm that the maximization of employment was
a priority of the elites."[87] Attitudes toward replacing technologies during
this period much resembled those of classical antiquity in that the
technologies were opposed by the political elites to avoid social unrest.
Relative to medieval times, the rise of strong nation–states between the
fifteenth and the seventeenth centuries also meant that governments
again had greater influence over technological progress. With the feudal
order of the Middle Ages, power was split among highly decentralized
feudal lords, who maintained their own armies. The territory of the
crown was thus merely a patchwork of scattered and largely independent domains, with no central administration. Over time, however,

growing competition between monarchs required more resources to be mobilized for warfare, and more centralized structures to achieve the pooling of those resources.[88] The military historian Quincy Wright estimates that fifteenth-century Europe consisted of some five thousand political units, but by the time of the Thirty Years' War (1618–48), these had been consolidated into some five hundred.[89] The emergence of infantry armies meant that the landed nobility lost ground as an efficient provider of military protection, and feudal oligarchies were replaced by centralized monarchies. According to the political scientist Charles Tilly, "war made the state and the state made war."[90] Between 1500 and 1800, Spain was at war with an enemy 81 percent of the time, while Britain and France were at war more than 50 percent of the time.[91] This, in turn, also spurred efforts to innovate. Indeed, as the economic historians Nathan Rosenberg and L. E. Birdzell Jr. argue, "In the West, the individual centers of competing political power had a great deal to gain from introducing technological changes that promised commercial or industrial advantage . . . and much to lose from allowing others to introduce them first. Once it was clear that one or another of these competing centers would always let the genie out of the bottle, the possibility of aligning political power with the economic status quo and against technological change more or less disappeared from the Western mind."[92]

The realization that political power was becoming harder to align with technological conservatism is suggested by the fact that governments began to subsidize engineers, grant patents to inventors, and created monopolies for key commercial interests. Famous examples of government-driven technological catch-up include Tsar Peter the Great's determination to modernize Russia, leading him to work at a Dutch shipyard under the pseudonym Pyotr Mikhailov, to learn about shipbuilding. But while governments clearly felt the need to promote technical progress, they were selective in the technologies they promoted, and as we have seen, they did their best to restrict the adoption of replacing technologies. Thus, overall, the technologies of the Renaissance were levers of Smithian growth rather than engines of Schumpeterian growth. Navigation, for example, became critical to the international trade that Europeans powers now engaged in. And for this, astronomical instruments and compasses constituted enabling technologies. Indeed, the Renaissance period has aptly been described as the "age of instruments" in technological terms. The telescope, barometer, microscope, and thermometer were among the prime technical achievements of the time, and they were adopted for a variety of purposes. While the telescope was used by Galileo to observe the moons of Jupiter, Prince Maurice of Nassau used it to look at the Spanish armies, while his captains employed it to spot enemy warships at sea. Even when trade and warfare were not among the intended applications of the inventions, they came to serve such purposes.[93]

The age of instruments came with important spillover effects, as the shops of instrument makers became meeting places for scientists, craftsmen, and amateurs and played a vital role in the dissemination of new ideas, facilitating the interaction between science and technology. As Cardwell has pointed out, "It is enough to record that by 1700 the foundations of modern technology had been laid. Appropriately, the word *technology* had been coined before the end of the century; and it seems likely that the word *inventor* was beginning to be used in the way in which it is understood today."[94] This, however, makes it even harder to explain why the Industrial Revolution did not arrive earlier.

2

PREINDUSTRIAL PROSPERITY

The frontiers of technology in Europe had expanded significantly by the eighteenth century. Yet the impact of this expansion on growth and prosperity remains controversial. Gregory Clark has gone as far as to suggesting that "had consumers in 8000 B.C. had access to more plentiful food, including meat, and more floor space, they could easily have enjoyed a lifestyle that English workers in 1800 would have preferred to their own."[1]

For anyone familiar with Jane Austen's writings about the eighteenth-century British upper classes, there can be no doubt that some enjoyed living standards far superior to those of hunter-gatherers. In Austen's *Sense and Sensibility*, first published in 1811, Colonel Brandon refers to a rectory that provided an annual income of £300: "This little rectory can do no more than make Mr. Ferrars comfortable as a bachelor; it cannot enable him to marry."[2] The average farm laborer at the time had an annual income of about a tenth of Mr. Ferrars, which still did not allow him to find a wife. To further put Mr. Ferrars's income in perspective, in the same year that *Sense and Sensibility* appeared, William Spencer Cavendish, marquess of Hartington and heir to the fifth duke of Devonshire, came of age. The sixth duke's inheritance included four country houses: Chatsworth House and Hardwick Hall in Derbyshire, Bolton Abbey in Yorkshire, and Lismore Castle in southern Ireland. He also had three London palaces to reside in: Chiswick House, Burlington House, and Devonshire House. And his estates were supported by land in Ireland and eight English counties, yielding an annual income of £70,000.[3] The extreme income disparities suggested by such anecdotal evidence are also borne out by the statistics. The top 5 percent of British income earners in 1801 captured more than one-third of total household income (in real terms), which had even increased slightly by 1867.[4] In that year, after a visit to the House of Lords, the historian Hippolyte Taine remarks that "the principal peers present were pointed out to me and named, with details of their enormous fortunes: the largest amount to £300,000 a year. The Duke of Bedford has £220,000 a year from land; the Duke of Richmond has 300,000 acres in a single holding. The Marquess of Westminster, landlord of a whole London quarter, will have an income of £1,000,000 a year when the present long leases run out."[5]

How did such inequalities come about? The first thing to note is that the incomes of wealthy noblemen, such as the Duke of Devonshire and the Marquess of Westminster, came from capital rather than labor. Capital was the predominant force behind the income disparities in Jane Austen's Britain. According to estimates by the economic historian Peter Lindert, the top 10 percent of the population had more than 80 percent of Britain's wealth in 1810.[6] Most of this wealth came from land. National wealth was roughly seven times the value of the national income, and agricultural land constituted about half of national wealth.[7] In other words, the fortunes of the landowning classes would not have been possible without one important technology: agriculture. Without it, the landed classes in eighteenth-century Britain would never have appeared. The fact that the gifts of the Neolithic revolution still shaped society in the eighteenth century, some ten thousand years later, suggests that despite millennia of technological change, economic life had not yet been fundamentally altered. Most people still worked on farms in the domestic system, which indicates that there had been little labor-replacing technical progress. Although there was an emerging middle class, social status and wealth was still derived from the land.

The Idiocy of Rural Life

For most of human history, there was no wealth and no inequality. The age of inequality began with the Neolithic revolution. The following period constituted only a brief episode of human history, relative to the forager era that preceded it. As noted, in the absence of any technology for storing meat, instant consumption was inevitable, and no significant food surplus was attainable. It was only after the invention of agriculture that food could be stored, land could be owned, and individuals could accumulate a surplus of significance—which in turn introduced the concept of property rights and a political structure to uphold those rights. Of course, prehistory does not provide any records of how the first political structures came about, but the rise of the feudal system in medieval Europe clearly constituted an exchange of peasant labor for knightly protection. The early beginnings of political authority are likely to have followed a similar pattern. The provision of a political structure provided some stability, but it came at the price of inequality.[8] Skeletons from Greek tombs at Mycenae of around 1500 B.C. show that royal skeletons were two or three inches taller and had substantially better teeth than those of commoners, suggesting that royals fared better in terms of nutrition. Further evidence is provided by Chilean mummies from around A.D. 1000, showing that the elite exhibited substantially lower rates of bone lesions caused by disease, in addition to other distinguishing features of wealth such as ornaments and gold hair clips.[9] The notion that political inequality stems from the invention of agriculture, as the philosopher Jean-Jacques Rousseau suggested, thus seems to hold.[10]

Of course, the price of inequality might be low if the commoner also benefited from the arrival of agriculture. One of the great questions in archaeology therefore concerns the impact of agriculture on the prosperity of ordinary people. Although data on living standards in preindustrial times remain sparse, food consumption clearly constitutes one important dimension. Building on the intuition that while an individual's height depends on genes, the heights of populations reflect patterns of food consumption, archaeologists often rely on heights to measure food intake.[11] Especially in societies where people are sufficiently poor for the demand for food to rise rapidly with their income, heights constitute a reasonable proxy for food consumption. Beyond height, anthropologists have looked at various indicators of well-being (including skeletal and dental features), which sometimes provide a somewhat different picture. Yet on balance, the available evidence suggests that the post-Neolithic rise in inequality was accompanied by a fall in average standards of living.

While it was long believed that the invention of agriculture dramatically improved the life of the commoner—relieving humanity of the burden of constant movement in the search for food—the body of data that has emerged since the 1960s shows that romantic views of the agricultural lifestyle are incorrect. Studies of societies that have shifted from foraging to agriculture for subsistence typically have found this transition to be associated with shorter people, deteriorating health, and an increase in nutritional deficiencies. For example, the anthropologists George Armelagos and Mark Cohen document declining health in nineteen out of twenty-one societies that underwent transformation to agriculture.[12] Reviewing the available evidence, Clark Spencer Larsen, another anthropologist, similarly concluded that the adoption of agriculture was accompanied by an overall decline in general health, as suggested by various skeletal and dental pathological conditions.[13] Although a number of studies have emerged since, Armelagos and coauthors recently revisited the question and found that the adoption of agriculture has been associated with a decrease in adult height and a reduction in general health. They further observe that the decreasing stature in populations holds across continents and the periods during which agriculture was adopted.[14] These findings are also consistent with evidence suggesting that hunter-gatherers had a much more diverse diet, and that the narrowing of the types of food consumed that is associated with agriculture led to growing deficiencies in some essential nutrients.[15]

The fact that living conditions deteriorated with the arrival of agriculture has left many economists, anthropologists, and archaeologists puzzled as to why hunter-gatherers would have voluntarily exchanged their lives for what the *Communist Manifesto* called the "idiocy of rural life."[16] One possibility, of course, is that the adoption of agriculture was

the result of population pressures and the increasing difficulty of foraging for food as population densities among hunter-gatherers gradually
increased toward the end of the Ice Age.[17] For example, the ecologist
Jared Diamond has suggested that "forced to choose between limiting
population or trying to increase food production, we chose the latter
and ended up with starvation, warfare, and tyranny."[18] But causality
might equally have run in the opposite direction. Another theory is
that higher productivity simply resulted in larger populations with no
per capita income gains. Agriculture was adopted because it was a better
technology, which initially generated higher incomes for the bulk of the
population. Nevertheless, with the arrival of agriculture, the cost of having more children fell as mothers no longer had to carry their babies in
search for food. And because higher incomes could feed more people,
population growth surged, offsetting any income gains in per capita
terms. Of course, there is no way of knowing in which direction causality ran. Most likely both explanations have some merit. What is clear is
that populations surged with the adoption of agriculture. Population
densities of hunter-gatherers were rarely over one person per square mile
and often substantially lower, while farmers averaged forty to sixty times
that density.[19]

The Population Curse

The idea that better technology results only in larger populations is an
appealing one because it also helps explain why growth was stagnant for
most of human history. Like the adoption of agriculture, the spread of
every new productivity-enhancing invention helped only grow the
population. The intellectual foundation for this intuition is the Malthusian model, put forward by Thomas Robert Malthus in 1798. This model
describes an organic society in which the laws that govern human economic activities are the same as those that govern all animal societies. The
sizes of both animal and human populations depend on the available resources for consumption. Over the long run, according to the
Malthusian model, people's incomes—and thus their resources for
consumption—are determined by fertility and mortality alone. So the

higher the fertility rate and the more people there are, the smaller share of these resources each individual has access to. Conversely, if the mortality rate increases for a reason such as disease or drought, those who are left will enjoy a larger proportion of the resources. Thus, even though the technological advances that took place in preindustrial times were cumulatively significant, slow technology adoption meant that no permanent income gains could be achieved. Because population adjustments take time, advances in technology had the potential of leading to higher incomes in the short run. But over the long run, growing incomes led to a reduction in death rates, and when birth rates started to exceed death rates, populations began to grow. In the end, the only effect achieved by moving toward a higher plateau of technology was a larger population, which eventually stopped growing when income returned to subsistence levels.[20]

Many historians have remarked that Malthus put forward his thesis just as the idea became irrelevant—at the onset of the Industrial Revolution, when England finally broke the iron law of wages and escaped the Malthusian trap.[21] Some economists and historians still believe that the preindustrial world was caught in a vicious cycle in which demographic negative feedbacks prevented per capita incomes from growing.[22] There is probably some truth to this belief, but to suggest that the Malthusian model applied to all preindustrial societies would be a stretch. First, empirical studies have shown that fluctuations in fertility and mortality in preindustrial societies were not primarily driven by variation in wages, at least from the sixteenth century onward.[23] Second, some places had already achieved sustained income growth prior to the Industrial Revolution.[24] While data on wages before the late Middle Ages remain scant, the Roman Emperor Diocletian issued an edict on maximum prices in A.D. 301, which included information on Roman wages. On the basis of Diocletian's wage schedules, the economic historian Robert Allen has estimated that a typical unskilled Roman worker earned just about enough to purchase a minimal subsistence basket, and that the workers' real wages were similar to those of their counterparts in eighteenth-century south-central Europe and Asia.[25] Yet by 1500, Britain and the Dutch Republic had already started to

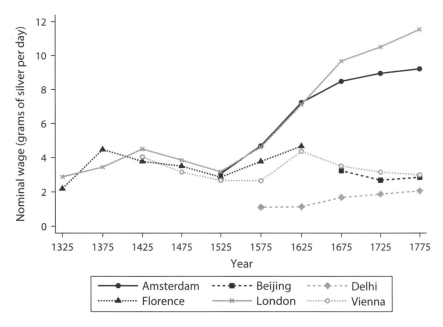

FIGURE 3: Nominal Wages in Grams of Silver per Day, 1325–1775

Sources: R. C. Allen, 2001, "The Great Divergence in European Wages and Prices from the Middle Ages to the First World War," Explorations in Economic History 38 (4): 411–47; R. C. Allen, J. P Bassino, D. Ma, C. Moll-Murata, and J. L. Van Zanden, 2011, "Wages, Prices, and Living Standards in China, 1738–1925: In Comparison with Europe, Japan, and India," Economic History Review 64 (January): 8–38.

experience a small divergence from the rest of Western Europe and the world, and by 1775, the wages of laborers in London and Amsterdam had pulled ahead of those of their peers elsewhere (figure 3).

The latest revisions to Angus Maddison's heroic gross domestic product (GDP) estimates point in a similar direction: per capita incomes were largely stagnant in most economies prior to 1500 but increased in Britain and the Dutch Republic thereafter.[26] In the seventeenth-century Ottoman Empire, per capita incomes (Int$700 in 1990 prices) were no higher than those in Byzantium and Egypt during the first century A.D., which at the time were slightly higher than in Britain, the Netherlands, and Spain (Int$600 in 1990 prices). Between the first and the eighteenth centuries, average incomes in Spain barely increased, stabilizing at roughly the same level as per capita incomes in Britain and the

Netherlands by the thirteenth century (around Int$900 in 1990 prices). But following the bubonic plague (the so-called Black Death) of 1348, which carried away 30–50 percent of Europe's population and caused a long period of population decline, average incomes in Britain and the Netherlands started to grow more rapidly.[27] However, such growth should not be overstated, as growing per capita incomes were largely the result of a shrinking population. As population growth rebounded in Britain, per capita incomes fell slightly in the period 1400–1500. Yet from 1500 onward, per capita incomes in Britain and the Dutch Republic almost doubled, reaching Int$2,200 and Int$2,609 (in 1990 prices) by 1800, respectively. Meanwhile, the rest of Europe—including Belgium, Germany, Portugal, Spain, and Sweden—witnessed no meaningful growth. Of course, there is no way of ascertaining that these estimates are correct, but wage data and GDP estimates alike suggest that the economies of Europe followed different trajectories in 1500–1800.

The Age of Discovery

Growing incomes after 1500 had little to do with replacing technology, however. If the Malthusian model provides a reasonable approximation of economic life before 1500, the next two centuries in Britain and the Dutch Republic are better characterized by the intuitions of Adam Smith. The great geographical discoveries—following the explorations of da Gama, Columbus, Magellan, and others—constituted the beginnings of an era of sustained Smithian growth. Trade emerged across continents, and new goods were discovered and consumed that had previously been unknown: colonial goods like sugar, spices, tea, tobacco, and rice—to name a few—were shipped distances that mankind had once not known existed. Though empirical evidence on the rise of international trade is sparse, data for the period 1622–1700 shows that British imports and exports doubled. The growing importance of trade is similarly suggested by the rapid expansion of shipping. Between 1470 and the early nineteenth century, the merchant fleet of Western Europe grew sevenfold.[28] As many of the colonial goods and other imports became attainable for a growing share of the population, people started to drink more tea, often sweetened with sugar; bought more luxurious clothing; and discovered new spices for their meals. The Industrial Revolution was thus preceded by a consumer revolution that created new desires and incentivized people to work harder to acquire the many newly available colonial goods.[29]

On the supply side, the surge in trade promoted industrial development. Many of the first factory builders in Britain were merchants taking advantage of the expansion of trade.[30] While the traditional crafts of the Middle Ages were largely practiced to make products for local markets, a growing entrepreneurial merchant class facilitated the emergence of rural industries to make items for export to other regions and foreign countries—a process for which the economic historian Franklin Mendels has coined the term "proto-industrialization."[31] The prominence of these industries is evident from statistical information. Following Gregory King's famous 1688 publication on the state and condition of England, historians long debated why Britain's trading economy was not reflected in the labor market statistics. According to King's estimates, only approximately 8 percent of the workforce were merchants or artisans. However, revisions of these estimates by Peter Lindert and Jeffrey Williamson, two economic historians, show that these numbers were in fact substantially higher: the number of traders, shopkeepers, and artisans amounted to 384,000, constituting roughly 28 percent of the labor force. Even though agriculture was still the predominant activity, Britain was dynamic for a preindustrial economy.[32]

While the age of discovery was not a time of economic wonders, most available evidence suggests that the British economy was growing. According to Maddison's estimates, growth rates in Britain between 1500 and 1800 averaged 0.22 percent per year.[33] Although estimates of preindustrial growth in 1990 prices inevitably rely on broad assumptions, different approaches relying on other data sources have yielded similar growth rates.[34] Skeptics may not be convinced by the assumptions underlying these estimates either, but it is telling that by the eighteenth century contemporary writers had no doubt that Britain was a relatively wealthy country. Daniel Defoe—best known for his novel *Robinson Crusoe*—wrote extensive accounts of his travels across preindustrial

Britain. In *A Tour through the Whole Island of Great Britain*, published in 1724, he observes that "labor is dear, wages high, no man works for bread and water now; our laborers do not work in the road, and drink in the brook; so that as rich as we are, it would exhaust the whole nation to build the edifices, the causeways, the aqueducts, lines, castles, fortifications, and other public works, which the Romans built with very little expense."[35] And he was not the only eighteenth-century observer left impressed by the wealth of Britain. Smith's descriptions of North America suggest that "labour is there so well rewarded, that a numerous family of children, instead of being a burden, is a source of opulence and prosperity to the parents [even though] North America is not yet so rich as England." The fact that the eighteenth-century Englishman was richer than people in previous generations is similarly suggested by Smith's assertion that "the annual produce of the land and labour of England . . . is certainly much greater than it was a little more than a century ago, at the restoration of Charles II[, and] was certainly much greater at the Restoration than we can suppose it to have been about a hundred years before, at the accession of Elizabeth."[36]

It is thus evident that although the picture painted of Britain in Jane Austen's writings mirrored economic reality in the sense that the wealth of the landed classes dwarfed industrial capital, economic life was changing.[37] Over the course of the eighteenth century, the share of land in total wealth had fallen significantly, and a wide range of new occupations had appeared, typically associated with the emergence of a new commercial and manufacturing class—referred to by Defoe as "the middle sort of mankind, grown wealthy by trade."[38] The economic structure of Britain was in many ways still a legacy of the Neolithic revolution, but the parallel rise of international trade meant that a growing share of the population benefited from growth. Moreover, because the commercial bourgeoisie for the first time had a higher fertility rate than the poor, the middle classes expanded rapidly, and any social mobility tended to be downward rather than upward.[39] This expansion was key to subsequent economic development. Middle-class families worked in occupations that required them to acquire skills rather than spend all of their time on costly leisure activities, while landed families could rely on income from capital to cultivate their refined taste for leisure and literature. This difference in mentality and ability is well captured by Smith, who suggests that "a merchant is accustomed to employ his money chiefly in profitable projects; whereas a mere country gentleman is accustomed to employ it chiefly in expense."[40] Because the investments parents make in their children's education and upbringing hinges upon the work they are expected to do, the bourgeoisie's work ethic was typically effectively transmitted to the next generation along with the "spirit of capitalism."[41]

The "bourgeois virtues," as the economic historian Deirdre McCloskey has called them, consisted of thrift, honesty, and diligence.[42] These virtues allowed them to accomplish the unprecedented. In *The Communist Manifesto*, even Karl Marx and Friedrich Engels alluded to the distinctiveness of this class, pointing out that the bourgeoisie "has been the first to show what man's activity can bring about. It has accomplished wonders far surpassing the Egyptian pyramids, Roman aqueducts, and Gothic cathedrals."[43] Indeed, the leading figures of the Industrial Revolution typically came from families that were already in some way involved in commercial and industrial activities. The historian François Crouzet's seminal compilation of information about 226 founders of large industrial undertakings whose fathers' occupations were known shows that some came from gentry and working-class backgrounds. Yet over 70 percent were born into middle-class families, of which a substantial share had made their fortunes in trade and commerce.[44] The intuition of Marx—that modern capitalism began during the Renaissance with the discovery of the New World—thus holds true.

However, the dynamism of preindustrial Britain should not be overstated. Though the agricultural share of employment had already fallen by 1700, as the emergence of small-scale industries led to an unprecedented expansion of Britain's trading economy, the distinction between agriculture and manufacturing before the Industrial Revolution was not clear-cut in practice. The rural industries that emerged were typically an off-season activity. Many of the workers living in the hinterland were both farmers and manufacturers. During the winter months, when agricultural work was not as plentiful, they engaged in spinning and weaving. Defoe characterizes the manufacturer as someone who had one horse to bring food and wool to the spinner and another to transport the clothing to market, while cows grazed on the land around his home.[45] While agriculture was not his main occupation, part of his living was derived from the land, which ensured his independence. In this so-called domestic system, there was no sharp distinction between the household, farm, and firm. Only around 30 percent of workers in early eighteenth-century Britain earned wages at some point. The vast majority remained self-employed, meaning that rural industries mainly consisted of small workshops, and even waged workers predominantly worked in their own homes. The prevalence of the "domestic system" meant that manufacturing—for the most part—remained "an industry without industrialists," in Crouzet's words.[46]

3

WHY MECHANIZATION FAILED

Why was Schumpeterian growth largely absent for so long? No single theory can explain why technological creativity failed to achieve higher standards of living for the commoner over thousands of years. The Malthusian trap is part of the explanation: productivity gains translated into larger populations and thus constrained income growth in per capita terms. But the world was not all Malthusian: from 1500 onward, standards of living did improve, and the age of discovery created sustained improvements in living standards for most people in Britain and the Dutch Republic. This would not have been possible without breakthroughs in technology. The rise of international trade was enabled by advances in shipbuilding and navigation, such as the three-master ship and the mariner's compass. Yet these technological advances did little to foster Schumpeterian growth—instead, they were levers of Smithian growth. Economic growth in the preindustrial world was thus not only quantitatively slower but also qualitatively different from what we regard as modern economic growth.[1] Growth in our own age relies heavily on technology adoption, creative destruction in employment, and new skills and knowledge that allow further innovations to emerge. Although the preindustrial world clearly did experience some of this kind of growth, it played only a secondary role in shaping the divergent economic trajectories in Europe. The real puzzle, therefore, is why technological creativity (which clearly flourished from time to time) did little to fundamentally alter economic life. Of course, the simple answer is that technological creativity is a prerequisite for growth, but not a

72

sufficient condition. Technical ideas need to be translated into reliable blueprints and prototypes, which in turn need to find an application in production, to have any impact on productivity and prosperity. The preindustrial era did not suffer from a shortage of imagination, it suffered from a shortage of realization. Leonardo da Vinci—the paradigmatic inventor of the preindustrial world—made drawings of hundreds of inventions, but he made hardly any effort to turn them into functioning prototypes. And numerous inventions that were turned into prototypes, such as Cornelis Drebbel's submarine, were not developed further. Even when applications were found, inventions often served political rather than economic purposes. Rulers of the Roman Empire, for example, directed technological efforts toward building grand structures to increase their popularity.

To be sure, for most of human history, technological advances did not take place within research and development departments focusing on finding a technical solution to a particular engineering problem. Technological development was organized very differently from how it is structured today, if it was organized at all. The importance of close collaboration between scientists and engineers for directing an idea toward the right application almost goes without saying today. But such collaboration was very rare in preindustrial times. The scientific revolution—which began with Galileo—assuredly facilitated more such interaction and later technological developments. In particular, the discovery of atmospheric pressure was essential for the development of the steam engine that eventually replaced water power as the engine of the Industrial Revolution. Yet other technologies of the Industrial Revolution could have been invented and put into widespread use without advances in science. So why were they not?

Broadly speaking, there are two strands of explanations. While some scholars have emphasized constraints on the supply of technology, others have pointed at limited demand. Joseph Schumpeter believed that for a given technology to be adopted, some kind of need must exist.[2] This was also the view of Thomas Malthus, who reckoned that "necessity has been with great truth called the mother of invention. Some of the noblest exertions of the human mind have been set in motion by the necessity of satisfying the wants of the body."[3] A number of examples of technological developments since the Industrial Revolution that conform to this view spring to mind, including the Manhattan Project, set up by the U.S. government to develop an atomic bomb before Nazi Germany could do so; the steam engine developed by Thomas Savery to pump water out of British coal mines; and the interchangeable parts pioneered by Eli Whitney to "substitute correct and effective operations of machinery for the skill of the artist which is acquired only by long practice and experience; a species of skill which is not possessed in this country to any considerable extent."[4]

To return to the preindustrial world, most demand-driven explanations of the lack of preindustrial growth tend to emphasize the fact that labor-saving technologies, which allow us to produce more with less, make economic sense only if capital is relatively cheap compared to labor. Possibly that was rarely the case in preindustrial times. In the context of classical civilizations, for example, the historian Samuel Lilley has argued that slaves were cheaper than machines, which provided few incentives to developing and adopt expensive machinery.[5] To push this argument slightly further, slaves were in many ways the robots of preindustrial times. In Hungary, unpaid serfs working on behalf of a feudal lord were called *robotnik*: this gave rise to the modern word "robot," which first appeared in Karel Čapek's famous play *R.U.R.* in 1921.[6] Slaves were able to perform just about any mundane manual task that one can think of, and they were certainly capable of performing a much wider set of physical tasks than can be accomplished by any robot technology today.

The view that slavery retarded technological development during the classical period is nonetheless highly controversial, and to extrapolate this intuition from classical times to all preindustrial societies would certainly be a stretch. Slavery in the Roman Empire had largely vanished by the end of the second century A.D. Yet the end of Roman slavery was the beginning of serfdom rather than freedom. Unlike slaves, serfs were allowed to retain some of the product of their labor, but like slaves, they were subject to many restrictions that ensured a steady supply of labor and exerted downward pressure on wages. Though the Black Death of 1348 produced labor shortages that ended serfdom in Britain, the government introduced legislation to prevent wages from rising, with long-lasting effects. And globally, slavery and serfdom persisted much longer. Even in 1772, four years before the American Declaration of Independence, estimates by Arthur Young suggest that only 4 percent of the world's population was free.[7] The remaining 96 percent were slaves, serfs, independent servants, or vassals.

While it is hard to say to what extent slavery might have retarded mechanization, the key question is not whether slavery (or serfdom) per se hindered the adoption of labor-replacing technology: incentives to mechanize hinge not on the freedom of the worker but on the price of labor, which remained low in preindustrial societies. The association between the abundance of cheap labor and slower mechanization is persuasively shown by a recent study, albeit in a modern setting.[8] In the American South, the long presence of slavery meant that agriculture had remained highly labor-intensive. Even though the slaves were emancipated during the Civil War, the wages of the black population remained low. The Mississippi's flooding in 1927 was the triggering event that put some counties in the American South on different trajectories from others, because many black families left the flooded areas in search of work elsewhere. Unable to prevent the loss of black labor, planters in flooded areas moved toward greater capital intensity and mechanization, relative to the unaffected areas—where cheap labor remained plentiful.

There are thus good reasons to believe that relatively cheap labor in preindustrial times created fewer incentives to put worker-replacing technologies into widespread use. In fact, Robert Allen has argued that the reason why the Industrial Revolution began in Britain is that at its onset, it was not economical anywhere else.[9] The path to the British Industrial Revolution, Allen suggests, began with the Black Death, which caused a long period of population decline and produced labor shortages that increased the bargaining power of workers.[10] As peasants demanded freedom instead of serfdom, wages eventually began to rise, even though legislation was put in place to hold the price of labor down.

Following Britain's success in trade during the age of discovery, wages began to grow faster. This success came with new challenges: given its high labor costs, how would Britain remain competitive in trade? The critical factor, Allen argues, was that British industrialists were fortunate enough to be sitting on a mountain of coal.[11] The early emergence of the coal industry was what distinguished Britain from other high-wage economies like the Dutch Republic. Facing low energy prices and high labor costs, British industry began to adopt machines that would not have been cost-effective elsewhere. But while such an explanation may seem appealing, newly collected data suggest that British wages did not grow as rapidly as previously thought.[12] What's more, even if we assume that wages in Britain were relatively high, early labor-saving technologies, like William Lee's stocking-frame knitting machine and the gig mill, were developed long before the Industrial Revolution but were vehemently opposed.

Examples of technological advances emerging from necessity are in fact seemingly few before the Industrial Revolution. Indeed, Joel Mokyr's magisterial review of technological developments in the preindustrial world suggests that "invention is the mother of necessity," provides a more accurate description of preindustrial inventive efforts.[13] Instead of technologies being developed in response to some preexisting demand, sporadic technological advances created previously unrecognized desires and new demand. Technological progress was often random and unpredictable, as was the demand that sometimes emerged from it. For example, Gutenberg's printing press created demand for books, education, and literacy—rather than a demand for books leading to the invention of the press. And other inventions were simply the result of serendipitous discoveries. When hunter-gatherers during the Ice Age first noticed residues of limestone and burned sand in their hearths, they could not possibly have foreseen how millennia of accidental discoveries would lead to the first Roman glass windows.[14] Similarly, when Evangelista Torricelli discovered that the atmosphere has weight, he could not have predicted the chain of events that would culminate in the invention of the steam engine.

The view that new technology creates its own demand implies that the lack of preindustrial growth was primarily a consequence of obstacles to the supply of technology. In support of a supply-driven explanation, a number of theories have pointed to different factors that are likely to have held the supply of technology back in the preindustrial era. For example, while it is widely acknowledged that entrepreneurial risk taking is critical to technological progress, it is rarely noted that innovation was riskier and less rewarding in preindustrial times. Before the age of mass production and the arrival of social safety nets, the upside of entrepreneurial risk taking was low, and the potential downside much greater. The fortunes that could be won by inventors in the nineteenth and twentieth centuries were previously unattainable because markets for new technology were typically local and thus substantially smaller, and entrepreneurial failure could at worst lead to starvation. In addition, because technological advances in preindustrial times often remained local, technologies used in one place that might have guided future technological advances in another location were often unknown there. Such path dependence sometimes put societies on technological dead-end trajectories. For example, in large parts of North Africa and the Middle East, the invention of the camel saddle (which occurred sometime between 500 and 100 B.C.) meant that the camel gradually replaced wheeled transportation, which reduced the resources available for building roads and bridges and led to poor infrastructure and lower incentives to innovate in other modes of transportation. However, as noted above, even though invention was a risky activity and it took time for technical knowledge to diffuse, pathbreaking technologies like the printing press were still developed and adopted.[15]

More importantly, although economists often are dismissive of culture as being an obstacle to economic development, there are good reasons to believe that beliefs long prohibited progress. An influential theory, championed by Mokyr, has maintained that the scientific revolution of the seventeenth century prepared the ground for a culture of growth.[16] There is probably much truth to this. A culture that replaced superstition with reason and a scientific attitude, which the sociologist

Max Weber deemed critical to technological progress, did not appear before the Enlightenment.[17] And superstition aside, most preindustrial intellectuals saw no virtue in mechanization. They shared the cultural attitudes of classical philosophers toward technological development, as aptly summarized by Bertrand Russell: "Plato, in common with most Greek philosophers, took the view that leisure is essential to wisdom, and is therefore not to be found among those who have to work for their living."[18] Indeed, in *Politics*, Aristotle wrote that "no man can practice virtue who is living the life of a mechanic or laborer."[19] In other words, work—especially the manual sort of work that was required for the construction of machines—was deemed unworthy by many of the greatest minds of classical times. Yet while the views of the upper classes in eighteenth-century Britain were little different from the nonprogressive beliefs of ancient philosophers, the beliefs of the middle or producing classes were subject to more profound change, and central to this change was the changing nature of religious beliefs. Although the relationship between technology and religion has always been a complex and tenuous one, it is indisputable that religious beliefs changed in preindustrial Europe, and with them attitudes toward technological progress. The Romans and the Greeks regarded nature as the domain of the gods: any manipulation of its forces by means of technology was considered sinful and even dangerous. This stands in contrast to medieval Christianity, which historians have argued paved the way for future technological progress as it embraced a more rational God. As Lynn White explains, "Christianity, in absolute contrast to ancient paganism and Asian religions . . . not only established a dualism of man and nature but also insisted that it is God's will that man exploit nature for his proper ends."[20]

There is no way to prove causality, but the notion that "if God would have wanted man to fly, he would have given him wings," was clearly rebelled against within the Latin Church. The writings of the Franciscan friar Roger Bacon in the thirteenth century envisioned the emergence of steamships, automobiles, and airplanes; and the monk Eilmer of Malmesbury similarly felt no sense of sin when attempting to fly using a glider.[21] In some ways, the clergy even facilitated technological development. The teachings of the Benedictine order, which had enormous

impact on medieval life, emphasized that work and production are virtuous and could provide a way to salvation. This is not to suggest that
Christianity was always favorable to progress. The famous controversy
surrounding Galileo's support of heliocentrism, which made him guilty
of heresy and led to his imprisonment, suggests the contrary. However,
while there can be no doubt that the oppression of science by the Latin
Church was an obstacle to some inventive pursuits, early industrialization had no scientific basis. The steam engine was a latecomer to the
industrialization process. Science became a pillar of economic progress
only in the nineteenth century. As Mokyr writes, "Many of the 'wave of
gadgets' that we associate with the classical Industrial Revolution—
steam power being the most notable exception—could have been easily
made with the knowledge available in 1600. What is beyond question is
that the relative importance of science to the productive economy kept
growing throughout the late eighteenth and nineteenth centuries, and
became indispensable after 1870, with the so-called second Industrial
Revolution."[22]

Another explanation for the time and place of the beginning of the
Industrial Revolution is that institutions in the preindustrial world did
more to prohibit innovation than they did to encourage it. Inspired by
the pioneering work of Douglass C. North, many economic historians
have argued that it was only after the Glorious Revolution of 1688–89,
when the English Parliament gained supremacy over the crown, that the
preconditions for the Industrial Revolution were established.[23] Before
then, rent-seeking monarchs and other so-called economic parasites
found it easier to extract revenue from others than to take part in productive activities, which required hard work. Article 4 of the Declaration of Rights of 1689 changed the rules of the game, as Britons could
no longer be taxed without their consent. Without the authorization of
Parliament, levying money for the use of the crown was deemed illegal.
But while this was surely an important event, explaining the long absence of any Industrial Revolution is not just a matter of identifying
variables that might have held back technical progress in general. As
noted, preindustrial culture and institutions did not hinder all progress.
Considerable technical progress was achieved before the eighteenth

century. The critical difference, it seems, was that governments before the Glorious Revolution frequently attempted to block worker-replacing technologies, and the key inventions of the Industrial Revolution were worker replacing.

Origins of the Industrial Revolution

So how did the necessary institutional change occur that would one day facilitate the Industrial Revolution? One compelling argument is that the road to industrialization began with the discovery of the New World. Daron Acemoglu, Simon Johnson, and James Robinson have demonstrated that where political institutions placed significant checks and balances on the monarchy, the growth of Atlantic trade strengthened merchant groups by constraining the power of the crown and helped them obtain institutional reform that favored industrial and technological progress.[24] Consequently, more rapid economic growth took place in economies with relatively nonabsolutist institutions—like those of Britain and the Dutch Republic—where commercial groups outside the royal circle were the key beneficiaries from trade. In Britain, for example, Parliament successfully prevented several attempts of both Tudor and Stuart monarchs to create royal monopolies, and consequently trade was typically carried out by merchants, either individually or in commercial partnerships. This stands in contrast to most of Europe, where royal trading monopolies prevailed. In the Portuguese case, trade with Africa and Asia was restricted to the royal trading house Casa da Índia. Casa de Contratación in Seville served the equivalent function for the Spanish empire, where colonial trade was a monopoly of the crown of Castille. And in France, the political influence of merchants diminished, if anything.[25] Though early Atlantic trade enriched some commercial groups outside the royal circle—especially the Protestant Huguenots—the siege of La Rochelle meant that the Protestant church was eventually banned by Louis XIV, leading most Huguenots to leave France.[26] In countries whose parliament failed to provide checks on the power of the executive, trade remained firmly under the control of the crown.

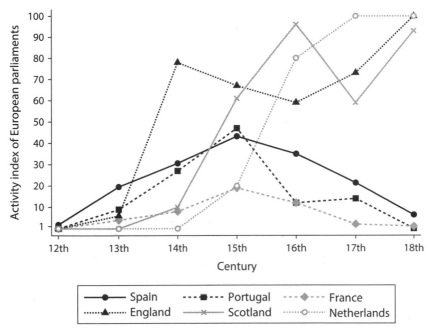

FIGURE 4: Activity Index of European Parliaments, 1188–1789
Source: J. L. Van Zanden, E. Buringh, and M. Bosker, 2012, "The Rise and Decline of European Parliaments, 1188–1789," Economic History Review 65 (3): 835–61.
Note: The activity index counts the number of years per century in which a parliament was assembled for official sessions. If the index is zero, no parliament was convened. If the value is 100, a parliament was convened in every year of the century.

The power struggle between parliaments and the crown was at the heart of many of the major sociopolitical conflicts, including the Dutch Revolt of the 1570s, the English Revolution of the 1640s, and the French Revolution of 1789.[27] The outcome of these conflicts in the North Sea countries, and the long absence of them in other parts of Europe, meant that the influence of parliaments declined in southern and central Europe, while parliaments gained in importance in the Dutch Republic and Britain. The economic historians Jan Luiten Van Zanden, Eltjo Buringh, and Maarten Bosker have found that colonial Europe experienced a sustained period of institutional divergence between 1500 and 1800.[28] As parliamentary activity in the North Sea countries surged, it declined elsewhere in Europe (figure 4). The political clout and activity of the parliament in France increased up until the mid-1500s but then decreased as the crown found ways to introduce taxes without approval of the Estates General. In Spain, the great discoveries associated with the New World (mainly silver and gold) resulted in new sources of income for the crown, which reduced the need to raise taxes—for which parliamentary approval was required. Convening Parliament was therefore no longer a necessity.

The increase in parliamentary activity in the North Sea countries can be explained by a series of events. In the Dutch case, commercial interests associated with the Atlantic trade led to conflict between Dutch merchants and the Habsburg monarchy—which ruled the Netherlands before the Dutch Revolt—and culminated in a war of independence in the 1570s. Merchants constituted the primary political force on the side of independence and naturally became the new ruling class as the Estates of Holland and the Estates-General of the northern Low Countries assumed sovereignty and created the Dutch Republic. In Britain, institutional change was shaped by the Civil War of 1642–49, when parliamentarian forces defeated the royalists, which led to the trial and execution of Charles I and subsequently the Glorious Revolution of 1688, when a coalition of English parliamentarians and Dutch military forces replaced James II with a constitutional monarchy led by the Dutch stadtholder William of Orange. The victory of Parliament in the Civil War meant that the fraction of members of Parliament who were sympathetic to industry increased dramatically.[29] What's more, the Glorious Revolution, which resulted in the Bill of Rights of 1689, restricted the ability of the crown to rule the country arbitrarily. The Mutiny Act, for example, forbade the crown to form and maintain a standing army without Parliamentary assent, which limited the ability of monarchs to overthrow Parliament militarily. By shorting the period for which taxes were granted, Parliament gained further political clout, as the crown had to convene Parliament regularly to cut a new deal. Also, to prevent the crown from controlling Parliament from within, new defenses were erected against seat and vote buying.[30]

The result was not only a shift of political power from the crown to Parliament, but also a shift of influence in favor of merchant manufacturers. Though merchants were hardly a majority group in Parliament, their interests were protected by the formation of the Whig coalition, which represented merchants and Protestant landowners.[31] Meanwhile, the landed aristocracy—whose members largely controlled the levers of political power before 1832—did not contribute very much to innovation and the mechanization of industry, but at least they didn't resist it.[32] This was in part because Britain's history as a trading nation had allowed them to diversify their wealth, so that some even stood to benefit from industrialization.[33] While the Glorious Revolution effectively marked the beginning of its slow decline, the House of Lords still looked after landed interests until the early twentieth century. Consequently, the British aristocracy were able to hold onto some of their powers and feel less threatened by the shifts in social and economic power that were taking place in the rest of society.[34]

All of this meant that Parliament increasingly acted to safeguard commercial and industrial interests. Contracts were enforced, and property rights were regarded above all. As Adam Smith observed in 1776, "security which the laws of Great Britain give to every man that he shall enjoy the fruits of his own labour is alone sufficient to make any country flourish."[35] Of course, not all institutions favored economic and technological development. The poor had limited access to education and could not serve on juries. Even with the Reform Acts of 1832 and 1867, most ordinary citizens did not have political rights. And the economic logic of many Parliamentary decisions was still guided by the flawed doctrine of mercantilism: the belief that trade is a zero-sum game. Some acts of Parliament prohibited the exportation of machinery and the emigration of artisans. Others were passed to protect British commerce and manufacturing from foreign competition. But even though Britain was by no means a modern democracy and lacked many of the characteristics of a laissez-faire economy, it had become a much more diverse, tolerant, and industrious society. Writing in 1689, John Locke observed that "toleration has now at last been established by law in our country."[36] People enjoyed freedom of expression and choice of occupation and were able to engage in just about any scientific and inventive activities they wanted, and merchants mixed with the landed classes. During his stay in Britain, Voltaire wrote that "the younger son of a peer will not look down upon business. Lord Townsend, a Cabinet Minister, has a brother who is content with leading a firm in the City."[37] And based on his travels in the early eighteenth century, Daniel Defoe writes about the English tradesman: "Trade is so far here from being inconsistent with a gentleman, that, in short, trade in England makes gentlemen: for, after a generation or two, the tradesmen's children, or at least their grandchildren, come to be as good gentlemen, statesmen, Parliament men, Privy Councillors, judges, bishops, and noblemen, as those of the highest birth and most ancient families."[38]

Such a state of affairs was unheard of outside the North Sea countries. Indeed, the relative influence of the merchant class in Britain is most evident when observing continental Europe, where—excepting the Dutch Republic—industry and commerce was controlled by the crown. In France, for example, Jean-Baptiste Colbert, who served as minister of finance under Louis XIV, was of the view that industrial development required state support, which he deemed critical to bringing the country back from the brink of bankruptcy. To grow the economy and make France self-sufficient in the production of luxury goods, Colbert established the Manufacture Royale de Glaces de Miroirs in 1665—among other state-run factories—to supplant the importation of Venetian glass, which was banned as soon as the French glass manufacturing industry was on solid footing. Statistical classifications drawn up by the inspectors of manufactures, which divided manufacturing industries into three categories, show the immense influence of the crown. First, there were the state factories funded by the royal treasury: the goods produced by these industries were mainly luxury articles enjoyed solely by the crown. The famous Manufacture des Gobelins, for example, employed legions of artisans who worked only at the crown's pleasure and embellished Versailles, Saint-Germain, and Marly. Second, the *manufactures royales* represented private enterprise producing for public consumption upon formal invitation of the crown in designated districts. And lastly there were the *manufactures privilégiées*, which enjoyed a royal monopoly for the production and sale of certain goods. All of these industries knew no competition, there was no mechanization, and manufacturing survived solely due to the support and patronage of the crown.[39]

European monarchs did not just fail to encourage industrial development, they actively blocked it. Francis I—the last emperor of the Holy Roman Empire, and then emperor of Austria-Hungary until 1835—clearly feared the political consequences of technological progress and did his utmost to keep the economy agrarian. The primary concern was that the establishment of factories would replace workers in the domestic system and concentrate the poor in cities, where they could organize and rebel against the government. To avoid the threat from below, Francis I blocked the construction of new factories in Vienna in 1802 and banned the importation and adoption of new machinery until 1811. When plans were put before him for the construction of a steam railroad, he responded: "No, no, I will have nothing to do with it, lest the revolution might come into the country."[40] Consequently, railroad carriages in the Habsburg Empire were long drawn by horses.

Tsar Nicholas I similarly feared that the spread of the mechanized factory in Russia could undermine his leadership. To slow down the pace of progress, industrial exhibitions were banned. And after a series of revolutionary outbursts across Europe in 1848, a new law was enacted to limit the number of factories in Moscow, explicitly banning any new textile mills and iron foundries.[41] As in the Holy Roman Empire, railroads were not considered just a revolutionary technology, but also an enabling technology for revolutions. Thus, the only railroad built before 1842 ran between Saint Petersburg and the imperial residences at Tsarskoe Selo and Pavlovsk; information about railroads was even censored in Russian newspapers. Worker mobility and the spread of information was not in the interest of the ruling classes. And the Russian elites were surely right to fear the mechanized factory. The *New York Times* correspondent in Saint Petersburg in 1895 reported: "The introduction of machinery in La Ferme cigarette factory led to a serious riot on Saturday. The employees, who believed that the use of machines would throw many of them out of work, smashed the machines and hurled the fragments out the windows."[42]

For a long time, as noted in chapter 1, British governments tried to block the spread of replacing technologies, too. Even in the seventeenth century, Charles I issued a proclamation against the diffusion of gig mills. But things changed after the Glorious Revolution. As Acemoglu and Robinson write, "In Tudor or Stuart England, Papin [whose steam digester was smashed by Fulda boatmen] might have received similar hostile treatment, but this all changed after 1688. Indeed, Papin was intending to sail his boat to London before it was destroyed."[43] It is undeniably noteworthy that even though examples of British monarchs blocking worker-replacing technologies were plentiful before 1688, such examples are hard to find thereafter. Part of the reason is that the guilds were weakened by Parliament and intensifying competition after the Glorious Revolution. Although they were abolished officially in England only by the Municipal Corporations Act of 1835, they had begun to lose members and power much earlier. As discussed above, the guilds did not resist technological progress when it enhanced their members' skills, but they did when it threatened to make their members obsolete. It was therefore a prerequisite for the Industrial Revolution, which rested upon worker-replacing machines, that the power of the guilds be reduced.

This happened naturally as markets became more integrated: the influence of the guilds didn't extend beyond their own cities, so as competition between cities grew, their political power declined. The shearers' guild, for example, was one of the strongest in the woollen industry and had been successful in securing good pay for its members. Through petitions and violence, they managed to block the introduction of the gig mill in the west of England for decades. But cascading competition changed the rules of the game. In Wiltshire and Somerset, where the guilds had violently opposed the gig mill for a long time, resistance ended as the region began to lose business to Gloucester—where shearers had figured out that they could produce at a lower cost and expand their business by using the mills.[44] New towns like Birmingham and Manchester, which emerged in formerly rural areas, were also free from guild regulations and naturally became the engines of the Industrial Revolution.[45] More broadly, a statistical analysis of 4,212 patents filed in the period 1620–1823 shows that areas in England that became more exposed to outside competition invested more in the invention of new technologies.[46] And crucially, the nature of technological progress changed as well. When the economic historian Christine MacLeod examined 505 patents filed in the period 1663–1750, she found that very few technologies were invented to replace workers. Forty-five percent of the patents were said to augment workers' skills. Another 37 percent were claimed to save capital. Only 2 percent were said to save labor. Between 1750 and 1800, however, the percentage of labor-saving technologies increased fourfold.[47] To be sure, inventors consistently underreported any labor-saving motives due to fear of opposition. But the upsurge in replacing technologies provides additional evidence of the fading power of the English craft guilds.

Contrast this with the experience of China, where the guilds (*gongsuo*) persisted much longer and had almost unrestrained control over their crafts.[48] They were more powerful than their European counterparts, and they used their power to forcefully restrain the introduction of worker-replacing technologies on a regular basis. One contemporary observer, Daniel J. Macgowan, wrote in 1886:

> Native merchants imported from Birmingham a quantity of thin sheet-brass for manufacturers of brass utensils at Fatshan, throwing out of employment a class of coppersmiths whose business consisted in hammering out the sheets heretofore imported in a thick form; but the trade struck to a man, would have none of the unclean thing, and to prevent a riot among the rowdiest class of the rowdiest city in the empire, the offending metal was returned to Hongkong. Further, a Chinese from America the other day imported thence some powerful sewing machines for sewing the felt soles of Chinese shoes to the uppers, but the native sons of St. Crispin destroyed the machines, preferring to go on as their fathers did, while the enterprising Chinaman returned to Hongkong, a poorer and sadder man. Again, some years ago a progressive Chinaman set up a steam-power cotton mill, only to be made useless by the very simple plan of the growers refusing to send in a pound of cotton. Filatures from France, effecting not

only a wonderful saving in time and money but improving the quantity and quality of the output of silk, succeeded at Canton for a while, and were introduced latterly by Chinese capitalists into the silk-rearing districts, only to be destroyed and wrecked by the country-folk.[49]

Fearing social unrest, Chinese authorities sided with the guilds. An 1876 report to the London Foreign Office highlights this:

> During the past year [1875–76] an attempt was made to launch a Steam Cotton-Mill Company at this port [Shanghai], for the purpose of manufacturing cotton piece-goods from native-grown cotton . . . similar . . . to the goods at present made by Chinese . . . but with the advantages of English machinery and steam-power. . . . When the enterprise came to be generally known to the Chinese newspapers, the attitude of the Cotton Cloth guild became so alarming that the native supporters [of the project] drew back. An idea was unfortunately circulated among the natives, and more particularly amongst the workers of native hand-made cloth, that the trade would be immediately put an end to if such a scheme were put into operation, whereupon the guild passed a resolution to the effect that no clothes made by machinery should be permitted to be purchased. . . . The local officials refused their support or [to] countenance the scheme through fear of causing riots amongst the people.[50]

Such opposition to replacing technologies and the long persistence of the *gongsuo* also help explain delayed industrialization in China. Chinese cities were much farther apart than their English counterparts, meaning that there was less competition between them and less threat to the power of the *gongsuo*. Thus, while competition between cities in England weakened the guilds in the eighteenth century, the economists Klaus Desmet, Avner Greif, and Stephen Parente argue that the lack of competition in China meant that industrialization had to wait another two hundred years, until China became integrated into the world economy. At the conclusion of the First Opium War in 1842, the British opened five so-called treaty ports for carrying out foreign trade in

China. By the closing years of World War I, their numbers had increased to almost one hundred. The competition introduced by foreign trade made China's technological backwardness only too apparent, and in the early twentieth century many labor-saving technologies from the West were imported.[51]

However, the weakening of the guilds in Britain was not merely because of competition between cities, which undercut the guilds' power. It was also a political choice, spurred by the rise of the new so-called chimney aristocracy and intensifying competition among nation-states. In eighteenth-century England, the polity and judiciary, which had previously supported the cause of workers and guilds and opposed replacing technologies, began to side with the innovators. Parliament ruled on a number of occasions against spinners, combers, and shearers who petitioned against cotton-spinning machinery, wool-combing machines, and gig mills. As mentioned above, the shift in the British government's stance on mechanization was due in part to the merchant manufacturers' becoming a more politically powerful force. Their fortunes depended on the success of the British Empire's trade, which in turn depended on mechanization to remain competitive internationally. And more broadly, Britain's dependence on trade made economic conservativism harder to align with the political status quo. The external threat of political replacement due to foreign invasion gradually became greater than the threat from below, as competition between nation-states intensified. The ruling elites were well aware that their military strength depended on their economic muscle.

The strong commitment of the government to supporting innovators is further underlined by legislation passed in 1769 that made the destruction of machinery punishable by death.[52] Of course, as we shall see in chapter 5, workers still did their best to oppose the introduction of labor-saving machinery. The Luddite riots in the period 1811–16 were due to the fear of replacing technological change among laborers, as Parliament revoked a 1551 law prohibiting the use of gig mills. However, the British government took an increasingly stern view of any attempt to halt the force of technology and deployed troops against the rioters. The sentiment of the government toward people's smashing of machines was made clear by a resolution passed after the Lancashire riots of 1779, which read as follows: "The sole cause of great riots was the new machines employed in cotton manufacture; the country notwithstanding has greatly benefited from their erection [and] destroying them in this country would only be the means of transferring them to another . . . to the detriment of the trade of Britain."[53]

Meanwhile, on the other side of the English Channel, matters unfolded very differently. As Britain was undergoing an Industrial Revolution, France was at the dawn of a political and social revolution. As the economic historian Jeff Horn has noted, the French Revolution made the threat from below very real for the French government.[54] Unlike British governments, which deployed massive levels of coercion to repress machine breaking, French governments feared that mechanization would exacerbate social upheaval. While English innovators and industrialists could count on government support against machine-breaking craftsmen, the turbulent state of politics on the other side of the Channel meant that French entrepreneurs were unable to rely on the protection of their government. As is well known, the classic work of E. P. Thompson suggested that political upheaval was inherent in Luddism.[55] But English machinery rioters were rebellious rather than revolutionary. In France, by contrast, the threat of revolution was real. The French machinery riots of 1789 had a much greater effect on delaying industrialization than their English counterparts. As Parisian crowds stormed the Bastille, angry woollen workers from the town of Darnetal broke through the line of royal troops guarding the bridges over the Seine. Arriving in the manufacturing suburb of Saint-Sever, they destroyed the machines that had been installed there. A long series of similar incidents followed, casting a long shadow over the country. At the newly established Calonne and Company, thirty machines were smashed by infuriated rioters. And in the suburbs of Rouen, more than seven hundred spinning jennies were destroyed. Some industrial pioneers, like George Garnett, tried to fight back, but the crowds were too large. And unlike in Britain, there were no troops to help. French industrialists and inventors could not put much faith in the willingness of the government to safeguard their interests, since it also feared that rebelling craftsmen would exacerbate the general state of unrest in the country.[56] Such political uncertainty undermined the willingness to invest in machines and industrial pursuits, which stifled economic progress in France. As Horn explains, "The possibility of a thoroughgoing social and economic revolution by the laboring classes ensured that neither the French state nor Continental entrepreneurs could safely maximize profits or innovate in response to labor militancy, as in Britain. . . . Because machine-breaking in 1789 was an aspect of the emergence of revolutionary politics, the supposedly assertive French state proved nearly powerless in clamping it down. Throughout the revolutionary decade (1789–1799), French industrial entrepreneurs could not rely on the state to repress working-class militancy."[57]

Conclusion

The slow rate of economic progress before 1750 cannot be explained by lack of inventiveness or curiosity. The preindustrial world gave rise to a host of important inventions, including the Antikythera mechanism, mechanical clock, printing press, telescope, barometer, and submarine. Some preindustrial inventions were arguably more sophisticated than the "wave of gadgets" that made the Industrial Revolution. The mere existence of technologically creative people, however, is evidently not a sufficient condition for economic progress. For that, technologies must find economic purpose and widespread use. As the economist Fritz Machlup has pointed out, "Hard work needs incentives, flashes of genius do not."[58] Flashes of genius clearly existed in preindustrial times, but incentives to invest in machinery were few.

Before the Industrial Revolution, political power was firmly held by the landed classes. The structure of power was shaped by the invention of agriculture, which meant that for the first time food could be stored, land could be owned, and individuals could accumulate a surplus of significance. This, in turn, led to the concept of property rights and a political structure to uphold those rights. The exchange of peasant labor for knightly protection created an unequal world, where rent seeking paid more handsomely than progress. The fear among the ruling classes that labor displacement would cause hardship, social unrest, and at worst a challenge to the political status quo meant that worker-replacing technologies frequently were resisted or even banned. This dynamic, in which the politically powerful had more to lose than they could gain from progress, kept the Western world in a technology trap where technologies that threatened people's skills were forcefully resisted.

A number of events tipped the balance in favor of innovators. The rise of nation-states and growing competition among monarchs meant that the cost of restraining technical progress increased significantly. Backward nations would soon find themselves overtaken—at the worst, conquered—by progressive ones, which made it harder to align economic conservatism with the political status quo. The external threat, in other words, became greater than the threat from below. The craft guilds, which did their utmost to resist replacing technologies, were weakened by growing competition between cities. Their weakening made it easier for governments to side with entrepreneurs and inventors, to the detriment of the guilds. Desmet, Greif, and Parente write:

> With less support from the judiciary and the polity, craft guilds started to resort to violent means when new technologies threatened jobs. These violent reactions, which took the form of riots, demonstrations and vandalism, became more frequent at the turn of the nineteenth century, culminating in the Luddite riots of 1811 to 1816. Rather than being a sign of a strength, these violent reactions were the death throes of a weakening guild system. . . . This is indeed what occurred, and with guilds becoming less effective in blocking the introduction of labor-saving technology, it was only a matter of time before England underwent a major industrialization and escaped its Malthusian trap.[59]

It was only after the ruling elites began to side with the innovators that British industry could mechanize.

PART II
THE GREAT DIVERGENCE

Between 1780 and 1850, in less than three generations, a far-reaching revolution, without precedent in the history of Mankind, changed the face of England. From then on, the world was no longer the same. . . . [N]o revolution has been as dramatically revolutionary as the Industrial Revolution, except perhaps the Neolithic Revolution.

—CARLO M. CIPOLLA, *THE FONTANA ECONOMIC HISTORY OF EUROPE*

Without this increased wealth appearing to benefit the bulk of the population in proportion to the effort it has supplied for its production; the opposition of two classes, of which the one increases in numbers and the other in wealth; of which the one earns, by increasing labour, only a precarious subsistence wage, whilst the other enjoys all the benefits of a refined civilization; these conditions are everywhere manifest, and are everywhere followed by the same movements of thought and feeling.

—PAUL MANTOUX, *THE INDUSTRIAL REVOLUTION IN THE EIGHTEENTH CENTURY*

The rise of the machines caused workers to rebel against technological progress. As we shall see, the technologies that made the Industrial Revolution were primarily worker replacing (chapter 4), which explains the widespread resistance to them (chapter 5). This time, however, political power was firmly with those who stood to gain from mechanization.

For the most part, workers lacked political power, so their case was hopeless.

Industrialization began with what to the modern eye seem to be a few minor inventions that enabled the establishment of the factory system and inaugurated an era of sustained industrial expansion that created the modern world. The story of the factory is much like that of science. Though it would be absurd to attribute modern science to Galileo Galilei, Francis Bacon, or René Descartes, they can justly be seen as founding fathers. In a similar fashion, it was only in the age of Richard Arkwright, Samuel Crompton, and James Watt that the technological foundations of the factory system emerged. Factories existed long before the Industrial Revolution, but they must be distinguished from the modern factory system—whose distinctive feature, as Karl Marx noted, was the introduction of machines.[1] The inventors of these machines can therefore equally be regarded as the inventors of modern industry.

Like the evolution of science, the rise of the factory system was a gradual and uneven process. The proposition of the economist Walt Rostow, that the Industrial Revolution constituted a "take-off" into self-sustained growth has been decisively disproved by subsequent empirical analyses, which indicates a more gradualist interpretation.[2] Not only was overall growth slow during the Industrial Revolution, but even industrial output didn't experience the kind of sudden surge that would suggest a revolution.[3] Per capita income growth between 1750 and 1800 was barely faster than in the early part of the century, but by 1870 per capita income in Britain was 82 percent higher than it had been in 1750. The corresponding annual growth rate of 0.53 percent was slow by modern standards. However, it was significantly faster than the growth rates achieved by preindustrial economies.

The macroeconomic impact of the Industrial Revolution was not large enough to be called an economic revolution, yet there are variables that suggest there was a technological revolution after 1750. The average annual number of granted patents more than doubled in the 1760s relative to the previous decade and continued to grow rapidly thereafter.[4] One might surely call into question the economic relevance of some patents, but the timing of the patent surge supports the historian T. S. Ashton's memorable phrase: "About 1760 a wave of gadgets swept over England."[5] Around that time, many of the defining inventions of the Industrial Revolution emerged, including Arkwright's water frame and Watt's separate condenser for the steam engine, both of which were patented in 1769.

The absence of an economic revolution is no mystery. The simple existence of better technology does not inevitably translate into faster economic growth. For that, widespread adoption is required, but the Industrial Revolution was initially confined to a small number of sectors that collectively constituted a fraction of the overall economy. Thus, in its early days, the Industrial Revolution was not an aggregate phenomenon. As the economic historian Michael Flinn explains, "The lesson to be learnt from the statistics appears to be one of the superimposition upon a steadily growing economy of a small group of extremely dynamic sectors. Statistically they represented, even by the end of the [eighteenth] century, a very small share of the national product, but the growth in them was sufficient to double the existing rate of overall growth in the economy."[6] The Industrial Revolution began in the textile industry, and that is where workers most keenly felt the force of the mechanized factory. This mechanization, as we shall see, set the wheels in motion for what economic historians have called the Great Divergence—the period after the Industrial Revolution, when the West grew much wealthier than the rest of the world. But in the early days of industrialization, a great divergence happened within Britain, too: wages stagnated, profits surged, and income inequality skyrocketed.

4

THE FACTORY ARRIVES

The annus mirabilis of 1769, as Donald Cardwell has called it, is often seen as the symbolic beginning of the Industrial Revolution.[1] As noted, it was the year when Richard Arkwright and James Watt patented their defining inventions. But the origins of the Industrial Revolution can actually be traced much farther back and largely occurred simultaneously with the evolution of the factory system. Exactly when the factory system first emerged is uncertain. It was first defined in Andrew Ure's 1835 *Philosophy of Manufactures* as something designating "the combined operations of many orders of workpeople, adult and young, in tending with assiduous skill a series of productive machines, continuously impelled by a central power."[2] The first legal definition dates from 1844, where it shall be taken to mean "all buildings and premises . . . where-in or within the close or curtilage of which steam or any other mechanical power shall be used to move or work any machinery."[3] Tracing the beginning of the factory system, in other words, requires tracing the application of machines powered by mechanical force in production. It was with the rise of worker-replacing machines that modern industry arrived at last.

The factory system is best understood in contrast to earlier modes of production. By the early eighteenth century, the domestic system was still predominant in Britain. The economic historian Paul Mantoux's descriptions of life and work before the ascent of the factory are instructive in illustrating just how transformative the Industrial Revolution was. In the domestic system, the typical artisan lived in a cottage with windows that were few and small. There was often only one room, which served as both living space and workshop. There was little furniture so that there was room, for example, for the loom of the weaver. The organization of labor was simple. If the artisan's family was large enough it did everything, with the minor operations divided up. For example, the wife and daughters might be at the spinning wheel, the boys carding the wool, and the husband working the shuttle. While some artisans employed other craftsmen, these workers ate and slept in the same house as their artisan master and did not regard him as belonging to a different social class. The artisan master controlled production and didn't depend on any financier, since he owned both the raw materials and the required tools. Part of his living was derived from the land, with industry often no more than an additional occupation. Production had barely changed since the Middle Ages.

Output growth in domestic industry was slow but steady. As markets became more integrated, the merchant became an indispensable middleman for some artisans, allowing them to sell their goods across Britain and abroad. Because the cloth produced by the artisan was typically not dressed or dyed, the merchant needed to take part in the process of finishing the goods before they could be sold in the market. For this purpose, the merchant had to employ workers, and thus he became a merchant manufacturer. These workers still lived in the countryside, where they were independent contractors, but their livelihoods increasingly came to depend on the merchant manufacturer. If the harvest was bad, they might lack the means to replace some of their tools and equipment. Aware of this dilemma, merchant manufacturers began to provide the tools for production. The independent contractors who had lived in the countryside now became employed and waged and were gathered under one roof in the town where the merchant manufacturer resided. In other words, people gradually lost ownership of the means of production and their autonomy over the pace of work, leading to the creation of what Karl Marx would call a working class. The slow but relentless divorce between capital and labor that characterizes the process of industrialization had begun. From the late seventeenth century onward, this process of alienation swept across the country, although in an uneven manner. In Yorkshire the independence of the artisan remained almost untouched, while in the district of Bradford, wealthy merchants controlled industry. But nowhere had the means of production changed. There was no mechanization.[4]

Why did the factory system emerge when it did? As noted above, the rise of international trade and growing competition among nation-states made it harder to align technological conservatism with political stability. In Britain, growth in wages meant that mechanization was necessary for the country to remain competitive in trade. Manufacturers selling abroad were incentivized to find ways of reducing labor costs due to growing market size, and increasing competition fueled political willingness to allow them to mechanize. What's more, manufacturers had the financial means to adopt expensive machines to replace workers, which for the most part, was economically and technically feasible only in a factory setting. Some equipment required large plants and thus was simply too large and complicated to fit into the living rooms of workers' cottages. Steam engines, iron-puddling furnaces, silk-throwing mills, and so on all required factories.[5] The development of the factory system was therefore a process of technological evolution, driven by economic and political incentives to mechanize. And while as indicated above its arrival is typically dated to the late 1760s, there were earlier factories: by 1718, the silk factory in Derby employed some three hundred workers in a five-story building.

The Rise of Machines

The silk industry began in Britain after a colony of skilled workers left France after the revocation of the Edict of Nantes and settled in the outskirts of London. In its early days, the British silk industry struggled, as smugglers brought cheap imported silk onto the market. The relatively high wages in Britain left domestic manufacturers unable to compete. Finding ways to reduce labor costs was thus a priority. Serious efforts were made to develop silk-throwing machines for this purpose, but they were unsuccessful. All the same, rumors spread that such machines already existed in Italy. In 1716, John Lombe undertook a risky journey to discover the Italian secret. Together with an Italian priest who was the confessor of the proprietor of a silk factory, Lombe devised a plan that gave him access to the machines, and he secretly managed to make drawings of them. These were then sent back to Britain hidden in pieces of silk. After his return a year later, John and his brother, Thomas—who supplied the necessary capital—set up the first silk factory near Derby. The silk-throwing machines were built based on the drawings from Italy, and Thomas Lombe made a fortune. Besides the wealth he accumulated through this piece of industrial espionage, he received a knighthood for his services to Britain. The mechanized silk factories of Derby were surely impressive. But even though vast industrial undertakings existed in Derby and Stockport, they were too small to have any meaningful impact on aggregate economic activity. The silk factories were "giants in an age of pygmies."[6]

The Industrial Revolution was heralded by developments in the silk industry, but it had its true beginnings in cotton. Only a marginal industry in 1750, the cotton industry rapidly expanded and eventually became the largest in Britain, accounting for as much as 8 percent of the gross domestic product (GDP) by 1830. The rise of industrial centers like the city of Manchester was just one consequence of its expansion, which came as British cotton manufactures outcompeted China and India— the leading cotton producers in the seventeenth century. Even in 1750, about eighty-five million pounds of cotton was spun annually in Bengal, while British producers churned out a mere three million.[7] One reason that Britain struggled to compete was that labor in Asia was cheaper. But its cost disadvantage was soon to become an advantage, as international competition spurred efforts to mechanize production.

Before the age of machines, cotton spinning was a laborious process. The whorl and spindle were still used to make fine yarn, while the spinning wheel made coarse yarn. At the dawn of the Industrial Revolution, cotton yarn was manufactured in three stages. In the first stage, packages of raw cotton were opened and any dirt removed. The cotton was then carded—a process in which strands of cotton were aligned into a roving. In the final stage, the roving was then spun into yarn. In a factory setting, each of these stages were mechanized. The pioneer of the modern cotton industry, and thus the Industrial Revolution, was Arkwright. As Mantoux writes, with "Arkwright machine industry ceased to belong solely to the realms of technical history and became an economic fact, in the widest sense of the word."[8] Though Arkwright was responsible for several inventions, his greatest achievement was surely the second Cromford mill, which was opened in 1776. In the mill, water-powered machines were set up in the sequence of production, which became a blueprint for other early cotton factories.[9]

To be sure, Arkwright did not single-handedly transform cotton. He just happened to be the first successful cotton industrialist. Decades earlier, in the 1740s and 1750s, Lewis Paul and John Wyatt developed a promising system of roller spinning. Wyatt was early to realize the potential of the factory system, but he failed to make it work in practice. According to his own estimates, roller spinning could reduce labor requirements by one-third, thereby increasing the profitability of British industry. Before the Glorious Revolution, he would probably have been more careful in masking any labor-saving effects, and the fact that he didn't suggests a greater acceptance for replacing technologies. Still, it would be a mistake to think that the subject had become uncontroversial. As will be discussed in chapter 5, workers in the eighteenth century often smashed machines that they perceived to threaten their jobs. This is probably why Wyatt felt the need to suggest that displaced workers would soon find new and better-paying jobs elsewhere: "An additional gain to the clothier's trade naturally excites his industry as well as enables him to extend his trade in proportion to his gain by the machines. By the extension of his trade he will likewise take in some men of the 33 per cent left unemployed. . . . Then he wants more hands in every other branch of the trade, viz. weavers, shearmen, scourers, combers, etc. . . . These workmen now having full employ will be able to get more money in their families than they all could before."[10] Worker-replacing machines, Wyatt argued, would not enrich only a few industrialists, but Britain as a whole. Though he was right in thinking that machines would eventually enrich Britain, his system of roller spinning did not even succeed in enriching himself and his companion. Lewis Paul was imprisoned for debt, and the machine was seized together with Paul's other belongings. After the two men went bankrupt in 1742, their invention was sold to Edward Cave—editor of the *Gentleman's Magazine*—who set up a small factory in Northampton with five water-powered machines, and eventually the Northampton factory ended up in the hands of Arkwright.[11]

Arkwright's success wasn't that of a brilliant innovator. His achievements rested instead in overcoming a number of engineering bottlenecks that allowed roller spinning to be put to practical use. Unlike the technologies of the Renaissance, the inventions that emerged in eighteenth-century Britain were actually 1 percent inspiration and 99 percent perspiration. And they were all developed with the same objective: cutting labor costs in production. Arkwright's water frame, which put roller spinning into practice, is estimated to have cut the cost of labor for spinning by two-thirds and overall costs of coarse cotton production by 20 percent. The economics of Arkwright's second invention, the carding machine, were similar. And like the water frame it was not an invention of staggering novelty. Indeed, the novelty of his invention was called into question, as his patent was challenged.[12]

The other key invention was James Hargreaves's spinning jenny. Hargreaves is said to have conceived it when he watched a spinning wheel fall to the floor and, while revolving, seem to do the spinning by itself. What is certain is that the machine was very simple. It was a rectangular frame on four legs with a row of vertical spindles on one end. Like the water frame, it was an invention that required no scientific breakthroughs. Its great advantage over the spinning wheel it replaced was that it allowed a single worker to spin several threads simultaneously. Although the spinning jenny was around seventy times more expensive than a spinning wheel, it was still much cheaper than building an Arkwright mill; it took up little space and did not require a factory setting.[13] The fact that it didn't require much alteration to the production process was probably one reason for its rapid adoption.

Though the spinning jenny did not facilitate the rise of the factory system directly, it did so indirectly. Samuel Crompton, who began spinning with a jenny as a boy, was among those who set out to improve it. The result was the Crompton mule, invented in 1779, which combined the draw bars of Hargreaves' jenny with the rollers of Arkwright's water frame. The mule was first adopted in domestic industry, but it was soon applied in a factory setting, where its original wooden rollers were replaced with steel rollers like those used by Arkwright.

As spinning machines ousted the spinning wheel, hand spinners were also ousted. It is thus unsurprising that few workers welcomed it. When rumors spread that Hargreaves had developed such a machine, residents of Blackburn broke into his house and smashed it. Indeed, incidents of workers smashing machinery regularly occurred during the classic years of British industrialization. Hence, even though political power had shifted to those who stood to gain from mechanization, inventors were still unlikely to describe their technologies as worker displacing or even labor saving. Jane Humphries, an economic historian, explains:

> Early eighteenth-century inventors rarely claimed that their innovations saved labour, inventors probably judging it unwise to publicise any adverse effects on local employment. Interestingly, they were more likely to promise employment creation, particularly of jobs for women and children, who by implication would otherwise be a burden on the rates. However, over time it became more acceptable to claim that an invention replaced labour, and by the 1790s patentees had lost all inhibition, with inventors in textiles, metal and leather trades, agriculture, ropemaking, docking and brewing all claiming such an advantage. Even then, savings were not of all labour but mainly the labour of skilled adults. Inventions were often advertised as reducing the need for strength or skill and so facilitating the substitution of unskilled women and children for adult trained operatives. The calculations by John Wyatt in defence of his (and Lewis Paul's) spinning engine are instructive, not least for the alertness shown to the interest of the poor law authorities in creating work for women and children. Wyatt claimed that a clothier who employed a hundred workers might turn off thirty "of the best of them" but take in ten children or disabled persons and thereby be 35 per cent richer, while the parish would save £5 in forgone poor relief. Since such substitution was at the heart of worker resistance to new technology, it

required a certain boldness to make such claims, and probably suggests that more inventions than announced were directed to this end.[14]

The extent to which spinning machines actually saved labor has been intensely debated. What is clear is that they saved labor costs and replaced workers who had done hand spinning. The adoption of the jenny, for example, was not merely a matter of substituting capital for labor but also one of replacing relatively expensive adult labor with that done by children. For example, Arkwright wrote about the ample supply of children in the southern Peak District, which probably explains his decision to locate production there. Indeed, early spinning machines were specifically designed to be tended by children (see chapter 5). As the sharp contemporary commentator Andrew Ure noted, "The constant aim and tendency of every improvement in machinery [is] to diminish the costs by substituting the industry of women and children for that of men."[15]

The benefits, of course, went beyond replacing expensive labor with machines and cheaper labor. Another motive was to gain greater control over the factory workforce, which went hand in hand with the employment of children. Many were pauper apprentices who worked in factories far from their families and friends. When they made up the bulk of the workforce, as they often did, they lacked the protection others were given by the mere presence of adult coworkers. They were often consigned to work without wages or rewards. So to control a large number of lawless children, many supervisors and managers resorted to using the stick rather than the carrot. Relative to adult workers, they had very little bargaining power and were easy to enforce the factory discipline on.[16] Clearly, as Humphries writes, manufacturers were well aware of the advantages of inventing in ways so "as to bypass artisan practices and controls and so sap resistance to change."[17]

All the same, while spinning was turned into a factory system in the late eighteenth century, weaving was still done with hand looms in domestic settings. Therefore, a concern was that after Arkwright's patent had expired, the number of spinning mills erected would surge to the extent that there would not be enough hands to weave all the cotton that was spun. This was discussed by Reverend Edmund Cartwright and a Manchester gentleman who deemed the construction of weaving mills unachievable. Cartwright set out to prove him wrong. The son of a country gentleman, who as a former Oxford student had until now been preoccupied with nothing but literature, Cartwright used the help of a carpenter and a blacksmith to prove his point. He invested decades and a fortune in constructing his power loom. Together with the Grimshaw brothers, Cartwright set up a factory containing four hundred looms powered by steam. Fearing that they would lose their jobs, however, weavers burned it to the ground. During the reign of Elizabeth I, Cartwright's loom would almost surely have been banned for fear that it could cause unrest. But at this time, British governments typically sided with innovators. Instead of banning the invention, the government helped fund it. Cartwright successfully petitioned Parliament for a grant in 1809, making the case that his machines were of great importance to Britain's competitiveness in trade.[18]

There can be no doubt that the power loom was a significant invention. As power looms improved over the course of the nineteenth century, so did productivity: the economic historian James Bessen has calculated that in 1800 it took a hand-loom weaver using a single loom nearly forty minutes to produce a yard of coarse cloth, while in 1902 a weaver could produce the same amount in less than a minute, operating eighteen automatic power looms.[19] But it did so at the expense of the hand-loom weavers it replaced. We shall return to the fate of the hand-loom weaver in chapter 5. For now, it is sufficient to note that with the introduction of the power loom, the triumph of textile mechanization was almost complete.

Iron, Railroads, and Steam

Most people think that the Industrial Revolution was powered by steam. There is surely some truth to this, but steam power was a latecomer to the industrialization process. While the shift from the muscular strength of people and animals to mechanical power was a defining characteristic of the rise of the factory system, the economic impacts of the steam engine became apparent only in the mid-nineteenth century. Without question, steam power had significant advantages over water power, whose use was always constrained by geography. As Marx writes, with the steam engine a prime mover finally arrived, "whose power was entirely under man's control, that was mobile and a means of locomotion, that was urban and not, like the water-wheels, rural, that permitted production to be concentrated in towns instead of, like the water-wheels, being scattered up and down the country."[20] But perhaps more importantly, its application was not confined to any single task or industry: unlike water power, it could be applied in land transportation as well. Like the computer and electricity, the steam engine was an example of what economists call a general purpose technology.

In contrast to other significant technologies of the eighteenth century, which were pure engineering efforts, steam power was a spin-off of the scientific revolution, building on the discovery that the atmosphere has weight. With the steam engine, science first took center stage in technological development, and its importance only continued to grow. Practical use of the discovery of atmospheric pressure began in the late seventeenth century with Thomas Savery, a British Army officer from Cornwall. In its early days, the steam engine—or the fire engine, as it was called—was nothing more than a pump, consisting of a boiler connected to a tank. The engine was developed specifically for the draining of copper mines, but Savery realized its general purpose nature. Beyond mining, he envisioned it being used to supply water to towns and houses, put out fires, and turn the wheels of mills. However, Savery's invention was not even fit for the purpose of draining mines. It worked only at a depth limited to about thirty feet. As soon as Thomas Newcomen's engine emerged in 1712, the fire engine was abandoned. But due to its inefficiencies, the Newcomen engine similarly failed to find widespread use. The vast amounts of energy required in production meant that few manufacturers adopted it. As late as 1770, it was almost exclusively used for draining coal mines and in places where coal was very cheap.

Steam power became economically viable only with James Watt's separate condensation chamber, which allowed condensation to take place without much loss of heat from the cylinder.[21] However, it took several decades for the Watt engine to become viable and required a partnership with Matthew Boulton for financial backing. Watt's steam engine was first used in 1784 in the Albion Flour Mill, in which the Boulton & Watt company had invested for promotional purposes. One year later, it was applied in cotton production and gradually spread to woollen spinning mills, sawmills, malt mills for breweries, pottery man-ufacturing, food processing, sugarcane mills, and iron and coal mining. Still, the immediate macroeconomic impacts of steam power were fairly limited. Calculating the so-called social savings of the steam engine, comparing it to the next best technology, the economic historian G. N. Von Tunzelmann has estimated that the national income of Britain in 1800 would have been reduced by only 0.1 percent if Watt had not in-vented the separate condenser.[22] Needless to say, such estimates are always only as good as the assumptions underlying them, but it is intui-tive that the aggregate economic impacts of the steam engine were neg-ligible before 1800. The available data suggest that a total of 2,400–2,500 steam engines were built in the eighteenth century.[23] And the impact of steam on the overall economy was still very slight as late as 1830, as the economic historian Nicholas Crafts has shown.[24] Though its productiv-ity contribution accelerated thereafter, especially in the period 1850–70, the economic impacts of steam were modest relative to those of later general purpose technologies like electricity and computers. Many sec-tors, including agriculture and construction, were left largely untouched. Nor did steam power enter people's homes. Similar to the economic benefits of electricity and computers, however, the productivity effects of steam were delayed. One reason is that adoption was slow because water power remained cheaper for a long time. There was no equivalent to Moore's Law operating in steam. Thus, most factories were driven by water power until the 1840s. Only around that time did the fuel consump-tion of steam engines drop sufficiently to make them economically viable.

The economic virtuosity of the steam engine became apparent as it revolutionized transportation during the mid-nineteenth century. Be-fore the railroad, the Industrial Revolution was largely local. Large parts of Britain were left unaffected by it. This is not to discount the advances made in transportation during the eighteenth century: the turnpike trusts, authorized by acts of Parliament to levy taxes and issue bonds for road construction, paved the way for a sizable road network in Britain.[25] The growth of the turnpike system during the eighteenth century greatly improved British roads, which together with better stagecoach technology dramatically reduced travel time. In the 1750s, it took some ten to twelve days to travel from London to Edinburgh. By the time of the first railroads in the 1830s, the same distance could be covered by stagecoach in about forty-five hours.[26] Still, at no previous point in history had people been able to travel faster than the speed of a horse, and horse travel was a luxury available to only a small percentage of the population. Most of the Britons who were to become train passengers had to walk to their destinations before the invention of the railroad. And relative to the stagecoach, trains were much faster and cheaper. The first trains traveled three times faster than a stagecoach could, and roughly ten times faster than the highest estimates of walking speed.[27] By the outbreak of World War I, it would have taken Britons an additional five billion hours to undertake all their train journeys, using only the means of transportation available to them before the railroad.

All the same, the arrival of the railroad was a long journey in itself. Not only did it require steam power, but cheap iron was another enabling technology for the railroad and indeed much of the Industrial Revolution. Iron went into the construction of factories, steam engines, machinery, bridges, and rails. Before the eighteenth century, the pig iron produced in blast furnaces was expensive and fragile. The first breakthrough was made in 1709—three years before the arrival of the Newcomen engine—when Abraham Darby developed a method of producing pig iron in furnaces using coke instead of charcoal. Though Darby cannot be credited with the invention of coke smelting, he made it economical. In the period 1709–1850, the average cost of pig iron is estimated to have declined by 63 percent.[28]

The path from coke smelting to railroads is best described by the evolution of the Coalbrookdale Iron Company, led by three generations of Darbys. Its story is an intriguing one because it illustrates the interconnectedness of the technologies that made the Industrial Revolution possible. Darby's method of iron production made cylinders in steam engines (of which Coalbrookdale became a leading producer) more accurate, and it also made steam engines more energy efficient—which helped reduce the cost of producing coke-smelted cast iron. A Bouton & Watt engine was installed in Coalbrookdale in 1774 and upgraded with a newer design in 1805. Over this period the rate of production tripled from fifteen to forty-five tons per day. For the transportation of the tons of materials, Coalbrookdale had a sixteen-mile railroad network by 1757. A decade later, the wooden rails were replaced with rails made of iron, creating the world's first iron railroad. Darby's method, in other words, was more than an enabling technology of the railroad. Coke smelting was how iron rails came to be used in the first place.[29]

While the road to steam-powered passenger rail travel began in Coalbrookdale, the completion of the journey took several decades. The first passenger railroad was opened in 1805, yet the carriages were drawn by horse. Before the railroad, many attempts were made to use steam power for land vehicles, but unpaved roads and tolls imposed by the turnpike acts meant that they failed to gain traction. Richard Trevithick, who built the London Steam Carriage in 1803, was one of the key figures behind the development of the steam-powered railroad. His achievement consisted in making the steam engine lighter and smaller by abandoning the separate condenser, which allowed it to be used more effectively in transportation. However, a number of other significant technologies were also required, including new and better gears, gauges, couplings, and so on. This series of inventions eventually culminated in George Stephenson's Rocket—the steam locomotive that would be used for travel on the first public and fully steam-powered railroad between Liverpool and Manchester.

The opening day of the Liverpool-Manchester Railway in 1830 was one of the major public events of the year, attended by Arthur Wellesley, Duke of Wellington and prime minister of Britain, among others. Although a triumph of British engineering, the introduction of steam-powered passenger travel was not without casualties. Against the advice given to passengers to remain on the train, William Huskisson, former cabinet minister and member of Parliament for Liverpool, who had resigned from the government after a disagreement over Parliamentary reform two years earlier, left the train during a scheduled stop and approached Wellington's carriage, seeking a reconciliation. Caught up in conversation with the prime minister, he saw one of the other locomotives approaching only when it was too late and stumbled on the tracks in front of the train. Because the Rocket was not equipped with brakes, the driver could only bring the train to a halt by shifting into reverse, which was a complex procedure. Huskisson died from his injuries that evening.

Unlike the Hindenburg disaster of the twentieth century, which spelled the end of airship technology, the widely reported Huskisson incident did not reduce railroad frenzy. As people across Britain became aware of this new form of long-distance transportation, there was nothing to hold back its diffusion. The railroad network in Britain expanded to 6,200 miles in 1850 and covered 15,600 miles by 1880. Around that time, the biographer Samuel Smiles described the railroad as "the most magnificent public enterprise yet accomplished in this country—far surpassing all that has been achieved by any Government, or by the combined efforts of society in any former age."[30] Much as the digital technologies of our age have enlarged the world, the railroad allowed people of the nineteenth century to travel beyond their previous horizons. With it, books, letters, newspapers, and people became more mobile, while inventions and ideas traveled at greater speed. Workers could more easily travel in the search for better jobs. And the decline in transportation costs meant expanding markets for manufacturers, which permitted regions to specialize in goods in whose production they held a comparative advantage. As ever-larger factories started to take advantage of economies of scale, local monopolies faced growing competition from outside industrialists. Factories of growing size also found it more economical to adopt steam power in production. In other words, the railroad spurred the adoption of steam in manufacturing, which in turn gave rise to a host of new labor-intensive occupations.

The contribution of the railroad to aggregate growth has been estimated by several economic historians who applied the concept of social savings, comparing the benefits of the railroad to the next best available technology. An early study by Gary Hawke puts the total savings associated with the railroads in the range of 6.0–10.0 percent of GDP in 1865; freight alone accounted for about 4.0 percent of GDP, whereas passenger travel is estimated to have accounted for 1.5–6.0 percent, depending on the value that passengers placed on comfort. However, Hawke's savings for passenger travel is downward biased, as it does not include any benefit for time savings.[31] Accounting for the time saved by passengers, Tim Leunig, an economic historian, estimates that the social savings from passenger travel amounted to some 5 percent of GDP in 1865 and reached 14 percent by 1912, and over that period, railways accounted for a full sixth of aggregate productivity growth.[32] To put these figures in perspective, it has been estimated that the social savings associated with the turnpike trusts at the eve of the railroad era added 1 percent to a much lower base of GDP.[33] However, the main benefits of the railroad came long after its invention. The economic significance of those benefits grew especially from 1870s onward, as the price of third-class travel was reduced sufficiently to allow people who had previously never been able to travel at all to do so for the first time.[34]

In other words, the full benefits of the Industrial Revolution took more than a century to be realized. Parts of Britain were largely unaffected, and the aggregate economic impacts of steam and railroads became significant only in the second half of the nineteenth century. But some places and industries felt the accelerating pace of change much earlier, and the expansion of some of those industries did have an effect on the aggregate statistics after 1800. To be sure, contemporaries took notice of the ascent of industry. In 1835, for example, Sir Edward Baines, a British journalist and member of Parliament, observed that "the causes of this unexampled extension of manufacturing industry are to be found in a series of splendid inventions and discoveries, by the combined effect of which a spinner now produces as much yarn in a day, as by the old processes he could have produced in a year; and cloth which formerly required six or eight months to bleach, is now bleached in a few hours."[35] And nowhere was this more evident than in the cotton city, Manchester.

5

THE INDUSTRIAL REVOLUTION AND ITS DISCONTENTS

In Benjamin Disraeli's *Coningsby*, published in 1844, a character struck by the technological capabilities of the time remarks: "I see cities peopled with machines. Certainly, Manchester is the most wonderful city of modern times."[1] By then, about two-thirds of British cotton production was carried out in factories; steam technology was substituting for muscular power; and the first general purpose railway between Liverpool and Manchester had opened more than a decade before. Modern industry was on the rise. But for all the glory surrounding the city of Manchester and other industrial powerhouses, there was far-reaching concern that the benefits of machines were not being widely shared. The same year that *Coningsby* appeared, Friedrich Engels published *The Condition of the Working Class in England*. The work was written during a stay in Manchester, yet unlike the contemporaries who were impressed by its armies of machines, Engels believed that machines served only to reduce the incomes of ordinary people, while benefiting a few industrialists: "The fact that improved machinery reduces wages has also been as violently disputed by the bourgeoisie, as it is constantly reiterated by the working-men. . . . The English middle classes prefer to ignore the distress of the workers and this is particularly true of the industrialists, who grow rich on the misery of the mass of wage earners."[2] The attitudes of laborers and the "middle sort" toward technological progress differed greatly.[3] As David Landes writes, the middle and upper classes were convinced that they were living in the best of all possible worlds. To them, technology was a new revelation, and the factory system provided the evidence to justify their new religion of progress. But the working poor, "especially those groups by-passed or squeezed by machine industry . . . were undoubtedly of another mind."[4] While industrialists marveled at the rise of machines, workers often resisted their introduction and expressed their fear of unemployment in verses like the following one:

> Mechanics and poor labourers
> Are wandering up and down
> There is nothing now but poverty
> In country and in town;
> Machinery and steam power has
> The poor man's hopes destroyed,
> Then pray behold the numbers of
> The suffering unemployed.[5]

The laboring poor surely had much to complain about. The conditions of the working class did not improve before the 1840s, and for many people, living standards were deteriorating. In rapidly growing manufacturing cities like Manchester and Glasgow, life expectancy at birth was some ten years shorter than the national average, which was only forty years anyway. A significant increase in incomes might have compensated for the undesirable side effects of working and living in factory cities, but such compensation was largely absent. Although some evidence suggests that wages in factory cities were higher than in rural regions, compensating in some measure for the dirty and unhealthy conditions, there was no urban wage premium in the north of Britain, when costs of living are factored in as well.[6] During the classic years of the Industrial Revolution, output experienced an unprecedented expansion, yet the gains from growth did not trickle down to labor.[7] In the period 1780–1840, output per worker grew by 46 percent. Real weekly wages, in contrast, rose by a mere 12 percent.[8] Taking into consideration that average working hours increased by 20 percent in the period 1760–1830, it is hardly an exaggeration to suggest that hourly earnings declined in real terms for a sizable share of the population.[9] The gains of the Industrial Revolution instead went to the pioneers of industry, as the rate of profit doubled.[10] As the capital share of national income expanded,

Peter Lindert has calculated that the income share captured by the top 5 percent almost doubled as well, increasing from 21 percent to 37 percent in the period 1759–1867.[11]

Various definitions and measurements of material standards support the view that many commoners fared worse during the early days of the industrialization process. There is a broad consensus that the average amount of food consumed in England did not increase until the 1840s.[12] Beyond food, households reduced their share of expenditures on non-essential manufactured goods. The retrenchment of consumption among the laboring classes meant that the growing demand for such goods came from the middle classes. Indeed, there was rapidly growing inequality in consumption over the classic years. While overall consumption among households increased, the percentage of households that could afford nonessentials declined among factory workers and those trapped in farming over the first half of the nineteenth century.[13] Though controversial, biological indicators similarly suggest that overall material living standards declined. Bearing in mind that all other things being equal, people who enjoy better nutrition grow to be taller, adult heights can be used as an indicator of people's material standards.[14] Building on this intuition, scholars have shown that the cohorts born in the early 1850s were shorter than any cohort born in the nineteenth century, and that the levels attained in the first decades of the century were not attained again before its last decade.[15] These studies show somewhat different temporal patterns, yet they concur that by 1850 men were shorter than they had been in 1760.

When discussing the decline in biological indicators of material standards, it is hard to separate nutrition from disease and general public health. Poor health was a critical issue during the Industrial Revolution, and its causes were intensely debated among contemporaries. Edwin Chadwick, whose 1842 *Report on the Sanitary Condition of the Labouring Population of Great Britain* investigated the matter, was of the view that public health was mainly an environmental issue. Industrialization spread disease among the poor because they lived in increasingly unhealthy environments. Solving the public health crisis was thus a matter of coping with the health challenges of industrial towns, such as garbage removal, drainage, and providing clean drinking water. In contrast, William Alison, a distinguished professor of medicine at Edinburgh University, insisted that low wages were a cause of poor public health. Unemployment, disappearing incomes, and poor nutrition, he argued, were critical factors in explaining the health conditions of ordinary people.[16]

Both explanations have some merit. One consequence of the Industrial Revolution was the rise of industrial centers, which were infamous not only for their lack of aesthetic appeal but also for their overcrowded and unhealthy environments. As incomes vanished and employment opportunities gradually declined in the countryside, workers increasingly moved to urban areas. In the period 1750–1850, the share of the population living in cities with more than five thousand inhabitants surged from 21 percent to 45 percent. Because of the horrendous living conditions in industrial towns, economic historians speak of an "urban penalty" associated with the rise of the factory system.[17] Even in 1850, life expectancy in Manchester and Liverpool has been estimated at thirty-two and thirty-one years, respectively—well below the national average of forty-one years.[18] But while the view of Chadwick can probably account for much of this difference, vaccination against smallpox was the most dramatic change in the disease environment of the period— which suggests that estimates of the decline in material standards should, if anything, be revised down even further. Moreover, the environmental perspective glosses over the fact that lower incomes also translated into poorer nutrition and shorter people. Even if incomes grew on average, many ordinary citizens saw their incomes vanish while middle-class incomes pulled away, as was also the case in America, where food prices rose more rapidly than the wages of working-class people in the early days of industrialization. A study by the economic historians John Komlos and Brian A'Hearn of the American path to industrialization concludes:

> The decrease in nutritional status of the American population during the structural change brought about by the onset of modern economic growth is inferred from the decline in average physical stature

for more than a generation beginning with the birth cohorts of the early 1830s. The decline occurred in a dynamic economy characterized by rapid population growth, urbanization, and industrialization. The decline in nutritional status was associated with a rise in both mortality and morbidity. These hitherto hidden negative aspects of rapid industrialization were brought about by rising inequality and a marked increase in real food prices, which induced dietary changes through the substitution away from edibles toward non-edibles. The implication is that the human biological system did not thrive as well as one would theoretically expect in a growing economy.[19]

The Conditions of England Question

What caused the misfortunes of the commoner? Even if wages in Britain were higher than in most other places, contemporaries worried that things were changing as machines were depriving people of their jobs. This belief was expressed long before Engels pondered the conditions of the working classes. For example, Sir Frederick Eden's famous inquiry into *The State of the Poor* in Britain in the 1790s expressed far-reaching concern that those living in workhouses—which provided employment and accommodation for the poor—were being made redundant by machines. Eden declared:

> Many persons complain of the introduction of machines into the woollen manufacture; and are of the opinion, that the engines for spinning, and carding wool, do not only deprive the industrious Poor, here, of employment, but are a great national disadvantage: I confess, that, to me, all the arguments I ever heard on the subject, would go to prove, that the land should be dug by labourers, and not cultivated by plows, and horses. . . . It is a great national misfortune that the woollen spinner can, by means of machines, do ten times the work he could perform without them.[20]

As mechanization picked up in industry and agriculture, concerns over the so-called machinery question intensified over the course of the early nineteenth century. Among economists, David Ricardo argued that "the

opinion entertained by the labouring class, that the employment of machinery is frequently detrimental to their interests, is not founded on
prejudice and error, but is conformable to the correct principles of political economy."[21] His famous chapter *On Machinery*, which asserted
that machines reduce the demand for undifferentiated labor and lead to
technological unemployment, prompted a number of theoretical approaches to prove that such unemployment was only a short-term problem.[22] Yet fear of machines, if anything, picked up over the following
decades. Victorian novelists like Charles Dickens and Elizabeth Gaskell,
whose writings capture many of the concerns of the time, frequently
echoed the sentiment of laborers toward machinery. In Gaskell's *Mary
Barton*, set in Manchester in the period 1839–42, a character remarks before a Parliamentary hearing in London: "Well, thou'lt speak at last.
Bless thee, lad, do ask 'em to make th' masters to break th' machines. There's
never been good times sin' spinning-jennies came up. Machines is th'
ruin of poor folk."[23] Contemporaries also worried about the effects of
machines on workers' wages, dignity, morality, independence, and social status. Dickens, who had visited several Manchester factories in 1839
and had himself suffered poverty and hardship, was appalled by the conditions in which people lived and worked. His novel *Hard Times* draws
upon those impressions. Similar to Marx, who contended that "the
worker makes use of a tool; in the factory, the machine makes use of
him," Dickens's fictional descriptions of the industrial landscape of
Coketown, where "the piston of the steam-engine worked monoto-
nously up and down, like the head of an elephant in a state of melan-
choly madness," stress the repetitive aspect of factory work, portraying
the worker as enslaved to the mechanical force of the factory.[24]

Beginning in the 1830s, the machinery question came to form part of
the broader debate on the "conditions of England question"—a term
first coined by Thomas Carlyle to refer to the conditions of ordinary
workers in Britain during the classic years of the Industrial Revolution.
Carlyle was a fierce critic of industrialization and believed that machines
served only to degrade workers. Other social reformers, like Peter Gaskell and Sir James Kay-Shuttleworth, similarly thought that the long
hours worked in the factories and the enforced focus of workers on the

repetitive motions of machines absorbed people's attention to such an extent that adverse effects on their moral and intellectual development were inevitable.[25] The domestic system was often described in idealized terms in contrast with the factory and was commonly referred to as the golden age of industry. Relative to people residing in industrial towns, it was argued, families in the countryside were protected from any outside influence that might injure their children's moral development, allowing their parents to guide their thoughts and feelings. And it was widely believed that domestic industry upheld the family structure, while the factory crowded together workers who had previously been scattered over various parts of the country, creating a new socially deprived class.[26]

Even if the contrasts were exaggerated, there can be no doubt that workers in the domestic system lived lives very different from those of their factory counterparts. In the domestic system, the absence of any clear separation between home and workshop meant that the artisan spent more of his time with his wife and children. The artisan worked according to his own needs, not the needs of any master. Although he had to work long hours, he decided for himself when a day's work began and when it ended. The repulsion felt by many workers toward the factory is thus easy to understand. To them, the factory, with its enforced hours and lack of freedom, closely resembled a prison. As Landes writes, the factory system "required and eventually created a new breed of worker, broken to the inexorable demands of the clock."[27]

Similar to the apocalyptic scenarios painted of the future of artificial intelligence today, people at the time of the Industrial Revolution saw a future in which technology would do more harm than good. Gaskell firmly believed that he was only witnessing the beginning, suggesting that in the future production would be almost entirely automated, with severe adverse consequences for employment:

> The adaptation of mechanical contrivances to nearly all the processes which have as yet wanted the delicate tact of the human hand, will soon either do away with the necessity for employing it, or it must be

employed at a price that will enable it to compete with mechanism. This cannot be: human power must ever be an expensive power; it cannot be carried beyond a certain point, neither will it permit a depression of payment below what is essential for its existence—and it is the fixing of this minimum in which lies the difficulty. . . .

The time, indeed, appears rapidly approaching . . . when manufactories will be filled with machinery, impelled by steam, so admirably constructed, as to perform nearly all the processes required in them, and when land will be tilled by the same means. Neither are these visionary anticipations; and these include but a fraction of the mighty alternations to which the next century will give birth. Well then, may the question be asked—what is to be done? Great calamities must be suffered.[28]

Gaskell was certainly no revolutionary, but it was his work that inspired Engels to ponder the conditions of the working classes, whose misery he attributed to the factory system. Engels's compatriot, Marx, with whom he wrote the *Communist Manifesto*, later expanded on Engels's work in an extensive chapter on machinery in *Das Kapital*, arguing that "machinery, when employed in some branches of industry, creates such a redundancy of labour in other branches that in these latter the fall of wages below the value of labour-power. . . . [N]owhere do we find a more shameful squandering of human labour-power for the most despicable purposes than in England, the land of machinery."[29]

Overall, machinery critics of the Victorian Age raised more questions than they answered. Yet they prompted defenders of mechanization—including Charles Babbage, Andrew Ure, and Edward Baines—to make a case for it. Babbage's *On the Economy of Machinery and Manufactures* presents machines as a helpful complement to the worker's labor, suggesting that "various operations occur in the arts in which an assistance of an additional hand would be a great convenience to the workman, and in these cases tools or machines of the simplest structure come to our aid. . . . The discovery of the expansive power of steam [has] already added to the population of this small island, millions of hands."[30] And in addition to making workers more productive, Ure declared that it was only with the spread of machines that new and better-paying jobs could be created, allowing ordinary people to climb the economic ladder:

> Instead of repining as they have done at the prosperity of their employers . . . good workmen would have advanced their condition to that of overlookers, managers, and partners in new mills, and have increased at the same time the demand for their companions' labour in the market. It is only by an undisturbed progression of this kind that the rate of wages can be permanently raised or upheld. Had it not been for the violent collisions and interruptions resulting from erroneous views among the operatives, the factory system would have been developed still more rapidly and beneficially for all concerned than it has been, and would have exhibited still more frequently gratifying examples of skillful workmen becoming opulent proprietors.[31]

This was also the view of Baines, who felt that powerful agitation among workers in industrial cities was chiefly motivated by "imagination and feeling much stronger than their judgement."[32] Like Babbage, Baines viewed machines as complements to labor rather than substitutes for it, and he argued that all classes of laborers employed in aid of machinery are well remunerated for their work. He added: "Instead of workmen being drudges, it is the steam-engine which is *their* drudge."[33] Examining data on 237,000 workers employed in cotton mills, Baines suggested that their wages were sufficient to buy not only necessities, but also many luxuries. Although his data showed that their nominal wages declined in the period 1814–32, he suggested that improvements in machinery allowed them to buy cheaper goods, compensating for that decline. Nonetheless, Baines observed that hand-loom weavers replaced by the power loom were in a "deplorable condition both in large towns and in villages; their wages are a miserable pittance, and they generally work in confined and unwholesome dwellings."[34]

The evidence on how workers fared economically as the mechanized factory displaced the domestic system is spotty. As we have seen, economic and biological indicators alike suggest that material standards during the classic years of the Industrial Revolution were stagnant or even declining for parts of the population. Such indicators are informative in that they capture the aggregate trajectories of material well-being. In its early days, however, the Industrial Revolution was not an aggregate phenomenon. The textile industry was the first to mechanize, and this is where the force of the factory was most keenly felt. Economic historians like Jane Humphries and Benjamin Schneider have recently drawn our attention to the personal tragedies the mechanized factory inflicted upon parts of the population. Hand spinning, which provided part-time work for hundreds of thousands of adults—mostly women—in rural Britain was the first trade to be affected. Humphries and Schneider show that hand spinning was condemned by mechanization in the late eighteenth century, and its demise came with prolonged agonies for families across rural Britain. As employment opportunities dried up for spinners, family incomes suffered a blow from which rural households struggled to recover.[35]

In the public imagination, however, it is the hand-loom weaver who remains the tragic hero of the Industrial Revolution. The autobiography of Walter Freer, born in 1846, recounted that, "before his birth hand-loom weavers had been labour's aristocrats."[36] Like hand spinners, the skills of weavers were rendered redundant by the onward march of mechanization. Examining the wages of weavers, Robert Allen has shown that poverty accompanied the spread of the power loom. Not only did wage inequality grow rapidly, but the earning potential of weavers was reduced to subsistence level.[37] The case of the hand-loom weaver sheds light on the conditions of England question more broadly: the incomes of many artisans vanished as the factory system spread. Humphries's seminal account of six hundred autobiographies of men who lived and worked during the Industrial Revolution provides many vivid descriptions of the personal tragedies that accompanied the disappearance of hand trades.[38] Their stories resonate with the view of Allen, who wrote: "The *standard of living issue* in the Industrial Revolution was the result of the destruction of hand loom weaving and other hand trades."[39] Indeed, even the detailed investigation into the case of the hand-loom weaver by Duncan Bythell—which is often cited to suggest

that their conditions were not as desperate as sometimes portrayed—
asserts that the power loom led to "the largest case of redundancy or
technological unemployment in our recent economic history."[40] In 1816,
the unemployment rate among weavers in the Stockport district was
60 percent. A decade later, 69 percent of hand-loom weavers in the
town of Darwen were still unemployed, some five thousand weavers in
Glasgow were out of work, and 84 percent of all looms in Lowertown
were standing empty.[41]

There can be no doubt that people suffered as their jobs vanished.
However, the extent to which unemployment was the result of mechanization during the Industrial Revolution is harder to judge, not just
because of sparse statistics but because unemployment has many
causes. The fact that the years of high unemployment among weavers
concurred with economic downturns suggests that unemployment in
part was cyclical rather than technological.[42] John Fielden's estimates,
which draw upon statistics from the relatively prosperous year of 1833,
probably come closer to an unemployment rate that can be deemed
technological: in a survey of the poor in thirty-three townships in Lancashire and Yorkshire, who chiefly worked as hand-loom weavers,
Fielden found that the rate of unemployment was around 9 percent.[43]
Whether such unemployment was permanent or temporary is yet
another matter. Even if the arrival of power-loom weaving caused unemployment among hand-loom weavers in certain locations, some
eventually migrated to jobs elsewhere.

The evidence on the mobility of workers is equally spotty. Around
1850, only a quarter of adults in major industrial cities like Manchester,
Glasgow, and Liverpool were actually born there, suggesting that Britons were highly mobile. However, more able and younger workers were
much more mobile than people in their thirties, who tended to remain
closely tied to their location and occupation and, if they did migrate,
typically moved to locations nearby. Few moved from the rural south
to the factory cities of the north.[44]

As mentioned above, the generational aspect of the Industrial Revolution was reinforced by the preference of factory owners for cheaper,
more docile children as workers and by the nature of work in the

factories. As Ure observed in 1835, "Even in the present day . . . it is found to be nearly impossible to convert persons past the age of puberty, whether drawn from rural or handicraft occupations, into useful factory hands."[45] With the aid of machines, spinning was quickly learned and needed little strength. During a Parliamentary hearing on child labor in 1833, a witness explained that "the discoveries of Arkwright, Watt, Crompton, and other great benefactors of mankind [transformed production in such a manner that] adults were superseded by children, whose wages were lower, and who soon acquired great dexterity."[46] Arkwright's first mills were almost entirely filled with young children, and Hargreaves's spinning jenny was brought to such perfection that a child was able to work 80 to 120 spindles.[47] As the number of spindles on cotton spinning mules rapidly increased, so did the number of children in the factories: the ratio of children to adults increased from 2:1 to around 9:1.[48] Wool combing was no different. Ure observed that a great "many self-acting machines have been contrived for performing the wool-combing operations. . . . After drying, the wool is removed to a machine called the plucker, which is always attended by a child, generally a boy of ten, twelve, or fourteen years."[49] These examples are crucially borne out by the statistics: by the 1830s, children constituted around half of the workforce in textiles, and about a third in coal mining.[50]

For factory owners, children provided cheaper substitutes for adult workers. The only cost of children was often their food and lodging. And when they were paid, their wage was between a third and a sixth of an adult's wage.[51] Not only was children cheaper workers, they were also easier to discipline. Alcoholism was a frequent problem among adult workers. When one engine man at the Boulton & Watt company received some money, he "drank so much the next day that he let the engine run wild, and it was thrown completely out of order."[52] And children could be made to work longer hours. Their working days lasted up to eighteen hours, and as machines could operate ceaselessly day and night, children were frequently forced into shift work, to take full advantage of the technological capabilities of the factory. For the only meal of the day, the children were often allocated no more than forty minutes and had to use part of their break to clean the machines.

Anybody failing to obey the factory discipline risked corporal punishment. Although the investigations made by factory commissioners who looked into the working conditions of children found that cases of extreme child abuse were the exception rather than the general rule, many children unquestionably suffered. In the Litton Mill, Ellice Needham pinched children's ears so that his nails met through the flesh, after hitting and kicking them. Robert Blincoe, a former child laborer, described ingenious methods of torture, including the filing down of children's teeth, the hanging of children by their wrists, and the tearing out of their hair with a cap of pitch. Of course, not every industrialist treated children poorly. But it is nonetheless appropriate here to quote Baines's reference to the factories as "hells upon earth."[53]

Adult workers were rarely treated with the same level of cruelty; their main concern was the threat to their incomes. Even assuming that machines did not temporarily reduce the overall demand for workers, there was no guarantee that the workers who were displaced would find better-paying or less hazardous jobs. Some artisans undoubtedly found jobs in the factories, but the cost of making the transition to a new job was often substantial, as it required occupational and geographical mobility. A recent study of Northamptonshire provides a case in point. As the worsted industry became increasingly mechanized in Britain, the domestic industry in Northamptonshire collapsed, and local producers were left unable to compete. The textile employment share of the Northamptonshire economy fell from 11 percent in 1777 to 1 percent in 1851, and the share of weavers and wool combers in the workforce declined at an even faster rate. The net population decline over the first two decades of the period suggests that some workers moved, perhaps to factory jobs in other regions. But the fact that agricultural employment in Northamptonshire surged as employment in textiles declined equally suggests that many textile workers moved into low-paying agricultural jobs. And because the influx of workers into agriculture could not be absorbed, it can be presumed that unemployment increased.[54]

As production processes kept changing, the skills of workers were becoming obsolete at an accelerating pace, putting pressure on the workforce to become more agile and adaptable to the rush of progress.

Even if unemployment was just temporary, people had to provide for themselves when they were without work by saving some of their limited earnings when they were employed: government-sponsored unemployment insurance was not available in Britain until 1911. Maxine Berg, an economic historian whose writings examine the machinery question in some detail, aptly sums things up:

> Working men and women felt keenly the unprecedented demands for mobility, both geographical and occupational. For them the machine meant, or at least threatened, unemployment, an unemployment which at best was transitional between and within sectors of the economy, and at worst affected the economy as a whole at times of scarce capital. For them the machine was accompanied by a change in the pattern of skills, and involved all too often the introduction of cheap and unskilled labour. . . . But the conceptual changes in political economy over the period are also very closely connected to class struggle. This shows in the very seriousness attached by political economists to the 1826 anti-machinery riots in Lancashire and to the 1830 agricultural riots.[55]

Defenders of mechanization, like Ure, were right in thinking that the factory would eventually create new and better-paying jobs. And clearly everyone might benefit from cheaper textiles. But this was little comfort to the workers who initially found their skills made redundant by worker-replacing technologies. The benefits of industrialization were rarely felt in workers' pockets before 1840s. Perhaps most telling is the reaction of the workers themselves. The Industrial Revolution created new factories and jobs, but it also created many Luddites. For many of the workers living through the Industrial Revolution, opposition was the rational response.

The Luddites

Any discussion of the machinery question must distinguish the short run from the long run. Although workers whose skills were made obsolete suffered initially, the Industrial Revolution eventually brought new goods that had been unattainable to previous generations within the reach of the poor, while creating new and better-paying jobs along the way. Nineteenth-century defenders of mechanization may have been right in thinking that the feelings of workers rebelling against machines were stronger than their judgment. Yet what does the long run matter to workers who lose their livelihoods, especially if they are unlikely to live long enough to see the benefits of the new technology? As the experience of the Industrial Revolution illustrates, the short run can be a lifetime, and of course in the long run we are all dead. Because the benefits and profits from mechanization came to the factory owners at the expense of workers, many reasoned that machinery threatened their livelihoods and must therefore be destroyed.

In fact, an economist might wonder why citizens would ever voluntarily agree to participate in the industrialization process if it reduced their own utility. One explanation, of course, is that the opportunity cost of taking on a factory job was reduced by people's steadily dwindling earning potential in domestic industry. Industrialization relentlessly reduced the price of manufactured goods, leaving rural industry uncompetitive, driving down the earnings of rural workers, and thereby forcing them to seek employment in the factories. What's more, the movement of people from the domestic into the factory system is only a puzzle if one believes that they had any other choice—but they didn't. Some workers did riot against the increasingly mechanized factory. But their efforts to halt the spread of machines were unsuccessful, as the British government sided with the pioneers of industry. As Paul Mantoux writes, "Whether [workers'] resistance was instinctive or considered, peaceful or violent, it obviously had no chance of success, as the whole trend of events was against it."[56]

Clashes between labor and the British government over the adoption of machine technology were not uncommon. On May 10, 1768, the first steam-powered sawmill in Limehouse—for which its founder Charles Dingley had been awarded the gold medal of the Society of Arts—was burned to the ground by some five hundred sawyers who claimed that it had deprived them of employment. Four days earlier, the sawyers had informed Dingley of their intent, but he was unwisely dismissive of their ability to put words into action. As the sawyers arrived, Christopher Richardson, one of Dingley's clerks, is reported to have confronted them, asking what they wanted: "They told me the saw-mill was at work when thousands of them were starving for want of bread."[57] The response of the British government to the Limehouse riot is even more telling than the sawyers' rage against the machines. In contrast to preindustrial monarchs who sought to halt worker-replacing technological progress for fear of social unrest, Parliament passed an act in 1769 that made the destruction of machines a felony punishable by death.[58]

However, the 1769 act did not prevent similar disturbances. In 1772, a factory using Cartwright's power loom in Manchester was burned down. And the riots of 1779 in Lancashire, where the use of machines had spread most rapidly, were every bit as dangerous. A letter by the industrialist Josiah Wedgwood, who was in the district at the time, describes the seriousness of the situation. He was met by a mob of several hundred workers in the road, and one of them told him that they had been destroying all the engines they could find and intended to do so throughout the country. The British government was decisive in its response, and prompt repression followed. Troops were sent from Liverpool, and the rioters were dispersed without much difficulty. A resolution passed after the Lancashire riots suggesting that restrictions on the use of machinery would deteriorate Britain's competitiveness in trade underlines not only the logic of the British government, but also the political clout gained by merchants relative to preindustrial times. Even if the diffusion of machines came at the expense of workers' utility and social unrest followed, Britain's competitive advantage in trade was not to be jeopardized. And the events in Limehouse and Lancashire can hardly be described as isolated examples. Further machinery riots occurred in West Riding, Yorkshire, and in Somerset, just to name a few places. Any violent attempt to compromise the spread of machines was quickly crushed by successive British governments.[59]

Working people also explored other ways to hinder the diffusion of machine technology. Petitions against various machines were laid before Parliament. Among the numerous appeals, wool combers petitioned against Cartwright's combing machine, journeymen petitioned against the use of machines in paper making, and cotton weavers petitioned against machinery they claimed to have driven them out of work.[60] But attempts to hinder the diffusion of machines by political means failed just as dismally. Again, the argument put forward by employers that machines were essential to trade, upon which the fortunes of Britain as a country rested, resonated more strongly with Parliament than did workers' complaints.

However, workers' concerns did not go unnoticed. A Parliamentary inquiry into the woollen industry was set up, which culminated in a famous 1806 report. Addressing the committee of the House of Commons, appointed to consider of the State of the Woollen Manufacture of England, on behalf of the clothiers, Randle Jackson tried a different line of argumentation, suggesting that mechanization deprived producers of their customers by putting them out of work.[61] Despite the argument that restrictions on the use of machines were also in the interest of industrialists, and the suggestion of many witnesses that machines had adverse impacts on people by diminishing the price of labor, the committee came to a much more optimistic conclusion, arguing that the use of "the machines has been gradually established, without, as it appears, impairing the comforts or lessening the numbers of the workmen."[62]

The conclusion of the inquiry into the state of the woollen industry underlines the hopelessness of the case of British workers. In addition to failing to hinder the spread of new technologies, they also failed to get Parliament to enforce legislation prohibiting the adoption of old replacing technologies. Despite having petitioned for a decade to enforce the prohibition on gig mills that dated back to the sixteenth century, Parliament repealed the old legislation in 1809. Further riots followed. During the Luddite risings in the period 1811–16, the Nottinghamshire rioters mainly targeted knitting frames, whereas in Yorkshire, the riots were led by the croppers who rebelled against the spread of gig mills—both old and established technologies. What the various machines that were smashed had in common is that they threatened jobs. And there were many incidents of rebellion. Jeff Horn explains:

Named after a supposed Leicester stockinger's apprentice named Ned Ludham who responded to his master's reprimand by taking a hammer to a stocking frame, the followers of "Ned Ludd," "Captain Ludd," or sometimes "General Ludd" targeted this machine for destruction. The movement began in the lace and hosiery trades early in February 1811 in the Midlands triangle formed by Nottingham, Leicester, and Derby. Protected by exceptional public support within their communities, Luddite hands conducted at least 100 separate attacks that destroyed about 1,000 frames (out of 25,000), valued at £6,000–10,000. Luddism in the Midlands died down in February 1812, but it had already inspired the woollens workers of Yorkshire to take action, beginning in January. A third outbreak took place in April among the cotton weavers of Lancashire. Factories were attacked by armed crowds. Thousands participated in these activities, including many whose livelihoods were not threatened directly by mechanization. Despite the diversity of the crowds, the Luddites generally destroyed only machines that were "innovations" or that threatened employment. They left other machines alone. The specific causes of these outbreaks varied not only according to region but also by sector. Collectively, these initial episodes of Luddism caused perhaps £100,000 of damage. Further waves of machine-breaking, in which a few hundred additional stocking frames were destroyed, came in the winter of 1812–13, in the summer and fall of 1814, and in the summer and fall of 1816 and the beginning of 1817.[63]

However, the Luddites were no more successful than their predecessors, except in forcing the British government to deploy ever more troops against them (see the introduction). Hence, the political situation of workers remained one of despair. When the Luddite riots broke out, another Parliamentary committee heard petitions for relief from the cotton workers and reported to Parliament in 1812. The report makes clear that the government would do nothing to put Britain at a disadvantage in international trade, even if it meant that workers suffered: "While the Committee fully acknowledge and most deeply lament the great distress of numbers of persons engaged in the cotton manufacture,

they are of opinion that [there should be] no interference of the legislature with the freedom of trade."[64] Lord Liverpool, who became prime
minister that year, was even of the view that any temporary aid to
workers who were made redundant would only impede their redeployment, to the detriment of the British economy. In a letter sent to Liverpool, Lord Kenyon explained that even though he did not expect the
conditions of the working classes to improve as a result of the acceleration in mechanization, the government should not seek to counteract the
force of technology.[65] Indeed, in 1812 and 1813, more than thirty Luddites
were hanged.[66]

As many historians have noted, the machinery riots of the Industrial
Revolution were not just the result of workers fearing displacement and
unemployment because of technological advances. The fact that incidents in which workers raged against machines became increasingly
common during the Napoleonic War and the continental blockade of
1806 suggests that other significant factors contributed to social unrest.
Beyond cyclical economic downturns caused by war and disruptions to
trade, the smashing of machinery was an expression of dissatisfaction
with deteriorating incomes; long working hours; and the lack of suffrage, freedom, and dignity. In some cases people rioted against the factory system as such, not just against the spread of machines. The relative
importance of these factors in explaining the many machinery riots is
hard to disentangle, especially since some of them are intimately inter-
twined. Nonetheless, there are clear cases when workers particularly
targeted machines they considered to be the cause of their misfortunes.
In Lancashire, spinners spared spinning jennies with twenty-four or
fewer spindles, while larger ones were destroyed. Moreover, rioters at
times smashed machines that had nothing to do with the factory system.
The "Captain Swing" riots that broke out in 1830 included more than
two thousand riots across Britain that solely targeted agricultural machines. Between September and the end of November 1830, 492 machines were destroyed, the vast majority of which were threshing machines.[67] Again, the British government took a stern line and ordered
the army as well as local militias to take action against any rioters; 252
death sentences were passed, though some sentenced to death were

instead deported to Australia or New Zealand.[68] While historians have long debated the causes of the Captain Swing disturbances, Bruno Caprettini and Hans-Joachim Voth, two economic historians, have shed new light on the matter, using newly compiled data on the diffusion of threshing machines. Their finding is intuitive: worker-replacing technology was the key determinant of the probability of unrest.[69] Where machines were adopted, the probability of riots was around 50 percent higher. Hence, although the smashing of machines in some cases might have reflected general economic and social dissatisfaction among workers, the statistical evidence on the matter suggests that machines themselves were the key cause of the workers' concerns.

Engels's Pause

The observation of Engels that industrialists "grow rich on the misery of the mass of wage earners" was largely accurate for the period he observed. While working people rioted against the mechanized factory, the British economy experienced a period of unprecedented growth. From the viewpoint of economic theory, it is a challenge to square stagnant or even falling real wages with a growing economy. But in the light of current economic trends, economists have developed models that show how wages and the labor share of income can fall as technology progresses.[70] As we shall see, these are also helpful in understanding the classic period of the Industrial Revolution. If technology replaces labor in existing tasks, wages and the share of national income accruing to labor may fall. If, in contrast, technological change is augmenting labor, it will make workers more productive in existing tasks or create entirely new labor-intensive activities, thereby increasing the demand for labor. The divergence between output and wages, in other words, is consistent with this being a period where technology was primarily replacing. Artisan workers in the domestic system were replaced by machines, often tended by children—who had very little bargaining power and often worked without wages. The growing capital share of income meant that the gains from technological progress were very unequally distributed: corporate profits were captured by industrialists, who reinvested them in

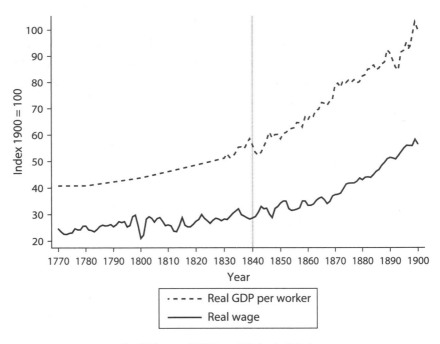

FIGURE 5: Real Wages and GDP per Worker in Britain, 1770–1900
Sources: See appendix, this volume.

factories and machines. This was the period that Allen has described as "Engels' pause," the time that Engels observed and wrote about.[71]

The classic period of the Industrial Revolution was an age of industrial capital. During the first four decades of the nineteenth century, the profit share of national income doubled, as both the share of land and labor declined. As noted above, in the classic years output grew almost four times faster than people's wages. Over the next sixty years, however, the situation changed (figure 5). In the period 1840–1900, output per worker increased by 90 percent and real wages by 123 percent: the great divergence between labor and capital income in Britain was followed by an episode of compression. In 1887, this was observed by the chief statistician of the British government, Robert Giffen. Using data on individual incomes that had been compiled since the introduction of the British income tax in 1843, Giffen showed that the total income of the wealthy had doubled since then, due to a doubling of the number of wealthy individuals. Not only had the number of wealthy people grown, but the total income of laborers had doubled as well, without their numbers increasing substantially. In other words, the wealthy had not become wealthier, there were simply more of them—while laborers were substantially better off.[72]

Giffen's analysis did not come as a complete surprise. The British government was already well aware of the surging levels of tax revenue coming from all sorts of laborers. Addressing the House of Commons more than two decades earlier, Prime Minister William Ewart Gladstone declared that "it is a matter of profound and inestimable consolation to reflect that while the rich have become richer, the poor have become less poor. . . . [I]f we look to the average condition of the British labourer, whether peasant or miner or operative or artisan, we know from varied and incontrovertible evidence that during the last twenty years such an addition has been made to his means of subsistence as we may almost pronounce to be without example in the history of any country and of any age."[73]

A critical question is why real wages eventually began to rise. The most convincing explanation is that technological change became increasingly labor-augmenting instead of labor-replacing, leading to the gradual replacement of physical capital by human capital as the main engine of growth. As with physical capital, the accumulation of human capital—in terms of skills, knowledge, and abilities—can be seen as an investment, because the costs of education and training may be offset by higher earnings at a later stage: productivity and the wages of workers are linked to skills. A well-known study by the economist Oded Galor makes the case that human capital became crucial only during the later nineteenth century, when technological progress increased demand for skills.[74] Although by no means perfect measures of the variety of skills demanded, rates of literacy and years of schooling provide frequently used indicators of human capital accumulation. In the early days of the industrialization process, investment in physical capital rapidly expanded, whereas human capital accumulation experienced little change.[75] In the period 1750–1830, literacy rates in Britain remained largely stagnant, but thereafter they increased rapidly.[76] The average years of schooling among male participants in the workforce equally showed no increase before the 1830s, but they had tripled by the beginning of the twentieth century. Meanwhile, the investment ratio in Britain almost doubled from 1760 to 1831 and then remained at roughly the same level until the outbreak of World War I.[77] Thus, before the 1830s, it seems that technological progress served to increase the demand for physical capital, while replacing workers and rendering their skills redundant. After that point, however, it led to a relative increase in the demand for human capital.

An explanation for the long absence of human capital accumulation is simply that there was little demand for it. As Landes has pointed out, much work during the early days of the Industrial Revolution could be performed without much or any formal education.[78] Literacy was rarely a requirement for employment in industry. Although workers in the factories clearly acquired new skills through on-the-job training, these jobs required less skill than the artisan jobs they were replacing. As Babbage noted at the time, before the factory, each worker had to be sufficiently skilled to perform every task in production—even the most difficult ones. There was no division of labor in the artisan workshop that allowed workers to specialize in a narrow set of tasks. In contrast, the division of labor that characterized the factory allowed only highly skilled workers to perform the most difficult tasks, while unskilled work could be left to unskilled laborers. The surge in child labor that accompanied the spread of the factory system bears witness to this view. As noted above, during the early Industrial Revolution, the share of children (those younger than fourteen) in the workforce grew rapidly and reached about half of the workforce employed in textiles and a third of coal miners during the 1830s.[79] This was not a purely British phenomenon: evidence from the northeastern United States shows that the share of children employed in manufacturing grew during the early stages of American industrialization but peaked in the 1840s.[80] In this regard, the American industrialization process followed a pattern similar to that in Britain. A recent study of the U.S. experience by the economists Lawrence Katz and

Robert Margo points out that "the machines were "special purpose" because they were designed to accomplish specific production tasks that had previously been performed with hand tools by skilled artisans. . . . Although special purpose, 'sequentially implemented' machinery displaced artisans from certain tasks in production, the machines could not run on their own—they required 'operatives.' Operatives were less skilled than the artisans they displaced in the sense that an artisan could fashion a product from start to finish, while the operative could perform a smaller set of tasks aided by machinery."[81] As the factory displaced the domestic system, Katz and Margo find, America experienced a hollowing out of middle-income artisan jobs—quite similar to the experience of today, when computers have caused middle-income jobs to be automated away (we shall return to the contemporary pattern in chapter 9). But instead of being replaced by computer-controlled machines, middle-income workers were replaced by machine-tending children.

However, this pattern becomes murkier over the course of the nineteenth century. To return to the British context, by the 1850s, the participation of children in the workforce had fallen dramatically. Quite possibly, the Factory Acts of the 1830s, which regulated working hours and improved the conditions of children in the factories, increased the cost of child labor and thus spurred the adoption of steam power, though causality might equally have run in the other direction. Regardless, the more widespread adoption of steam power from the 1830s onward, and the subsequent arrival of machines of greater size, meant that more-skilled operatives were required: the complementarity between factory equipment and the human capital necessary to operate it grew stronger as machines became more complex. Contemporaries like Peter Gaskell had already observed this tendency in the 1830s: Gaskell asserted that "since steam-weaving became so general as to supersede the hand-loom, the number of adults engaged in the mills have been progressively advancing; inasmuch that very young children are no longer competent to take charge of a steam-loom."[82]

Determining when technological progress became augmenting is hard. Real wages started to grow after 1840, suggesting that there was an inflection point around that time. But the process was naturally as gradual as the adoption of new technologies. Steam power started to have a meaningful impact on aggregate growth only in the 1830s, around the time when child labor reached its peak. The first railroads were also built in the 1830s, and their growth over the later part of the nineteenth century exacerbated the growing demand for human capital. With the railroads, the Industrial Revolution went from local to national, as larger factories took advantage of economies of scale to serve expanding markets. Because steam power was more rapidly adopted by larger factories, the transportation revolution made production more skill intensive. Accelerating productivity growth, which accompanied the adoption of steam, also helped offset some of the negative consequences of displacement for labor by creating additional demand in the economy. But the reason that real wages grew even more rapidly than productivity is that labor was benefiting from the creation of new jobs more broadly. As factories grew in numbers and size, entirely new skilled occupations emerged. Factories needed managers, accountants, clerks, salespeople, mechanical engineers, machinists, and so on. The growing prominence of skilled occupations was probably a contributing factor to the surge in literacy over the second half of the nineteenth century: workers in skilled occupations were more literate than those in unskilled ones.[83]

Did the growing importance of human capital cause wages to grow after 1840? Skeptics have pointed out that evidence on the evolution of the skill premium, as economists call it, is sparse for the nineteenth century.[84] One study has found that there was no return on human capital, but this is no surprise since it focused on the return on old skills in construction, which were not affected by mechanization.[85] However, mechanization required new skills, which were eventually reflected in workers' wages. In the American context, James Bessen has traced the wage trajectories of factory weavers over the course of the nineteenth century, as the power loom and steam came into use. Similar to the macroeconomic trends in wages in Britain, growth in the wages of factory weavers in America followed mechanization only after a delay of several decades. The reason, Bessen argues, is simple: power-loom weavers needed new skills that took time to acquire, and their skills took even longer to be reflected in their wages. Because the new technologies were initially not standardized across factories, which often used different types of looms, the skills of the weavers were not of much use in factories other than the one in which they worked. Consequently, it was only after machines had become more standardized that factory workers could threaten to leave their jobs if they were not paid for their skills.[86]

Of course, factors other than education and skills might have affected long-run trends in wages as well. Government regulation, such as minimum wage requirements, and the bargaining power of labor unions are other significant variables, but they cannot account for the surge in wages relative to output in Britain around 1840. The first minimum wage in Britain was introduced only in 1909. Moreover, Owenism and Chartism (the most significant ideologies before the mid-nineteenth century) did not establish any significant national labor movement: "There seems to be little evidence . . . that the movement was either extensive or coordinated enough to make much impact on the distribution of income in Britain before 1850."[87] Even as late as the 1890s, when the first comprehensive statistics on union density were published, unionization rates across Britain were low: about 4 percent of the workforce were members of a union.[88] Rising wages are best explained by the industrialization process itself.

Conclusion

It is often suggested that the British Industrial Revolution marked the beginnings of a great divergence between the West and the rest. But just as important, early mechanization was accompanied by a great divergence within Britain as well. This period, which has been called Engels's pause, saw stagnant or even deteriorating living standards for many citizens. It took some seven decades for common people to see the benefits of technological progress trickle down into their pockets. The earnings of hand-loom weavers, for example, rapidly diminished in response to the spread of the power loom. In the early days of industrialization, the gains from growth overwhelmingly went to owners of capital.

In preindustrial times, monarchs frequently held technology back to reduce the risk of political upheaval, as they had little gain and much to lose from creative destruction. By the eighteenth century, however, a new industrial class had become a strong political force in Britain. Since machines were critical to Britain's competitive advantage in trade, and thus the fortunes of industrialists, political leaders were determined to facilitate the diffusion of machine technology—even if it came at the expense of workers' utility. As discussed above, perhaps more important was the growing competition among nation-states and the erosion of the political power of craft guilds, which meant that the ruling classes suddenly had less to lose and more to gain from mechanization. Thus, governments began to side with innovators and pioneers of industry rather than with angry workers. Though it may seem illogical that workers willingly accepted the rise of the factory if it reduced their well-being, this incorrectly presumes the absence of coercion. As the mechanized factory displaced the domestic system, causing the incomes of artisan workers to vanish, many raged against the machine. The Luddites did their utmost to bring progress to a halt, but their case was hopeless, as they lacked political power. That was now held by those who stood to gain from progress, to the detriment of many other people.

The short run, however, must be distinguished from the long run. During the closing decades of the Industrial Revolution in Britain, a new growth pattern emerged: as productivity growth accelerated with the adoption of steam, real wages began to rise in tandem. This happened largely in the absence of organized labor or any significant government intervention to boost wages. The reason is straightforward. During the classic years of industrialization, technology took the form of capital that substituted for skilled workers in existing tasks, as the mechanized factory displaced the domestic system. While new tasks also emerged in the early factories, they required a different breed of worker: spinning machines were designed to be tended by children who cost little to employ, had no bargaining power, and were relatively easy to control. Much like advanced robotics today, machine-tending children replaced middle-income workers. In contrast, in the later stages the arrival of more complex machines required more skilled workers in the factories, who found their skills augmented by technology. And ever-larger factories required more engineers and more skilled people in management and administration. Technical change turned from replacing to enabling, which served to increase the bargaining power of labor as workers' skills became more valuable. It is hardly a coincidence that the arrival of the modern growth pattern marked the end of widespread resistance to machinery. Attitudes toward technological change, as we shall see, are shaped by whether people can expect to benefit from it.

PART III

THE GREAT LEVELING

The Luddites, who opposed technological change, proved very wrong, insofar as new, higher-paying opportunities for work opened up to replace the ones they lost. Henry Ford's invention of the assembly line for producing automobiles in his Highland Park, Michigan, facility actually lowered the average skill levels required to build an automobile, breaking apart the complex operations of the earlier carriage craft industry into simple, repeatable steps that a person with a fifth-grade education could accomplish. This was the economic order that supported the rise of a broad middle class and the democratic politics that rested on it.

—FRANCIS FUKUYAMA, *POLITICAL ORDER AND POLITICAL DECAY*

Fears that technological advances will wipe out jobs aren't new. During the Depression in the 1930s, Charles Beard and other leading American thinkers blamed engineers and scientists for creating the conditions for mass unemployment. In the early 1960s, fears of automation returned as businesses began heavily relying on computers for the first time and machine tools slowed job growth on the shop floor. Even Woody Allen, then a rising standup comedian, took note of the automation hysteria in a routine about how an automated elevator destroyed his father's job.

—GREGG PASCAL ZACHARY, "DOES TECHNOLOGY CREATE JOBS, DESTROY JOBS, OR SOME OF BOTH?"

Could mechanization have progressed uninterrupted if people's standard of living had continued to deteriorate and Engels's pause had persisted? Running the counterfactual is of course impossible, but workers in the early nineteenth century clearly didn't meekly accept market outcomes, and those who found their livelihoods threatened by machines did their utmost to resist them. Why Western countries in the twentieth century rarely saw Luddites opposing the introduction of machinery is a question to which historians have unfortunately paid little attention. The reason, however, is evidently not that the pace of change decelerated. On the contrary, with the introduction of steam power in the second half of the nineteenth century, mechanization accelerated. And following electrification and the arrival of the internal combustion engine—known as the Second Industrial Revolution—mechanization increased even further in the twentieth century.

While other European countries took different approaches to industrialization, what they have in common is that they industrialized later than Britain. They were able to play industrial catch-up by adopting technologies already invented in Britain, which allowed them to take different paths to industrialization. As discussed above, in France, the imminent threat from below during the revolutionary era meant that the government could not repress worker agitation against machinery, as the ruling elites did in Britain. Consequently, as Jeff Horn has argued, industrialization in France was not only delayed but was also fundamentally different, as it was characterized by greater state intervention—which mediated the different interests between labor and capital.[1] And in Prussia, like in Britain, institutional reforms that removed guilds' restrictions on trades were fundamental in facilitating industrialization.[2] But unlike in Britain, in Prussia education played a much greater role in industrialization from the beginning. We saw in chapter 5 that education became important in Britain only in the later stages of industrialization when more skill-intensive technologies, like steam power, came into play. In Prussia, technologies already invented in Britain could simply be put to use with the necessary skills, so that education played a greater role in industrialization from the start.[3]

The focus here, however, is not on explaining divergent paths to industrialization. Catch-up growth will always be different from growth that stems from expanding the frontiers of technology into the unknown.[4] With the Second Industrial Revolution, which began in the 1870s, the United States took over technological leadership from Britain—which means that to trace the frontiers of technology, we shall henceforth have to focus on the American experience. The key question is why resistance to machinery ended. To be sure, the rise of the welfare state made the experience of losing one's job less harsh. But as late as 1930, welfare spending in America (including unemployment benefits, pensions, and health insurance and housing subsidies) accounted for a mere 0.56 percent of the gross domestic product.[5] It took the Great Depression and World War II to spawn the rise of the welfare state. Of course, the relative absence of Luddite sentiment might also reflect the fact that workers joined labor unions to fight for better pay and working conditions. But unlike the craft guilds of the preindustrial era, which vehemently opposed the technologies they perceived to threaten their members' skills, unionized workers didn't focus their anger on machines. While the United States may have had the most violent labor history of the industrial world, after the 1870s, workers rarely if ever targeted machinery. Why? The reason, I shall argue in the following chapters, is that people began to see technology as working in their own interest. Though it is hard to prove that this caused the relative absence of Luddite sentiment throughout the twentieth century, that absence is even harder to explain in isolation from what actually happened to working people as a result of technological progress.

We know that new technology can destroy jobs, create entirely new ones, or radically transform the nature of jobs that on paper appear to be the same. As noted above, if technological change is of the replacing sort, productivity growth alone might not offset its negative impacts on employment and wages. Enabling technologies, in contrast, not only increase productivity but also reinstate labor in entirely new tasks, occupations, and industries more broadly. In a major study, the economists Michelle Alexopoulos and Jon Cohen found that America's great inventions of the period 1909–49 were predominantly of the enabling sort. Some jobs were clearly destroyed as new ones appeared, but overall, new technologies boosted job opportunities enormously. Indeed, gigantic new industries emerged, producing automobiles, aircrafts, tractors, electrical machinery, telephones, household appliances, and so on, which created an abundance of new jobs. Vacancies rose and unemployment fell as the mysterious force of technology progressed.[6] The technologies that Alexopoulos and Cohen examined were those of the Second Industrial Revolution. The authors demonstrate that the internal combustion engine and electricity did more to create jobs than other technologies. Labor-saving machinery had similar effects on productivity, but it did not boost employment by as much—which suggests that electricity and the internal combustion engine also placed workers in previously unimaginable jobs. Thus, economists have come to conclude that this was a period when technology was working in the interest of labor. As Daron Acemoglu and Pascual Restrepo write: "The importance of . . . new tasks is well illustrated by the technological and organizational changes during the Second Industrial Revolution, which [led to] the creation of new labor-intensive tasks. These tasks generated jobs for a new class of engineers, machinists, repairmen, conductors, back-office workers and managers involved with the introduction and operation of new technologies."[7]

In a world where enabling technologies create an abundance of new and better-paying jobs, even replacing technologies are not too bad for labor. While the twentieth century was a period of unprecedented churn in the labor market, it was also one in which most workers could still expect to come out ahead. The ever-growing number of semiskilled jobs created in America's factories provided abundant opportunity even for those who found themselves displaced. Men were able to leave the drudgery of working in the fields for more pleasant and better-paying factory jobs. Indeed, rather than being pushed out of the farms by replacing technologies, most people were pulled into the smokestack cities of the Second Industrial Revolution, which offered better pay and working conditions. At the same time, the mechanization of the household allowed women to leave unpaid housework behind for paid office jobs (chapter 6). Some farm laborers, railroad telegraphers, elevator operators, longshoremen, and so on surely lost out—especially during the 1930s, when the Great Depression meant that there were fewer alternative job options, prompting incidents of machinery angst. Yet even then, there was not any worker resistance to the introduction of machinery comparable to the sort we saw in the nineteenth century (chapter 7). For labor, the benefits of mechanization were simply too large. The continued expansion of manufacturing and rising educational attainment allowed the vast majority to switch into better-paying and less hazardous jobs, making ordinary Americans the prime beneficiaries of progress (chapter 8). True, labor-management relations is likely to have played a role in easing the transition, along with increasing workers' wages and improving working conditions in general. True, the emergence of the welfare state made losing one's job less harsh. The point is not to downplay the importance of social inventions. The point is that technology itself made everyone better off, to the point where members of Karl Marx's proletariat became firmly middle class. Consequently, the rational response of labor was to allow mechanization to progress while minimizing the adjustment costs imposed on working people.

6

FROM MASS PRODUCTION
TO MASS FLOURISHING

When Thomas Jefferson visited Britain in 1786, America was a young republic and a technological backwater. James Watt's steam engine was the technological wonder of the time and proof of Britain's relative technological progressiveness. It is "simple, great, and likely to have extensive consequences," Jefferson remarked.[1] Those consequences would eventually become apparent in America as well. During his travels across North America in 1831, Alexis de Tocqueville marveled that "no people in the world have made such rapid progress in trade and manufactures as the Americans."[2] The once-lagging United States was catching up in some sectors and would soon take the technological lead in others. By the time of the Paris Universal Exposition of 1867, American technological progressiveness was widely acknowledged: Americans received prizes and medals for a wide variety of new technologies, ranging from telegraphs, locomotives, and sewing machines to reaping and mowing machines. Over the next half century, annual patenting almost quadrupled, having expanded thirteenfold after the Crystal Palace Exhibition of 1851. Thus, in 1900, when Edward W. Byrn surveyed recent technological progress at the Patent Office, he observed:

> a gigantic tidal wave of human ingenuity and resource, so stupendous in its magnitude, so complex in its diversity, so profound in its thought, so fruitful in its wealth, so beneficent in its results, that the mind is strained and embarrassed in its effort to expand to a full appreciation of it. . . . With the advent of the dynamo electricity has

taken a new and very much larger place in the commercial activities of the world. It runs and warms our cars, it furnishes our light, it plates our metals, it runs our elevators, it electrocutes our criminals; and a thousand other things it performs for us with secrecy and dispatch in its silent and forceful way.[3]

Electricity and the internal combustion engine were the general purpose technologies of the century, not only affecting every aspect of industry but also transforming the lives of average citizens. One of the great coincidences in economic history is that Karl Benz's successful trial of his gas engine on New Year's Eve 1879 occurred just ten weeks after Thomas Edison's invention of the electric light bulb.[4] Thus, if the annus mirabilis of the Industrial Revolution was 1769, when Richard Arkwright and Watt both patented their defining inventions, 1879 can be seen as the symbolic beginning of the Second Industrial Revolution. The individual contributions of Edison and Benz should not be overstated, however. They were part of the tidal wave of innovation that transformed industry, culminating in the age of mass production. As the American businessman Edward Filene noted, mass production was to the Second Industrial Revolution what the factory system was to the first.[5]

Electrifying the Factory

Mass production will always be associated with the Ford Motor Company. The achievement of Henry Ford and his engineers was not just to develop a revolutionary new vehicle: they successfully harnessed electricity to devise an advanced system of production as well. Before the arrival of the Model T, "mass production" was not even part of our vocabulary. By 1928, when Ford opened its complex in River Rouge, Michigan, the term had become universal. Although it was first properly defined in an article published under Henry Ford's name—but written by Ford's spokesman, William J. Cameron—in the *Encyclopaedia Britannica*, it gained traction before the article appeared due to a Sunday feature in the *New York Times* in 1925 titled "Henry Ford Expounds Mass Production." Ford's ghostwritten article argued that mass production was an American invention. According to his own definition, which entailed the complete elimination of manual labor in the fitting of parts, he was right in thinking so.[6] And like the factory system of the British Industrial Revolution, it was a technological event—it required a machine-tool industry capable of producing interchangeable parts and electric motors to drive the machines. The flood of new goods and gadgets demanded by ordinary Americans could not possibly have been produced in large numbers at sufficiently low cost without these two developments.

In some ways, mass production was an extension of the factory system with new and better technologies. As the historian David Hounshell has argued, the road to mass production began in antebellum America. Eli Whitney, Samuel Colt, Isaac Singer, and Cyrus McCormick are often viewed as the pioneers of the so-called American system of manufacturing, in which complex products are assembled from mass-produced individual and interchangeable parts. The superiority of this system was widely recognized during the 1851 Crystal Palace Exhibition in London. As one visitor observed, "Nearly all American machines did things that the world earnestly wished machines to do. . . . Most exciting was Samuel Colt's repeat-action revolver, which was not only marvelously lethal but made from interchangeable parts, a method so distinctive that it became known as the American system."[7]

Yet the concept of interchangeable parts was not an American invention, if it can be regarded an invention at all. Christopher Polhem, a Swedish engineer, produced a wooden clock using interchangeable parts in the 1720s. The achievement of American industry was to devise sufficiently accurate machine tools to allow uniform parts to be massproduced. For parts to be interchangeable, they had to be identical. The ability to produce identical parts in large numbers only followed successive improvements in machine tools. Much of the machine technology that eventually found its way into Ford's factories had its origins in the production of firearms, from which the machine-tool industry emerged.[8] Colt's dictum that "there is nothing that cannot be produced by machinery" outlined the principle that Ford later would turn into practice.[9] Colt's dictum rested on his faith in machine tools, a faith that was not widely shared at the time. In 1854, during the Crimean War, the British Parliament had formed a select committee to investigate ways of producing arms in the cheapest possible manner. To explore the question of whether small arms could be produced by machines, the America system provided a natural staring point. Joseph Whitworth, a machine-tool manufacturer from Manchester and one of the experts called upon to give evidence to the committee, visited manufacturing establishments in some fifteen American cities. Whitworth was clearly impressed with what he witnessed. In his report, he noted that "whenever it [machinery] can be introduced as a substitute for manual labour, it is universally and willingly resorted to." Whitworth did not share Colt's view that everything could be mechanized. Skilled hand labor would always be required, he argued before the committee. The key point of disagreement concerned the interchangeability of parts. Colt was of the view that machines were capable of churning out identical parts, requiring no manual labor to make them sufficiently uniform. In contrast, Whitworth maintained that sufficient uniformity would be impossible to achieve, so that hand labor would always be needed for fitting purposes.[10]

Great technological leaps are by definition rare, and it was another half century before Ford proved Colt right. Ford was the first to demonstrate that minimizing costs and maximizing production was a profit-maximizing strategy, allowing companies to tap into a seemingly unlimited consumer market. But as Hounshell has noted, this strategy required better machines that made it possible to produce uniform parts. Well aware of the assembly problems associated with interchangeable parts not being identical, Ford's engineers made accuracy the prime machine-tool requirement. Special machines were built for this purpose. "Ford's machinery was the best in the world, everybody knew it," one contemporary authority on the subject remarked.[11] No hand labor for fitting was required in any of Ford's assembly departments. In 1908, when the Model T left the factory, it was the first product to meet these standards.

The remaining challenge was assembling the parts. The solution was found in continuous flow production, which allowed workers to remain stationary as parts were moved to them. A prerequisite for the moving assembly line was the diffusion of electric power throughout the factory to provide light and power machines. Electricity triggered a complete reorganization of production. The moving assembly line introduced at Highland Park, just north of Detroit, Michigan, in 1913 successfully harnessed all of this new technology. Electric motors permitted the use of machines of greater accuracy and speed; electric craneways reduced labor requirements in handling and hauling; electric light facilitated precision work; and electric fans made the factories healthier and temperatures more bearable. And most importantly, the flexibility offered by electricity allowed for constant reconfiguration of the factory to speed up production. The assembly of a Model T took around twelve man-hours in 1913. A year later the same car could be assembled in one and a half hours, while electrification allowed for similar time savings in the production of individual components.

To be sure, many factories were electrified before 1900, yet electricity in its early days was mainly used for lighting. Between the opening of Edison's Pearl Street, New York City, station in 1882 and World War I, the cost of household lighting declined by 90 percent due to the availability of better bulbs and improvements in power generation and transmission. Though the benefits of electric light are hard to quantify, it clearly had significant technical advantages over gas. Electrification made working conditions healthier by reducing the level of air pollution in the factories. It made the workplace safer by reducing the risk of fire, which lowered fire insurance costs. And, brighter light improved accuracy, allowing interchangeable parts to be made sufficiently uniform to eliminate hand fitting. In short, for a variety of reasons it benefited businesses and workers alike. Reports suggest that worker absence due to illness decreased by 50 percent following the introduction of electric light. It is therefore hardly surprising that when the U.S. Government Printing Office allowed workers to choose between electric light and gas, all chose electricity.[12] The superiority of electricity was simply overwhelming in every conceivable way. As an electrician at one printing office pointed out: "The advantage to be gained from changing over from belted steam driving to individual electric motor for printing-press work is not alone in power saved, but better grade of work, less spoiled sheets, cleaner, healthier rooms for employees, less repairs to machinery, and most of all, an increased product without a corresponding decrease in value of the presses by running at too high speed. There has never been a hitch in the motive power; not a motor has given out. In fact, such a freedom from interruption of power has never been known in the history of the office."[13]

The use of electric motors long remained confined to traction, however. At the beginning of the twentieth century, steam and water still provided more than 95 percent of the mechanical power in American factories. But during the early part of the century, supply-side changes propelled factory electrification, and electric motors supplied 80 percent of the mechanical drive in 1929.[14] From Chicago to the Gulf of Mexico and from the Atlantic Coast to the Great Plains, contemporaries observed that America was experiencing a "giant power transformation . . . comparable only with the industrial revolution that began a hundred years ago."[15] This new power transformation, a reporter for the *New York Times* wrote in 1925, was "bringing a Second Industrial Revolution."[16]

One reason for the delay is that electric drive required motors that were sufficiently reliable and efficient to outperform mechanical systems driven by steam or water. Such motors gradually arrived after 1884, when Frank J. Sprague developed the first practical DC motor. Tests soon showed the virtues of the new electrical system: energy loss from friction created by the gears, shafts, and belts of mechanical systems made the benefits of the electric motor all the more apparent. The shift to electric motors was gradual but relentless. As their capacity grew almost sixtyfold over the first half of the twentieth century, their use surged—making electricity by some margin the chief prime mover of industry. Another contributing factor was the arrival of the AC motor, developed by Nikola Tesla, which could be adapted to drive just about any machine: "Tesla's contributions to the introduction and rapid diffusion of AC electric motors was [thus] no less important than Edison's efforts to commercialize incandescent light."[17] In fact, the contribution of electric motors to American productivity was far greater than the productivity effects of electric light. Electricity could no longer just light things, it could power them as well.

But the more important reason why the prime contribution of electricity to productivity took so long to take effect was that realizing the full benefits of factory electrification required experimentation with reconfiguring the factory: "Just as electrifying the city was not merely a matter of substituting streetlights and trolleys for gaslight and horsecars, electrifying the factory was more than a simple substitution of motors for water wheels and steam engines."[18] Electrification, reorganization, and modern management were all part of the same process. As Paul David has noted, the main boost to American manufacturing productivity was delivered only in the 1920s, two generations after the first factories were electrified.[19] This was in large part due to the relatively late transition to unit drive, which has been outlined in some detail by the economic historian Warren Devine Jr.[20] Before 1900, direct drive—by which machines were connected to a centrally located power source, most often a steam engine or water wheel, through a mechanical link—had been the predominant production system. In this system, steam engines and water wheels were simply replaced by electric motors as the source of power, but there was no reorganization of production. The entire network of line shafts and countershafts still operated continuously once the central power source was started, regardless of the number of machines in use. And if the central power source broke down, all of the machines ceased to work, meaning that production was down until the power source had been repaired. Like in steam- and water-powered factories, the power source was often housed in its own room, and thus a jungle of leather belts, pulleys, and rotating shafts was required to distribute power throughout the factory. The basic design of factories had barely changed since the days of water power, when the distribution of power dictated the organization of production.

Dispensing with the apparatus for mechanically distributing power throughout the factory was a critical step in harnessing the flexibility of electricity. Yet fully appreciating the virtues of electric motors as a means of driving machinery took a long time, as mentioned above. Group drive was an intermediate stage in the evolution of the factory that allowed medium-size motors with shorter shafts to drive groups of machines. Then electrical engineers found that they could get rid of the shafts altogether by equipping each machine with smaller electric motors. The switch to unit drive unleashed a revolution in factory design. The flexibility of unit drive allowed factory workflows to be reconfigured to accommodate assembly line techniques, as machinery could now be arranged according to the natural sequence of manufacturing operations. To take full advantage of assembly-line techniques, new factories like Ford's in River Rouge were single-storied, which had further benefits—including substantially lower construction costs per square foot. And the elimination of the system for mechanically distributing power made it easier to install overhead cranes for the hauling of interchangeable parts, as rotating shafts were no longer hanging from the ceiling. Many of these changes saved more capital than labor, and most labor savings were related to construction and maintenance tasks rather than operations. Harry Jerome noted this at the time:

> Changes in industrial technique not altering materially the amount of labor involved in the operation immediately affected, may, nevertheless, alter substantially the labor required in other processes; for example, through a reduction in the floor space required, in reduced waste of materials or damage to product, or through savings in fuel, power, supplies, or wear and tear on machinery. All these—floor space, materials, equipment—require labor in their construction, and any economies in their use have an indirect effect on the demand for labor. Thus the electrification of the power department of a factory may reduce maintenance labor owing to the absence of belts and shafting.[21]

How did workers fare as a result of electrification? We shall return to this question in chapter 8, but apart from the health benefits described above, it is noteworthy that working Americans also saw their incomes grow rapidly. Mass production not only put an array of new goods within the reach of average American households but also put labor on a virtuous cycle in which the explosive growth of manufacturing required an ever-growing number of operators whose skills were made more valuable by more capital being tied up in machines. Factory jobs were simple compared to the emerging jobs in today's tech industries, and workers could learn most tasks swiftly on the job. As the historian David Nye has pointed out, "One advantage of making the tasks brief was that every job could be learned quickly. Not only could virtually anyone work at Ford; workers could be moved around."[22] To be sure, as will be discussed in chapter 7, the speedy churn in the labor market brought some adjustment problems. But on the whole, in the period up until the 1970s, most people could expect to see their wages rise. As the economist Frederick C. Mills observed in the 1930s, "Under the pressure of mechanization men have had to learn to do new things in new ways."[23] In the glass industry, for example, Jerome notes that the "potential displacement of hand blowers [had been] met successfully . . . by the conversion of jar blowers into blowers of other forms of ware unaffected by the machine; or by placing hand blowers in positions as machine workers."[24] In many industries, not just glass, hand labor shifted to machine-aided operations. As factories were electrified, some workers were replaced in maintenance and hauling tasks, but the enlargement of machine operations meant that more productive and better-paying jobs emerged for them (chapter 8). The greatest virtue of the Second Industrial Revolution was that it created entirely new jobs for average people at the same time as making new goods available to them. The flood of electric appliances that entered American households benefited people in their capacity as consumers and producers alike.

Machines of Liberation

If the prime feature of the Industrial Revolution was the mechanization of industry, the defining characteristic of the Second Industrial Revolution was the mechanization of the household. While steam power transformed the factory in the nineteenth century, it left the home untouched. Electricity, in contrast, revolutionized the home as well. Companies like General Electric and Westinghouse led the way in expanding the array of electrical appliances available to the average citizen, such as the iron (first introduced in the market in 1893), vacuum cleaner (1907), washing machine (1907), toaster (1909), refrigerator (1916), dishwasher (1929), and dryer (1938), to name just a few. All of these inventions were by no means American, but America was their largest market and the American housewife their greatest beneficiary. The so-called household revolution did not just make the home more comfortable and enjoyable. It also replaced the housewife in an array of unpaid tasks, which allowed women to take on paid jobs in industry, contributed to a rapid expansion of the American labor force, and delivered a boost to household income. Pioneers of electrification foresaw this development. In 1912, Edison told *Good Housekeeping*: "The housewife of the future will be neither a slave to servants nor herself a drudge. She will give less attention to the home, because the home will need less; she will be rather a domestic engineer than a domestic labourer, with the greatest of all handmaidens, electricity, at her service. This and other mechanical forces will so revolutionize the woman's world that a large portion of the aggregate of woman's energy will be conserved for use in broader, more constructive fields."[25]

Consider the American home of 1900. Most homes still lacked running water, and very few had electricity and central heating (figure 6). In the absence of electricity, light was typically provided either by candles or kerosene lamps. The danger of fire was part of everyday life, and in the worst cases sparks from open-flame lamps or open hearths could set entire homes ablaze. The discovery of fire during the Stone Age was an invention that the household still relied upon. Before the days of central heating, an open hearth provided most of the heat. Wood or coal had to be carried into the home, and removing the ashes and making up the fire each day was tedious work. And despite the considerable effort involved in keeping the dwelling warm, most rooms were as cold as the outside during the winter: "Rags stuffed into cracks provided the only insulation. Most rooms were hotter near the ceiling, floors almost universally chilly."[26] In American bedrooms, iron ingots or ceramic bricks (which had to be heated in the kitchen stove) placed in the bed provided the main source of heat on cold nights. Moreover, the lack of running water meant that for nearly every American the pleasure of taking a bath entailed carrying a heavy tub made of wood or tin into the kitchen, where it was filled with water heated on the stove: "Even in the early twentieth century, working-class housewives had to haul water from hydrants in the street, a task little different from centuries when farm housewives had brought water from the nearest creek or well. All the water for cooking, dishwashing, bathing, laundry, and housecleaning had to be carried in—and then hauled back out after use."[27]

For women in agriculture, who devoted their labor to both the farm and the household, life was particularly harsh. In addition to taking care of the home, nearly all poultry flocks were cared for by women, who were also responsible for feeding the livestock and frequently helped in the fields. A 1920 report by the Department of Agriculture found that women on average worked 11.3 hours per day during the year and 13.1 hours in summer. They had 1.6 hours of leisure time during a summer day and an extra 0.8 hours in the winter. Half of the women surveyed were up at 5:00 A.M. As the majority of farms lacked running water, the day typically began with a walk to the spring or the pump from which water was hauled for the preparation of breakfast—which was made without the help of any electric kitchen appliances. Rural electrification, the report argued, was part of the solution: "As power on the farm is the greatest of time savers for the farmer, so power in the home is the greatest boon to the housewife."[28] Yet rural electrification took off only after President Franklin D. Roosevelt signed the national law establishing the Rural Electrification Administration on May 2, 1936, which provided funds to local cooperatives that private power companies had neglected.

As more homes electrified, American companies made a push to expand the use of electric appliances by seeking to appeal directly to housewives (figure 6). During the 1930s, a pamphlet handed out in Muncie, Indiana, aptly featured the phrase, "Electricity, the Silent Servant in the Home." A General Electric advertisement read: "A Man's Castle is Woman's Factory."[29] The message was clear: hiring the silent electric servant would free up time spent on household chores. Many tasks that are thought of as simple today were not so simple back then. Take, for example, the task of washing. In 1900, 98 percent of households used a scrub board. Hand washing entailed carrying wood or coal to the stove, where the water was heated. Then, the clean clothes had to be wrung out, mostly by hand, and hung on clotheslines to dry. After that,

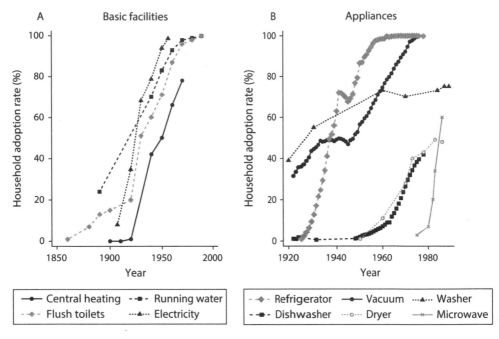

FIGURE 6: The Diffusion of Basic Facilities and Electrical Appliances in the U.S.
Source: J. Greenwood, A. Seshadri, and M. Yorukoglu, 2005, "Engines of Liberation,"
Review of Economic Studies 72 (1): 109–33.

the equally laborious task of ironing began. Instead of an electric iron, heavy flatirons were used, which required continuous heating on the stove. According to one study carried out in the mid-1940s, the electric washing machine saved three hours and nineteen minutes relative to hand washing for every wash. A woman doing laundry by hand walked 3,181 feet to complete the task, but only 332 feet if she used electricity. In similar fashion, the time required for ironing was reduced from 4.50 to 1.75 hours, and the amount of walking required was cut by almost 90 percent.[30]

It is worth noting that electricity became the servant of wealthy and poor alike. New technologies were evidently first adopted by more affluent Americans: even on the same street, some women might be scrubbing clothes by hand as in medieval times, while others had an electric washing machine. Over time, however, along with access to basic household facilities, the relative reduction in the price of appliances made them available to all. When the National Electric Light Association (NELA) surveyed the adoption of electric appliances among Philadelphia households in 1921, it found that only the iron and the vacuum cleaner had reached half of the respondents. Electric refrigerators had a slower start in part because iceboxes provided a cheaper alternative, but also because refrigerators long remained unaffordable to most Americans even if they were willing to replace their iceboxes. In 1928, a refrigerator cost $568, but its price fell rapidly so that the cost was only $137 in 1931—and refrigerator sales skyrocketed in response. Most other appliances were already affordable to most Americans. The cheapest washing machine in 1928 was sold at a price equivalent to three weeks of income; an electric vacuum cleaner cost about a workweek's pay; and the cheapest electric iron could be bought for less than one day's pay.[31]

The average American home soon had "everything electric except the canary bird and the janitor."[32] As prices came down, low- and middle-income households were the greatest beneficiaries. Wealthy households previously had human servants to do the most unappealing tasks and hard physical work. It was only with the household revolution that the rest eventually were able to afford servants, albeit mechanized ones. By 1940, modern conveniences had started to trickle down to a sizable share of the population. From then on, the wealthy and poor alike tapped into the networks of electricity, gas, water, and sewers to which all citizens gradually got equal access. The price of technologies like washing machines, refrigerators, and dishwashers meant that they reached the poor later, but mass production and installment credit soon made them affordable to the majority of people.

Yet the impact of the household revolution on the overall time spent on in-home production remains controversial. An early study by the economist Stanley Lebergott intuitively showed that hours fell sharply.[33] According to his estimates, the workweek of the housewife was reduced by forty-two hours between 1900 and 1966—a staggering finding. However, as suggested by another economist, Valerie Ramey, it seems that Lebergott inadvertently included hours spent on in-home production by all family members as well as domestic servants, rather than only those spent by the housewife.[34] Ramey's estimates, in contrast, reveal a much more modest decline in time spent on in-home production by prime-age women: a reduction of eighteen hours between 1900 and 2005. But strikingly, this decline was offset by men spending more time on work in the home. How can this be reconciled with the obvious fact that household appliances reduced labor requirements in in-home production? One explanation offered by Ruth Schwartz Cowan, a historian of technology, is that the number of hours did not decline, as technology replaced only the domestic labor of servants. To produce the middle-class standard of health and cleanliness of the 1950s, according to Cowan, the American housewife of the 1850s would have needed three or four servants. But with the help of her new electric servants, the American housewife of the 1950s could do so single-handedly.[35]

It is certainly true that over this period domestic service gradually disappeared, and it was not without reason that inventions like French's Conical Washing-Machine were looked upon with distrust, even though in 1860 a writer for the *New York Times* argued that washerwomen had nothing to fear: "This machine will lighten the labor, save the hands, and relieve many of the wearing and disagreeable features of hand-washing, but is not designed to, and will not, take the place of a single young woman at service, we feel confident. If young women would improve their condition, they would do well to avail themselves of all such aids for performing household work."[36] With the benefit of hindsight, we know that laundresses were right that their skills would become obsolete, yet it was only half a century later, when the electric washing machine finally arrived, that their occupation began to wane. At a meeting of the Housekeepers Club of Pittsburgh in 1921, housekeepers complained about washerwomen "playing the phonograph instead of the washing machine" and insisted on new practices where servants will have to "get in tune with the mistress or take the air."[37]

While this lends some support to the view of Cowan, it would be a mistake to think that technology didn't affect household work beyond displacing domestic servants. First, as electric appliances greatly reduced the need for hard physical labor in the home, many women began to spend more of their time on less tedious domestic work, such as teaching and looking after their children. Second, standards of cleanliness and nutrition increased as new technologies emerged.[38] As household tasks were made simpler, they were performed more often. People changed their clothes more regularly, took more frequent baths, and cleaned their homes more routinely: "Instead of dragging rugs outside a few times a year and beating them, the whole house could be vacuumed once a week."[39] Third, Ramey's estimates make more sense when converting hours per person into hours per household: hours spent on in-home production per household fell by a staggering 38 percent in the period 1900–2005. Of course, it is true that the size of the American household also declined over this period, but to suggest that hours per person of in-home production did not fall ignores economies of scale.[40] It doesn't make much difference if dinner is prepared for a family of two or five. Research showing that female labor force participation expanded much more rapidly in areas where electric appliances were used more extensively provides evidence in favor of this view. Figure 7 shows the staggering increase in female labor force participation over the course of the twentieth century. In the period 1900–80, the female workforce expanded by 51 percentage points. An influential study by the economists Jeremy Greenwood, Ananth Seshadri, and Mehmet Yorukoglu, published in the *Quarterly Journal of Economics*, estimates that the household revolution alone can account for 55 percent of this increase.[41] As the home became increasingly mechanized, more women entered the labor market to take on paid and often more fulfilling jobs, and more families suddenly had two incomes, making many American households richer in the process.

Clearly, the increase in the numbers of women entering the workforce was not just due to the labor-saving impact of technology. Cultural and social factors also played an enormous role, but they are beyond the scope of this book. What's evident is that many women entered the workforce despite the continuing pressure on women to stay at home, and technology made it easier for them to do so. Just as the mechanization of in-home production increased the supply of women able and willing to enter the labor market, office machines increased the demand for them. Like the typewriter, which first appeared in 1874, office machines spawned large offices and sparked an early ascent of women in

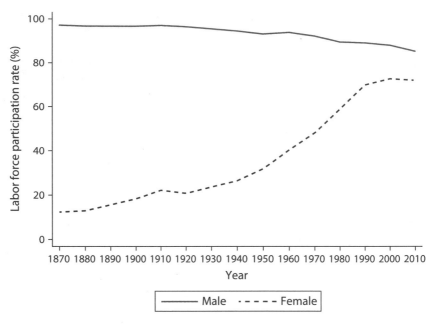

FIGURE 7: U.S. Labor Force Participation Rates for People Ages 25–64 by Sex, 1870–2010
Sources: 1870–1990: Historical Statistics of the United States (HSUS), Millennial Edition
Online, 2006, ed. S. B. Carter, S. Gartner, M. R. Haines, A. L. Olmstead, R. Sutch, and G. Wright
(Cambridge: Cambridge University Press), Table Ba393-400, Ba406-413, Aa226-237, Aa260-271,
http://hsus.cambridge.org/HSUSWeb/HSUS EntryServlet; 2000–2010: Statistical Abstract of
the United States 2012 (SAUS) (Washington, DC: Government Printing Office), Table 7 and 587.
See also Gordon, 2016, figure 8-1.

the clerical workforce. Writing in *Scientific American*, Vincent E. Giuliano explained:

> With the typewriter came an increase in the size of offices and in their number, in the number of people employed in them and in the variety of their jobs. There were also changes in the social structure of the office. For example, office work had remained a male occupation even after some women had been recruited into factories. (Consider the staffing of Scrooge's office in Charles Dickens' "A Christmas Carol.") Office mechanization was a force powerful enough to overcome a longstanding reluctance to have women work in a male environment. Large numbers of women were employed in offices as a direct result of the introduction of the typewriter.[42]

But as we shall see in chapter 8, most of the growth of the clerical work-force happened after 1900, when the proliferation of office machines, the mechanization of in-home production, and the desire to boost family income allowed women to make a great leap forward. In the period 1950–70 in particular, about 11.4 million women newly took up clerical occupations, while only 1.5 million men did so. The term "pink collar," which became increasingly common in the 1970s, referred to the growth in the female, machine-tending clerical workforce.[43] Much of the rise in labor force participation over the twentieth century, in other words, was due to mechanization, not just despite it.

The Ride to Modernity

As essential a part of this story as the transformation of the home and the factory was the revolution in the movement of goods and people. Indeed, the economic historian Alexander Field has argued that pro-ductivity growth in the period 1919–73 can be thought of as "a tale of two transitions."[44] The first involved the redesign of the factory to take advantage of the virtues of electricity, whereas the second constituted a shift toward the horseless age, as motorized vehicles revolutionized trans-portation and distribution. The second transition gradually eclipsed the first, beginning in the 1930s. But the road to motorized transportation began much earlier.

Despite the rapid expansion of the railroad industry, Americans before the twentieth century were still subject to the tyranny of the horse. The railroads allowed goods and people to move faster and more cheaply from railhead to railhead, but horses were needed to transport them to their final destination.[45] One reason horse technology predomi-nated long after the Industrial Revolution is that steam power failed to revolutionize intracity transportation: "Steam engines could not be used on city streets because of fear of fires started by sparks, deafening noise, thick smoke, and heavy weight that shook foundations and cracked street pavements."[46] One solution was to take transportation underground. The Metropolitan Railway, which opened in London in 1863, was initially powered by steam, but the discomfort caused by the smoke in the tunnel made it an unappealing mode of transportation. The steam-powered subway never reached America. Yet intracity transportation experienced continuous transformation during the second half of the nineteenth century, with horse-driven omnibuses, cable cars, and electrified streetcars. When the New York City subway finally opened in 1904, it was powered by electricity and traveled more than ten times faster than the horse-driven omnibus. Together, cable cars, streetcars, and subways gave rise to suburban America, allowing average citizens to escape the city. But for all their benefits, these modes of urban transport left personal transportation untouched. People were still enslaved by the network and timetables of public transportation and relied on travel by horse, which was more flexible and the equivalent of the modern taxi. Its drawbacks were significant, though. First and foremost, horse travel was slow: horse-powered carriages traveled at a speed of no more than six miles per hour, and for longer journeys the horses regularly had to be replaced. Second, horses were too expensive for most Americans. Urban homes rarely had enough space for horses, and most workers lacked the financial means to buy the supplies required to feed them. Consequently, less than a fifth of the working population relied on horsecars to commute between home and work. The vast majority had to walk—which was hardly made more pleasant by the remaining horses.[47] In the late nineteenth century, horses in urban areas are estimated to have dropped 5 to 10 tons of manure per square mile, and carcasses of dead horses could be left on the street for days.[48] The occupations dealing with the removal of horse manure and carcasses are probably not missed even by the most ardently nostalgic people.

In November 1895, a new periodical appeared in New York City called *The Horseless Age*. It was created in response to the advent of the automobile. Despite the many drawbacks of horse technology, a horseless age looked unlikely to most contemporaries. The automotive industry was in an embryonic state.[49] Four automobiles were produced in the whole of America in 1895. The only serious alternative to the horse as a flexible means of personal transportation at the time was the bicycle. However, the "ride to modernity," as it has been called, can only aptly be described as such in that it paved the way for the automobile.[50] Riding a bicycle was, for the most part, a risky undertaking. In *Taming the Bicycle*, Mark Twain describes his attempt to ride a high wheeler in the 1880s. The adventurous experience is best captured by the concluding words of his essay: "Get a bicycle. You will not regret it, if you live."[51] The arrival of the safety bicycle with its smaller wheels, and the subsequent invention of the pneumatic bicycle tire, eventually brought about the golden age of cycling in the mid-1890s: "People went cycle mad; the bicycle industry appeared to be an El Dorado, and there was a rush to engage in it."[52] But cycling in America soon went out of fashion. And as the bicycle industry declined, many bicycle companies instead became producers of automobiles.

The bicycle was in many ways the bridge to the automobile.[53] The father of the American bicycle industry, Albert A. Pope, did not just predict the rise of the motor carriage; he also employed Hiram Percy Maxim to realize his prediction. The Pope Manufacturing Company never became a leading producer of automobiles—Pope declared bankruptcy in 1907—but the industry as a whole contributed to solving many of the mundane engineering problems that needed to be addressed for the later mass production of automobiles, such as accurately machined gears and pneumatic tires. Of perhaps even greater importance was that with the arrival of the bicycle, Americans first experienced the freedom of horseless personal transportation. Maxim claimed that he had first conceived of the benefits of the automobile when riding his bicycle. Looking back in 1937, he recalled: "It carried me over a lonely country road in the middle of the night, covering the distance in considerably less than an hour. A horse and carriage would require nearly two hours. A railroad train would require half an hour, and it would carry me only from station to station. And I must conform to its time-table, which was not always convenient."[54] According to Maxim, it was the experience that many people gained from cycling that created demand for convenient and cheap personal transport—and thus the automobile.

Efforts to develop motor carriages had already begun in the eighteenth century, using steam engines. However, despite decades of experimentation, steam cars never reached the mass market. Steam engines were too heavy, unsafe, and inefficient to revolutionize personal transportation. The automobile revolution would have to await the development of the internal combustion engine. The first gas engine was patented by Nikolaus Otto in 1864, but it was unsuitable for road transportation. Practical designs of automobiles began with gasoline engines and carriages built by Gottlieb Daimler and Wilhelm Maybach and, independently, by Karl Benz. They were the first to successfully power a road vehicle with a gas engine. All the same, for some time it looked like the electric motor might be the one to propel the horseless carriage: "In 1900 there appeared great likelihood that the electric automobile might advance to perfection as rapidly as other branches of electrical appliances had done."[55] However, a number of events tipped the balance in favor of the internal combustion engine. Charles Kettering's invention of the electric starter made gasoline-powered cars easier to operate; the expansion of the intercity road network meant that the greater range and speed of gasoline-powered cars could be taken advantage of; large petroleum discoveries contributed to a decline in the price of fuel; and as a result of advances in mass production in Ford's factories, the price of gasoline-powered automobiles fell rapidly, while the price of electric cars did not.

As Pope noted, the days of the horse were already beginning to wane, but for the automobile industry to take off, good roads were required, "not only in and about cities, but throughout the entire country."[56] Such complementary infrastructure had to be built from scratch. The two million roads that existed in America at the beginning of the twentieth century are best described as a network of dirt tracks. For the few people who could afford to travel by motor carriage, flat tires and blowouts were the norm rather than the exception. In the early twentieth century, a Vermont doctor and his chauffeur were among the first to drive across the country. Their trip from San Francisco to New York took them sixty-three days.[57] Days have since become hours: the same journey would now take some forty hours by car, in the absence of traffic jams, according to Google Maps. The rapid expansion of the road network during the 1920s and 1930s allowed drivers to cross America from coast to coast on highways without even having to drive through unpaved stretches on major routes. Growing traffic flows in and out of cities were helped by new displays of American engineering excellence, including the Benjamin Franklin, George Washington, Golden Gate, and Bronx-Whitestone Bridges. And as the number of gasoline stations increased, roadside commerce started to flourish as well, creating a whole set of new jobs.[58] Before 1920, there was virtually no commerce along American highways. A colorful description from 1928 reveals a marked contrast: "Every few hundred yards there is a . . . filling station, half a dozen colored pumps before it. In connection with the stations and between them are huts carrying the sign 'Hot Dogs.' Where there is neither hut nor filling station there are huge hoardings covered with posters."[59]

Scholars disagree on whether road construction paved the way for the automobile industry or vice versa. It is probably fair to say that causality ran in both directions.[60] Without the growing demand for automobiles, there would have been little incentive for both business and governments to invest in expensive infrastructure. The automobile revolution could not have happened without better production technology to make automobiles affordable to ordinary citizens. The 1901 Mercedes—the "first modern car in all essentials" and the holder of the world speed record, having reached 40.2 miles per hour—was sold on the American market at a price of $12,450, roughly twelve times the annual per capita income at the time.[61] Consequently, automobile ownership was at first only attainable for a fraction of the population. Things changed markedly only with the appearance of Henry Ford's revolutionary Model T. When its production began in 1908, it was priced at $950; by the time production ceased in 1927, its price had fallen to $263. Expressed as a ratio to annual disposable income per person, its purchase price fell from 316 percent in 1910 to 43 percent in 1923. That year, the dominance of the Model T reached its peak: over half of the cars sold in America were Model Ts. The expansion of installment credit throughout the 1920s further helped reduce the annual share of disposable income citizens had to spend to own a car, making automobiles affordable to all but the poorest Americans. The share of households with a registered motor vehicle exploded in response, growing from 2.3 percent in 1910 to 89.8 percent in 1930.[62]

For most Americans, the automobile did not just replace the street-car as a mean of transportation between the home and the factory. People began using their cars to go shopping, visit friends and relatives, and drive out to the countryside on the weekends to escape the noise of the city. By changing the way people worked and lived, the automobile changed the face of the North American continent. With better and cheaper transportation, cities were no longer merely a great congregation of people. Cities developed special areas for factory work, shopping districts, and suburban neighborhoods for living. They were subdivided into areas of work, consumption, and living. And many of those working in the city no longer had to live within its limits. In the words of Ralph Epstein, writing in 1927: "The countryside is not only brought nearer the city; the city itself becomes, in all but its corporate name, indeed a part of the surrounding country. New York is no longer merely Manhattan, Brooklyn, and the Bronx; it is also Long Island, Rye, New Rochelle, and indeed a part of Connecticut and New Jersey."[63]

In the same way motor vehicles transformed the city, they revolutionized farming. The transition from horse to motor power was without question the greatest transformation of agriculture since the domestication of animals to substitute for human muscle. During the nineteenth century, the mechanization of farming lagged behind that of manufacturing simply because steam engines were unsuitable for unstructured environments and too expensive for most farmers.[64] Even breakthrough inventions of the nineteenth century like Cyrus McCormick's reaper were pulled by teams of horses.

As automobiles replaced horses in transportation, tractors replaced them in agriculture. The share of farms with tractors surged from 3.6 percent in 1920 to 80 percent in 1960. Over the same period, the numbers of horses and mules on farms declined from twenty-five million to a mere three million (figure 8).[65] And while causing a mass redundancy of horses, they delivered a significant boost to economic growth: the economist William White estimates that the direct social

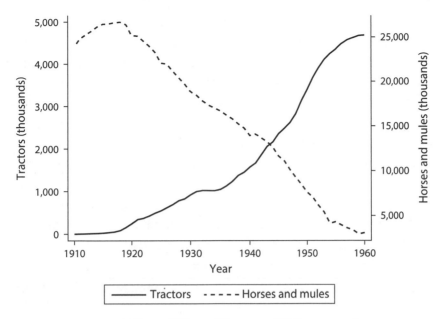

FIGURE 8: Horses, Mules, and Tractors on U.S. Farms, 1910–60
Source: R. E. Manuelli and A. Seshadri, 2014, "Frictionless Technology Diffusion: The Case
of Tractors," American Economic Review 104 (4): 1368–91.

savings of the tractor were in excess of 8 percent of the gross national product in 1954. Tractors eliminated huge inefficiencies, such as the fact that horses consumed a fifth of farm output.[66] But while the tractor is responsible for most of the downfall of the farm horse population, motor trucks and passenger cars contributed as well by making possible new forms of swift delivery and distribution of goods. With automobiles and motor trucks, distances that had formerly taken a day by horse could be covered in an hour. Consequently, the cost of hauling fell dramatically, allowing American farms to serve larger markets.[67] The widening radius of farming operations is documented by a 1921 study of Corn Belt farmers by the Department of Agriculture, which showed that after the introduction of motor trucks, many farms changed the markets for some or all of their products.[68]

Motor vehicles powered economic growth well beyond agriculture. In the 1930s, transportation and public utilities together with wholesaling and retailing accounted for nearly half of economy-wide productivity growth. And trucking and warehousing accounted for about a third of the growth in the transport and public utilities sector.[69] Significant investment in roads meant that by the time of the Great Depression, truckers could drive across the continent without having to go through any unpaved stretches. In response, truck registrations increased by 45 percent in the period 1929–41. During the Great Depression, businesses were able to take full advantage of the alternative and more flexible methods of distribution offered by trucking. In the cities, department stores began to employ trucks to deliver packages to surrounding rural areas, allowing consumers to just place a telephone call instead of having to drive into the city. And in short-haul transportation, trucks provided a flexible alternative to the horse in moving goods between railheads, farms, factories, wholesalers, and retailers.

The productivity trends of the interwar years reasserted themselves after the Paris Peace Treaties of 1947.[70] The greatest boost to productivity was the continuation of the second transition observed by Field, enabled by motorized vehicles. But like the productivity gains of electrification, the full impact on growth of the internal combustion engine was delayed. The reason was not just the war but also that the infrastructure required to support motorized transportation was lagging. It was only after infrastructure spending resumed and accelerated with the Federal Aid Highway Act of 1956 that the full benefits of motorized vehicles could be realized.[71] Until then, the railroad was deemed more effective than highway freight, as is evident from President Roosevelt's report on interregional highways, submitted to Congress in 1944. The report argued that "all the evidence amassed by the highway-planning surveys points to the fact that the range of motor truck hauls is comparatively short. There is nothing to indicate the probability of an increasing range of such movements in the future."[72] As absurd as this prediction might seem retrospectively, the experience of World War II lent it some support. During the war, the percentage of ton-miles freight traveled by truck declined to 5.6 percent in 1943, at which time the railroads carried 72 percent of intercity freight. All the same, the relative importance of the railroads declined thereafter. By 1958, trucks carried 20 percent of all ton-miles; a figure that rapidly increased after the completion of the interstate highway system. Economists have affirmed the contributions of the highway system to postwar productivity, finding that spending on interstate highways was responsible for over a quarter of the increase in American productivity in the 1950s and 1960s, while during the 1980s it accounted for a mere 7 percent increase.[73] Indeed, the heyday of trucker culture came toward the end of the golden age (1947–73) of productivity growth. In the 1970s truck drivers became the new American cowboys and were frequently romanticized in blockbuster films like *Smokey and the Bandit*.

While the trucking industry itself fueled American productivity, it also had significant spillover effects on transportation and trade more generally. In conjunction with the rise of the trucking industry, the container revolution was an engine of postwar growth. And containerization emerged directly from trucking. Malcom McLean, a trucking entrepreneur, invented the container as a mean of integrating the segmented industries of shipping, trucking, and railroads. The first successful container shipment dates from April 26, 1956, when McLean's Ideal-X made its maiden voyage from Port Newark to Houston, Texas. This seemingly unspectacular shipment was, together with Christopher Columbus's discovery of the New World, one of the key events in the history of trade. In the same way that railroads and steamships paved the way for the first wave of globalization—which abruptly ended with World War I—containerization was the technology that underlay the second wave of globalization, beginning in the postwar years. According to a recent study, the container boosted bilateral trade by 320 percent over the first five years of its adoption.[74]

The container did not just change the world of trade. Containerization was a driver of Smithian and Schumpeterian growth alike. Contemporaries hailed it as "an extension of our mass production techniques into the carrying of overseas trade."[75] Besides eliminating twelve separate handling steps in the moving of goods between the manufacturer and the consumer, container terminals are estimated to have increased the volume a dock laborer was able to handle from 1.7 to 30 tons per hour.[76] Although the construction of such terminals was capital intensive, faster rates of throughput saved substantial amounts of capital—not to mention the capital savings associated with the associated decline in theft. A well-known joke at the New York wharves before the age of containers was that the dockmen's wages were "twenty dollars a day and all the Scotch you could carry home."[77] Containerization put the lid on the Scotch, reducing the cost of insuring cargo in the process.

With the advent of the container, winds of change swept through American harbors. These winds meant stormy seas for longshoremen. "Just how far it will go, no one knows, but the idea of moving domestic and overseas cargo in boxes or containers is swelling like a tidal wave," the *New York Times* noted in 1958.[78] Like many transformative technologies, containerization was not welcomed by everyone. Before the age of containers, ports were places crowded with thousands of longshoremen loading and unloading ships. After containerization, large crews of longshoremen were replaced by machines. All handling was gradually done with cranes and special forklift trucks. But the longshoremen were not passive bystanders. In 1958, the president of the New York district of the International Longshoremen's Association made it clear that dockmen would not handle containers, claiming that they deprived too many longshoremen of work. "We do not propose to be drowned in this wave," one negotiator for the longshoremen union added.[79] Labor-management disputes surrounding containerization were a recurring theme through the 1960s. Yet even the unions underestimated the dramatic impact the container would have on their members' jobs. In 1968, Thomas W. Gleason, president of the International Longshoremen's Association, reported that the port of New York had provided 40.7 million man-hours of work, down 3.0 million man-hours from the previous year. Under pressure from containers, he predicted, total man-hours could be reduced to 28.0 million. Eight years later, when a federal court dismissed a labor-management agreement designed to protect the jobs of remaining longshoremen, the port accounted for a mere 19.0 million man-hours.[80]

But while the twentieth century clearly saw the spread of some replacing technologies, most progress was of the enabling sort. One reason that the horseless age was not accompanied by a jobless age is that human workers, unlike horses, have the means of acquiring new skills, which allows them to take on tasks outside the realm of machines. The automobile, the motor truck, and the tractor reduced the comparative advantage of the horse as a prime mover in agriculture and a means of moving goods and people around. The result was a gradual reduction in the horse population, not of the working population. For example, people employed by the street railroads "engaged in a struggle for existence against the competition of the private automobile and the motorbus."[81] But employment in the operation, service, and maintenance of motorized vehicles increased greatly: the occupation of truck driver is now the largest single occupation in many American states (see figure 20). In addition, vast employment opportunities were created in the production of motorized vehicles. As we shall see in chapter 8, the reduction in the demand for farm laborers was accompanied by the expansion of industry, which meant that farmworkers had good alternative job options. The automobile industry, for instance, soon grew to eclipse the railroad as the leading employer of American workers. From being an industry so unimportant that it was not even reported separately in the 1900 census, automobiles became the largest manufacturing industry in 1940. Employment in automobiles grew 765 percent faster than total manufacturing employment over the first three decades after the industry emerged.[82] To put this figure in perspective, employment in semiconductors grew 121 percent faster than the overall manufacturing employment in the three decades after its invention in 1958.[83] Research by Alexopoulos and Cohen affirms the general perception that motor vehicles boosted employment more than other technologies.[84] And their employment contributions extended far beyond the auto industry. They also fueled job creation in supplier industries, construction, transportation, tourism, car services, and road commerce. As the historian David L. Lewis wrote in 1986, the auto industry's glory days of the 1950s and 1960s will surely not be repeated, but it still "directly employs some 1.2 million people, while the payrolls of auto dealers, service stations, and other related businesses are several times larger. All told, the industry provides jobs for one in every six American."[85]

7

THE RETURN OF THE MACHINERY QUESTION

In 1930, William Green, then president of the American Federation of Labor, wrote an article for the *New York Times*. The story line was a familiar one:

> Today our captains of industry recount with pride increases in productivity, installation of machines. . . . They glory in things—in technical progress, in management, in the progress of science—but what thought do they give to musicians displaced by music reproductions; to the art of the actor, forgotten in the latest movietone; to the Morse operator displaced by the teletype; to the steel worker displaced by a new process; to the carpenter watching a house assembled by units; to the printer turned out by the teletypesetter? Such workers in thousands have been turned out without jobs, and without future employment in the craft in which they have invested their all.[1]

Indeed, when reading Green's article, it is hard not to think of Friedrich Engels's assertion that industrialists "grow rich on the misery of the mass of wage earners."[2] Yet such parallels can be taken too far. True, episodes of machinery angst emerged from time to time in the twentieth century. But while some workers struggled to adjust to mechanization, there was no Engels's pause. As we will see in chapter 8, wages rose in tandem with productivity, working conditions improved, and America became more equal as technology progressed. It is somewhat telling that not even union leaders advocated slowing down the pace of change. Unlike Britain in the classic period of the Industrial Revolution, America in the twentieth century didn't experience outright resistance to technological progress because, as Green went on to argue in his article, mechanization brought improved material well-being for the majority of the working population. The Industrial Revolution had shown that society as a whole could gain from technological progress over the long run but that mechanization could bring a painful period of transition for some. Green also pointed to its adverse effects on those members of the workforce whose skills were made redundant. To support workers who went through pain for the sake of society's gain, he proposed a dismissal wage for workers to help them cope with adjustment, a shorter workweek so that leisure would become more widely shared, a system of federal employment agencies to make job matching more efficient, the provision of vocational training to update workers' skills, and higher wages to stimulate demand and bring industries to work at full capacity.

Green was by no means alone in being concerned about the machinery question. In the early 1930s, discussions of machines stealing citizens' jobs were featured in radio talk shows, films, and academic conferences, and the Committee on Labor of the House of Representatives even held several hearings on the subject.[3] The return of machinery angst cannot be explained in complete isolation from the Great Depression, which certainly exacerbated and prolonged concerns about technological unemployment. Yet the latter was not the cause of the former. As the economic historian Gregory Woirol has pointed out, "The honor of starting the technological unemployment debates belongs to Secretary of Labor James J. Davis."[4] In a 1927 speech, two years before the outbreak of the Great Depression, Davis was the first to take note of the technological challenges facing labor:

> For a long time it was thought impossible to turn out machines capable of replacing human skill in the making of glass. Now practically all forms of glassware are being made by machinery, some of the machines being extraordinarily efficient. Thus, in the case of one type of bottle, automatic machinery produces forty-one times as much per worker as the old hand processes, and the machine production requires no skilled glass blowers. In other words, one man now does

what 41 men formerly did. . . . The glass industry is only one of many industries that have been revolutionized in this manner. I began my working life as an iron puddler, and sweated and toiled before the furnace. In the iron and steel industry, too, it was long thought that no machinery could ever take the place of the human touch; yet last week I witnessed the inauguration of a new mechanical sheet-rolling process with six times the capacity of the former method.[5]

But like Green, Davis was no Luddite. Technological progress, he added, must continue:

If you take the long view, there is nothing in sight to give us grave concern. I am no more concerned over the men once needed to blow bottles than I am over the seamstresses that we once were afraid would starve when the sewing machine came in. We know that thousands more seamstresses than before earn a living that would be impossible without the sewing machine. In the end, every device that lightens human toil and increases production is a boon to humanity. It is only the period of adjustment, when machines turn workers out of their old jobs into new ones, that we must learn to handle them so as to reduce distress to the minimum. . . . Please understand me, there must be no limits to that progress. We must not in any way restrict new means of pouring out wealth. Labor must not loaf on the job or cut down output. Capital must not, after building up its great industrial organization shut down its mills. That way lies dry rot. We must ever go on, fearlessly scrapping old methods and old machines as fast as we find them obsolete.[6]

Hardly any serious commentator argued in favor of slowing down the pace of mechanization, despite the sense that manufacturing employment was beginning to wane. Two new sources of productivity data, published in May 1927 and suggesting that manufacturing employment had fallen between 1919 and 1925, had sparked the technological unemployment debates. During the 1927 December meetings of the American Economic Association, the newly compiled data naturally became a subject of intense debate. As the economist John D. Black remarked,

"It is hard to believe that there was an actual decrease of 7 per cent in the number of workers in agriculture, manufacturing, mining, and railway transportation in this short period."[7] Most analysts had been assuming that the exodus from agriculture had primarily been absorbed in manufacturing.

All the same, a series of studies made it increasingly hard to dispute the fact that some workers struggled to adjust. In 1928, before the Great Depression, the Senate Committee on Education and Labor asked the Brookings Institution to inquire into "how many dispossessed laborers were being absorbed by American industry."[8] The study traced the fates of 754 workers who lost their jobs to mechanization across a variety of industries. While 11.5 percent had found a new job within a month and 5.0 percent were still looking for work after a year, the vast majority were out of work for more than three months but eventually found employment elsewhere. Other research similarly showed that transitional costs could be significant. The economist Robert Myers studied 370 displaced cutters in Chicago's clothing industry during the period 1921–25 and found that the average duration of unemployment was 5.6 months, although after a year 12.9 percent of the former cutters were still without work.[9] And workers' ability to adjust was highly age dependent. Fully 90 percent of the displaced cutters over the age of forty-five either failed to find work or were forced into lower-paying jobs. In contrast, for the most part young adults managed to switch into better-paying jobs. Both studies found that roughly half of displaced workers found new jobs that paid as well as their former ones.[10] As in Britain in the Industrial Revolution, in America older people found it especially hard to adjust to new technology. And many workers who found new employment were left worse off economically, at least in the short run.

Unsurprisingly, workers with highly specialized skills faced the most serious adjustment problems. Many musicians, for example, struggled to adapt as the motion-picture branch of the entertainment industry underwent exceptionally rapid technological change. The adoption of sound-producing machines made unnecessary the employment of living musicians in the projecting theater, so that the number of employed musicians declined. In Washington, D.C., the union of musicians and the owners of the motion-picture theaters agreed on a reduction in employment of 60 percent. Part of the decline was offset by the growing demand for musicians by local radio stations. But only a few musicians could make a living in radio broadcasting. Like other displaced worker surveys, a study of a hundred displaced musicians in Washington's motion-picture theaters showed that the majority experienced a subsequent fall in earnings. On the bright side, the loss of the theater musician was the gain of the motion-picture machine operator. The switch from the silent to the sound movie was "accompanied by an improvement in the status of the motion-picture machine operator, both by replacing the customary boy helper with a licensed operator, and by increasing the average earnings of the projectionists." According to representatives of five leading moving-picture theaters, more operators were added than the approximately ten thousand musicians who lost their jobs to the sound picture. But even if new types of jobs were added, that provided little relief to the musicians whose skills were not applicable to other employment.[11]

Displaced worker surveys, it is true, say nothing about the aggregate. The National Research Project on Reemployment Opportunities and Recent Changes in Industrial Techniques, set up to investigate the role of technology in unemployment, unfortunately failed to provide much conclusive evidence. In a 1932 article, David Weintraub, director of the project, reached an optimistic conclusion about the impacts of technical change on employment during the 1920s. However, his later analysis suggested the opposite: mechanization, he found, had been a key factor in unemployment.[12] As economists today still struggle to isolate the share of nonemployment attributable to technology, it should be no surprise that research efforts of the 1930s faced similar challenges. At the time, Leo Wolman, who served in the National Recovery Administration during the Great Depression, pointed at several empirical issues limiting progress in studies of technological unemployment, of which many seemed hard to overcome.[13]

Despite statistical challenges, the emerging consensus among contemporary economists was that there was technological unemployment, albeit of temporary nature. The likes of Paul H. Douglas, Alvin

Hansen, and Rexford G. Tugwell all argued that labor market rigidities were impeding the process by which workers would be reabsorbed into new jobs: the expense of moving between locations, the human drags of retraining, and the psychological pressures of job loss made adjustment costly and hard. To ease transitions, Douglas argued against home ownership and too narrow specialization in education and for some form of unemployment insurance and a federal employment agency. In the absence of such policies, he suggested, it will be "almost inevitable that labor will resist and oppose most attempts to raise efficiency of industry."[14]

To what extent economists shaped the public discourse is hard to say. Few economists, if any, thought that policies to slow the pace of progress were a good idea. Yet displaced worker studies and the depth of the depression did prompt the adoption of unconventional policies by twentieth-century standards. The great exception of twentieth-century America was the administration of President Franklin D. Roosevelt, which actually tried to slow the pace of mechanization. Of the 280 regulations in the National Recovery Administration, thirty-six included restrictions on the installation of new machines.[15] By focusing too much on worker-replacing technologies, the administration missed many of the enabling technological advances of the time. Michelle Alexopoulos and Jon Cohen write:

> In this respect, it is quite possible that the Roosevelt administration's preoccupation with the labor displacing effects of new technologies was largely a consequence of its focus on innovations in manufacturing. Had it been able to take a broader perspective that encompassed the rapid growth of new products linked to automotive and electrical advances, the administration may have been more sanguine about the employment impact of these new technologies. It may even have embraced the idea that the great wave of gadgets that swept over the U.S. from the mid-1930s actually prevented a bad situation from becoming a good deal worse.[16]

Instead, the debate stopped only when unemployment ceased to be a concern. As late as 1940, Roosevelt warned in his State of the Union address that America had to begin "finding jobs faster than invention can take them away."[17] It took the attack on Pearl Harbor and America's entry into World War II for the machinery question to ebb. Beating the Axis powers required everyone to work at full capacity, and they did.

The machinery question, however, only faded temporarily. The first few introductions of electronic computers into the workplace sparked a panic about the threat of automation to jobs in the news media, and the upsurge in unemployment that came with the three post–Korean War recessions led people to connect the two. Looking back in 1965, Robert Solow noted: "Whenever there is both rapid technological change and high unemployment the two will inevitably be connected in people's minds. So it is not surprising that technological unemployment was a live subject during the depression of the 1930s, nor that the debate has now revived."[18] As noted above, the technological unemployment debate actually predated the Great Depression. However, machinery angst in the twentieth century was clearly cyclical, and this time it followed an upswing in unemployment after the Korean War. Though it is hard to detect much progression in the nature of the debate, our vocabulary bears witness to the progression of technology: the discussions of the 1950s and 1960s centered on the new popular term "automation."[19] Just like "technological unemployment" in the 1930s, "automation" and its discontents became one of the defining themes of the postwar years.

In America, the first comprehensive inquiry into the employment effects of automation was undertaken in 1955, when twenty-six leaders of labor, industry, and government testified before a Congressional subcommittee.[20] The subcommittee concluded that "all elements in the American economy" accept and welcome progress, change, and increasing productivity," but that "no one dare overlook or deny the fact that many individuals will suffer personal, mental, and physical hardships as the adjustments go forward."[21] During the hearings, nobody proposed measures to restrict the use of machines or even disputed the desirability of automation. Rather, the witnesses urged that greater attention be given to the social problems emerging from dislocation and noted that there was concern that older workers in particular would face difficulty in finding new and better jobs. Union representatives voiced the traditional demands for labor to get a larger share of the nation's increasing productivity in the form of higher wages, shorter hours, and a lower retirement age. But Secretary of Labor James P. Mitchell responded: "I repeat, there is no reason to believe that this new phase of technology will result in overwhelming problems of readjustment. Science and invention are constantly opening up new areas of industrial expansion. While older and declining industries may show reducing opportunity, new and vibrant industries are pushing out our horizons."[22]

The automation debate extended far beyond America's borders. When the International Labour Organization (ILO) convened the fortieth annual ILO Conference in 1957, automation was a topic on everyone's mind. In connection with the conference, David A. Morse, the ILO's director general, wrote an article on the subject for the *New York Times*. Just as William Green had argued that mechanization was nothing new in the 1930s but was now progressing faster than at any time in history, Morse suggested that "no one could claim that automation is new. The productivity of machines has been increasing the productivity of man for many centuries. Perhaps the *newest* thing about automation is its tendency to speed up the rate of technological change and thus both to multiply the opportunities for social progress and to pile up the social problems likely to go with it."[23] As in the 1930s, the 1950s was held to be different because of the growing pace of technological change. Morse also pointed to the human tragedies that emerged when automation meant displaced workers failed to find work elsewhere. But on the whole, he was optimistic about the opportunities arising for "a better living, and for a better world society."[24]

In the public debate, those worker-replacing technologies that reduced the bargaining power of labor unsurprisingly received the bulk of the attention, but things were not quite as straightforward as they might seem. Take the case of the elevator operators, for instance. During a general strike on September 24, 1945, elevator operators left more than fifteen hundred office buildings across Manhattan empty. Workers swarmed in the lobbies and crowded the sidewalks, while a few brave ones tried to climb the endless stairways of the highest skyscrapers. Such disruption was immensely costly and bad for business, and the arrival of the automatic elevator seemed to be the best way to ensure that such a thing would never happen again.

However, substituting automatic elevators for manual ones required public acceptance. Many citizens were first terrified by the idea that automatic elevators might leave them hanging on a cable hundreds of meters in the air, without any operator being responsible for their safety. Such concerns seem familiar today, given the present discussions about the adoption of autonomous vehicles. But like human drivers today, elevator operators were not infallible. Injuries were frequent, and several operators in New York City are reported to have been involved in fatal accidents. One operator on Seventh Avenue was killed when the elevator "shot up and pinned him at the top of the door," and another was found "wedged between the lift and the door" in the Bronx.[25] Indeed, a report filed by the Elevator Industry Association in 1952, in response to attempts seeking to restrict the introduction of automatic elevators, came to the conclusion that automatic elevators were fully five times safer than manually operated ones.[26]

When the jobs of truck and taxi drivers will go remains to be seen. But people in the 1950s were clearly right in thinking that the occupation of the elevator operator would soon become a distant memory. In 1956, the *New York Times* predicted that "the elevator operator may be joining the coachman and trolley car motorman on the road to oblivion."[27] That year in New York City alone, 43,440 elevators (about one-fifth of those in operation in America) were estimated to have carried 17.5 million passengers a combined distance of halfway to the moon. But by 1963, it was reported that although 35,000 elevator operators were employed in New York City in 1950, only 10,000 of their jobs were left. While the Empire State Building was still one of those with manually operated elevators, one article reported that a $2 million investment would serve to cut operating costs associated with operators' salaries, pension plans, and sick leaves. At the Chrysler Building, forty-eight of its fifty-two elevators had already been converted to automatic operation. Two-thirds of elevator operators had been reassigned as porters and handymen, while the remaining third found themselves looking for new work.[28]

For the most part, however, automation anxiety concerned computers. As Abraham Raskin wrote in 1961, "The ultimate horror in labor's book was the announcement that a computer . . . will do the work of seventy-five of the biggest computers now in use. . . . When computers start creating unemployment among computers it is really time to start worrying."[29] As we will see in chapter 9, much like artificial intelligence today, the first computers did not have any meaningful impact on labor markets even in the 1960s. In fact, the effect of computers on employment was not felt before the 1980s. Even so, most commentary on the early introduction of computers focused on the fear that they would leave many Americans unemployed. Such worries reached the heart of the government, with some congressmen fearing that the displacement of government employees by electronic computers would imperil the practice of giving politicians government jobs in recognition of their service. As statistical work could be better accomplished by machines, politicians were torn between the "desire for increased efficiency and the fear that the patronage system will be wrecked," in the words of C. P. Trussell. The matter was treated most seriously. In 1960, a subcommittee of the House of Representatives, chaired by Congressman John Leskinki of Michigan, recommended that workers who might experience displacement should be given sufficient notice to be retrained to handle new machines, equipping them with the skills to keep their jobs. In the case of a net reduction in employment, the subcommittee also recommended a hiring freeze to allow displaced workers to take vacant jobs. However, it did not advocate any policies to slow the pace of computer adoption.[30]

During the 1960 presidential campaign, Senator John F. Kennedy gave a spirited speech on the automation dilemma at a rally in Detroit that was quite similar to what Secretary of Labor Davis had said in 1927. The message was straightforward. The coming automation revolution, Kennedy suggested, is "a revolution bright with hope of a new prosperity for labor and abundance for America—but it is also a revolution which carries that dark menace of industrial dislocation, increasing unemployment, and deepening poverty."[31] After Kennedy became president, the first formal report by his Advisory Committee on Labor Management Policy, published in 1962, claimed that "it is clear that unemployment has resulted from displacement due to automation and technological change." But it added that "it is impossible, with presently available data, to isolate that portion of unemployment resulting from these causes."[32] Yet such caution was not enough for Kennedy to hold back on the issue. When asked, "How urgent do you view this problem, automation?" during a news conference in 1962, he responded:

> Well, it is a fact that we have to find, over a 10-year period, 25,000 new jobs every week to take care of those who are displaced by machines, and those who are coming into the labor market, so that this places a major burden upon our economy and on our society. . . . But if our economy is progressing as we hope it will, then we can absorb a good many of these men and women. But I regard it as the major domestic challenge, really, of the '60s, to maintain full employment at a time when automation, of course, is replacing men.[33]

The tragedy of Kennedy's assassination the following year didn't end the automation debate. Soon after taking office, President Lyndon Johnson set up a National Commission on Technology, Automation, and Economic Progress. Like Kennedy, Johnson was not opposed to automation. When signing the bill to establish the commission, he noted that "technology is creating both new opportunities and new obligations for us. . . ." He saw both an opportunity for faster productivity growth and the obligation to make sure that no worker and family "must pay an unjust price for progress." Automation, he argued, could be the "ally of our prosperity if we will just look ahead, if we will understand what is to come, and if we will set our course wisely after proper planning for the future."[34] Large parts of the commission's report, published in 1966, was dedicated to investigating "the belief that technological change is a major source of unemployment" and the fear that technology would eventually "eliminate all but a few jobs, with the major portion of what we now call work being performed automatically by machines."[35] But unlike Kennedy's Advisory Committee on Labor-Management Policy, the commission concluded that persistent unemployment in the period 1954–65 was not due to automation, and the conclusion was supported by some actual analysis. "The persistence of a high general level of unemployment in the years following the Korean war," the commission asserted, "was not the result of accelerated technological progress. Its cause was interaction between rising productivity, labor force growth, and an inadequate response of aggregate demand."[36] Yet despite this conclusion, the commission considered automation sufficiently disruptive to recommend making the government the employer of last resort, expanding free education, and introducing a guaranteed minimum income.

The parallels with the technological unemployment debates of the 1920s and 1930s are many. Like the National Research Project of the 1930s, the commission of the 1960s was set up to investigate the role of technology in unemployment. Though the findings of the commission were more conclusive, both bodies failed to settle the debate and, as in World War II, ended the technological unemployment concerns in 1940; another war effectively stopped the automation debate in 1965. As Woirol writes, "Automation then remained a major popular issue through the mid-1960s. Only after unemployment fell below 4 percent during the Vietnam War did automation begin to disappear as an everyday topic in popular publications."[37]

What is missing from most commentary and scholarship, however, is an understanding of how workers felt about technical progress. As we have seen, in Britain during the classic years of the Industrial Revolution, workers made their voices heard in one way or another. They petitioned Parliament, urging it to block the spread of worker-replacing technologies. They expressed their frustration in novels and poems. And they rioted against the spread of machinery. While the sentiment of union leaders like William Green suggests that workers were not interested in blocking the progress of technology in the twentieth century, there is little direct evidence of workers' attitudes toward mechanization during the days of the technological unemployment debates. One rare source of information is letters written to the Roosevelt administration during the Great Depression. These letters include policy proposals by ordinary citizens that provide some insight into the concerns of the American public. Using a sample of eight hundred letters, Woirol recently classified a small percentage of their proposals.[38] Most common were plans intended to increase consumer purchasing power through the implementation of minimum wages, price controls, government loans, pension or unemployment insurance programs, and direct job-creation plans. Other plans favored the expansion of public employment in work projects of various kinds. But some citizens also favored policies to stop job displacement forces: 5 percent of the letters argued for restrictions on labor-saving machinery.

While Woirol's sample might not be representative of the American public, it does suggest that even during the years of most extreme hardship, few people believed that restrictions on machinery were a good idea. However, this somewhat limited evidence does not shed any light on how workers whose jobs were directly affected by technical progress perceived it. But during the 1950s and 1960s, sociologists made serious efforts to study workers' attitudes toward mechanization. Their findings underline the intuition underpinning this book—that attitudes depended very much on how workers adjusted to the technology. In the context of the installation of an IBM 705 computer in a large public utility, for example, William Faunce, Einar Hardin, and Eugene Jacobson found that "for many individuals this was a period of growth; for others a period of failure and disillusionment. The change severely tested marginal employees and supervisors, while at the same time giving the more experienced and able ones the opportunity to develop and to demonstrate their work potential. The dislocation and the loss of duties and jobs was a serious problem for some employees."[39]

Similar studies were also conducted in a factory setting. Surveying workers in two power plants, Floyd Mann and Lawrence Williams found that operators in the more automated plant on average liked their current jobs more.[40] They had to spend less time doing dirty work, felt they had more responsibility, and had more contact with other employees. This, of course, tells us little about how they perceived the transition associated with mechanization. To that end, Faunce, a sociologist, investigated how people fared as they were transferred to an automated automobile-engine plant in 1958.[41] He found that workers overwhelmingly preferred the automated factories to the old ones, primarily due to the reduction in the handling of heavy materials, which made their jobs less physically demanding. But attitudes toward mechanization were not always favorable from the onset. In a study of an automated steel mill, Charles Walker found that job satisfaction varied significantly throughout the process of adjustment: "The same job characteristics, all stemming from the automatic or semiautomatic operations of the mill which had at first been feared and hated, were later the source of satisfaction."[42] Once the new became the familiar, attitudes shifted.

Thus, in a review of the literature, Faunce and coauthors aptly summarize the matter as follows:

Field research suggests that the impact of office automation upon job satisfaction varies depending on ... whether the employees are in electronic data-processing departments which gain work tasks or in other affected departments that lose tasks, whether the computer is of large or medium size, and on several other circumstances. Office employees think the broad impact of office automation is to eliminate jobs and regard the methods changes as temporarily disruptive, but they often welcome change and rarely reject mechanization as such. Attitudes toward change appear to depend on the ability of the individual to deal effectively with change and on the skill with which the organization manages the change. Studies of factory automation suggest that automated plants are preferred as work places to less advanced plants, although they provide important sources of dissatisfaction. The sources of satisfaction and dissatisfaction vary over the course of adjustment to automation.[43]

To be sure, none of these studies surveyed displaced workers, and it stands to reason that those who found their jobs taken over by machines had less favorable attitudes toward automation. Indeed, even if no displacement took place, attitudes toward technological change were evidently shaped by how workers were affected in their current roles. When workers found part of their jobs transferred to machines, the loss of duties was more likely to be accompanied by fears of losing their jobs. In contrast, when new tasks and duties were created, workers often felt a sense of increased responsibility, although they sometimes worried about inadequate training. While their perception of technological changes was clearly always contextual, attitudes in large part depended on whether the technology augmented or replaced workers' skills. For the most part, as we shall see, technical progress did the former. Mechanization made workers' skills more valuable in existing tasks and created many entirely new ones, thereby increasing the bargaining power of labor and allowing workers to earn better wages. This also helps explain why there were few Luddites in the twentieth century.

8

THE TRIUMPH OF THE MIDDLE CLASS

How did the accelerating pace of change affect the majority of working Americans? Despite rapid mechanization, twentieth-century America never experienced machinery riots on a comparable scale to those in Britain during the classic period of the Industrial Revolution. In the nineteenth century, in contrast, there were some incidents of workers rebelling against machines. In 1879, the year that Thomas Edison invented the light bulb, the *New York Times* reported the story of Elias Grove, whose wheat-threshing machine was destroyed in a fire. Ten days later a letter arrived with a warning: "Mr. Grove: You will stop your other machine or next will be your life. We intend to stop steam threshing. We do not get enough work through the Winter and Summer."[1] The article noted that a number of farmers had received similar threatening letters. However, more recent examples of American workers destroying machinery for fear of losing their jobs are hard to come by. The historian Daniel Nelson writes:

> The mechanization of agriculture was not painless. As early as the 1830s, the advent of threshing machines provoked protests, occasionally violent, from men who devoted their winters to flailing wheat. More serious opposition appeared in the mid-1870s, when recession coincided with the appearance of the labor-saving twine-binder, a machine that substantially reduced the size of harvest crews. Strikes and terrorist acts punctuated the normally placid midwestern summer of 1878. Ohio, still an important wheat-producing state, was the

center of violence. . . . As the economy improved in 1879 and fewer city workers turned to the countryside for employment, the crisis passed. Thereafter there was little or no overt resistance to mechanization.[2]

This is not to suggest that U.S. labor history was otherwise peaceful. The renowned labor historian Philip Taft and Philip Ross, a professor of industrial relations, have argued that "the United States has had the bloodiest and most violent labor history of any industrial nation in the world."[3] But labor violence in twentieth-century America rarely targeted machinery. It is indeed somewhat telling that Taft and Ross's detailed review of the causes of strikes and incidents of labor violence does not provide a single reference to the introduction of labor-saving technology as the cause of such incidents. Other studies examining the determinants of strikes in the period 1900–70 do not even consider mechanization as a potential cause of workers' decisions to strike.[4] One reason for the absence of any machinery riots might be that white Americans in the twentieth century had other means of expressing their frustration. Instead of casting their votes with sticks and stones, they could simply show up at the ballot box. However, despite the privilege of being able to vote, workers frequently showed their frustration about their wages and working conditions in general through violence. So why did they not violently oppose mechanization? The most convincing explanation is the simplest one: labor, for the most part, greatly benefited from the steady flow of new technologies.

The emergence of trade unions, it is true, also provided a mechanism for settling disputes that was not available to British workers in the early nineteenth century: labor unions became legal in Britain in 1825, when a very small percentage of workers joined them, and their members obtained the legal right to strike only in the 1870s.[5] And the approach taken to mechanization by the unions in the twentieth century sheds light on its benefits for their members. Union leaders were well aware that the pay a worker takes out of his envelope at the end of the week depends on the amount of mechanical power standing behind him. Factory electrification allowed workers to produce more and thus earn more. Instead of raging against the machine, workers and trade unions battled to maximize their gains from progress. From the perspective of trade unions, mechanization was a way of achieving many of the benefits their members demanded, including higher wages, shorter hours, and earlier retirement. Walter Reuther, who had spent a large part of his career leading the union of American automobile workers, was evidently not opposed to mechanization. His attitude was simply that people's purchasing power must grow in tandem with the productive capacity of American industry. Reuther was also a vocal proponent of a guaranteed annual income. In an interview, he said that he looked forward to "the day when the worker would spend less time at his job and more time working on a concerto, a painting or in scientific research." He confidently predicted: "Technological advances will make that possible. . . . In the future an auto worker may only work 10 hours at the factory. Culture will become his main preoccupation. Working for a living will be sort of a hobby."[6] Technology, he believed, would turn the backyard of the American worker into a Garden of Eden.

Most trade union officials may not have shared Reuther's utopian vision, but as long as their members gained from progress, mechanization was equally in their interest. A series of separate case studies by the Bureau of Labor Statistics (BLS) illustrates this point by showing that unions frequently took an active role in the mechanization process. In one bakery, the introduction of semiautomatic production techniques was handled through collective bargaining, which allowed managers and union officials to resolve issues of displacement, downgrading, and compensation: "The consensus of the workers, as expressed by the local union president, was that the results of the changes on the whole were advantageous to them. . . . The local union president believed the workers have shared in the greater productivity of the plant through the wage increases and fringe benefits obtained in the past few years."[7] Naturally, the change to increased automation meant some shifting of people from jobs in reduced activities to jobs in expanding ones. And in some cases, the shift meant a downgrading in skill level, while in others upgrading took place. When informing the union business agent about the company's plans, including estimated displacement, management guaranteed that any workers downgraded into jobs of lower pay would continue to be paid at their current rate. This announcement was perceived to reduce automation anxiety and was later formalized into a union contract. In every case that a union was involved, the BLS shows, its approach was to make sure that its members reaped the benefits of mechanization rather than stood in the way of it.[8] The benefits of technology to labor as a whole were simply too large, even if some workers lost out in the process. And any transitions were helped by the fact that companies often compensated workers who lost their jobs to machines. Unlike in the nineteenth century, the focus of labor in the twentieth century was on managing the transition instead of blocking technological progress. And the unions, acting in interest of their members, were for the most part a helpful facilitator in this regard. Looking back in 1984, the Congressional Office of Technology Assessment noted that "labor-management relations play an important role in the introduction of new technology. Using collective bargaining, organizing, and political strategies, unions in the United States have attempted to minimize what are perceived to be the socially harmful effects of new technologies on the labor force. Their efforts have generally been directed toward easing the adjustment process rather than retarding the process of change."[9]

Labor had good reason to praise progress. As the winds of technological change swept through the workplace, jobs became more pleasant, less hazardous, and better paid. Mechanization made it possible for people to move away from sweat and drudgery to well-paid jobs that were less physically demanding. The outflow of workers from agriculture to blue- and white-collar jobs laid the foundations for the emergence of a growing and increasingly prosperous middle class. The American experience before the 1980s thus contrasts markedly with that of the British in the classic period of the Industrial Revolution, where the human costs of displacement were high because workers were left little choice but to take lower-paying jobs. As the factory replaced the domestic system, few had the means to acquire costly human capital to become managers, accountants, clerks, mechanical engineers, and so on. Instead, they were left competing for low-skilled production jobs that were simple enough to be performed by children. But if workers are able to shift into less hazardous, more enjoyable, and better-paying jobs, any distress will be short-lived. In the twentieth century, the abundance of semiskilled jobs in America's offices and factories, brought by the Second Industrial Revolution, provided the best reassurance for those who feared unemployment. Some Americans, it is true, failed to climb the economic ladder. As noted, older workers with highly specialized skills living in isolated areas often struggled to adjust and could be forced to shift into low-skilled jobs at lower wages and reduce their standards of living, at least temporarily. But while mechanization made a few workers worse off individually in the short run, the expectation that they would benefit in the medium term seemed justifiable to the vast majority of ordinary people.

The End of Drudgery

Perhaps the greatest contribution of mechanization was that it made the workplace safer and less physically demanding.[10] Consider for a moment the stark contrast between the air-conditioned offices in which most Americans work today and the environment in which most citizens worked a century ago. In 1870, almost half of working Americans were still employed in agriculture. The occupation of farming was not just hard work, it was also economically risky. Because most people derived their subsistence from the natural world, they had to contend with heavy rains, droughts, forest fires, swarms of insects, and so on. The farmer was thus left coping with many variables beyond his control that could have devastating economic consequences. The dust storms in the 1930s, for example, blew vast quantities of topsoil off the farmland. On "Black Sunday" in 1935, "one such storm blanketed East Coast cities in a haze."[11] By the 1940s, many Great Plains farms had lost more than 75 percent of their topsoil, and farmers had lost around 30 percent of the per-acre value of their land as a result.[12] And if the outdoor nature of the job made things unpredictable, the uncertainty of maintaining a steady income was made even greater by the constant risk of injury. Long working days with animal power as the only help meant a constant strain on muscles. Danger and drudgery were part of everyday working life.

Miners were hardly better off. Workers could spend several days underground without sunlight. Before electrification, kerosene lamps provided the only light miners would have. In addition, mine workers were constantly exposed to cave-ins, explosions were always a risk, and lung disease often came as part of the work package. In the late nineteenth century, roof collapses, flooding, and accidental explosions meant that coal-mining deaths occurred on a daily basis.[13] And although the introduction of machines in the factories meant that exceedingly hard and laborious tasks were replaced, they rarely made the workplace safer. Statistics on industrial accidents associated specifically with machines remain sparse, but such accidents were evidently frequent enough for the *New York Times* to run out of phrases to describe such incidents: "killed by machinery," "killed in the machinery," "mangled by machinery," "terribly mangled by machinery," "crushed by machinery," and "scalped by machinery" appeared in the many headlines of reports on deaths from industrial accidents during the 1870s and 1880s. Among the many casualties, one proprietor of a large paper mill in Lambertville, New Jersey, got his clothing caught in the shafts and "was thrown violently to the floor and the top of his head was torn off."[14] Another engineer in Newark, New Jersey, was "crushed to a pulp" after he was trapped in the shafts of the engine. Beyond machinery accidents, explosions and fires were a constant threat. The Triangle Shirtwaist Factory fire of 1911 in New York City, described by the news media as the "the worst calamity that has befallen us since the burning of the Slocum," cost the lives of 148 workers, most of them young women.[15] As fire ravaged the factory, many people jumped out of the windows—only to be picked up either squashed or fearfully injured. Few of the workers who escaped death but were left seriously injured or disabled received any meaningful compensation to support themselves and their families.

For the most part, electrification was a blessing to workers, making factories brighter, more pleasant, and safer, although electricity also brought the previously unknown dangers of shock and electrocution into the factory. Immigrants who came in contact with the force of electricity for the first time were the prime victims: "A newly landed Croatian lad of seventeen was killed by fooling with a switch with wet gloves on, watching the sparks fly."[16] Overall, however, electrification and safety went hand in hand. Belts, gears, and shafts were the main sources of factory accidents, posing a constant danger to workers' fingers, arms, and lives. The switch to unit drive eliminated the jungle of belts and shafts and the accidents associated with them. Electrical machines also stirred up less dust, making the air cleaner and working conditions healthier. The displacement of gas by electric light further reduced humidity and improved oxygen in factories, while making acid fumes a matter of the past. And, increasingly automatic machines eventually lightened toil. It is thus no wonder that factory electrification for the most part was welcomed by workers (chapter 6). Indeed, in the period 1926–56, when the first comprehensive statistics on injury frequency rates were collected, the average number of disabling injuries in manufacturing was halved, as was also the case in mining.[17] As one factory worker at Henry Ford's River Rouge plant marveled in 1955, "Automation has saved me. . . . If I had to lug those heavy blocks into position like I used to I could not last till I was 65. Now I expect to be working till I am 80."[18] His only complaint was having put on thirty-three pounds since being aided by machinery.

As work gradually became less physical demanding, industrial hygienists of the 1950s and 1960s speculated that many injuries from the lifting, handling, and unloading of goods would become historical concerns. In textile weaving, safety devices were introduced on looms, automatically disconnecting the machine in case of any accident. And one Ford plant is reported to have experienced an 85 percent reduction in cases of hernia after the installment of automatic machines.[19] Automation, according to the BLS, meant that man no longer had "to pace himself to the rhythm of the machine, a rhythm which may be an unnatural one and result in tension and possible accidents."[20]

Machines meant the end of the most hazardous, dirty, and backbreaking jobs. For millennia, agriculture had been most people's preoccupation. Now, in less than a century, technology had shifted the bulk

TABLE 1: The Changing Composition of the U.S. Workforce, 1870–2015

		1870	1900	1940	1980	2015
Farmers and farm laborers		45.9%	33.7%	17.3%	2.2%	1.0%
Blue-collar workers	Total	33.5	38.0	38.7	31.1	21.5
	Craft workers	11.4	11.4	11.5	12.0	8.4
	Operatives	12.7	13.9	18.0	14.7	8.9
	Laborers	9.4	12.7	9.2	4.3	4.3
White-collar workers	Total	12.6	18.3	28.1	38.9	37.3
	Clerical workers	1.1	3.8	10.4	19.2	15.5
	Sales workers	2.3	3.6	6.2	6.7	6.2
	Domestic service workers	7.8	7.6	4.4	0.6	0.0
	Other service workers	1.4	3.2	7.1	12.3	15.6
Managers and professionals	Total	8.0	10.0	15.1	27.8	40.1
	Managers and proprietors	5.0	5.9	7.9	10.4	14.7
	Professionals	3.0	4.1	7.1	17.5	25.4

Source: 1870 to 1980 from Historical Statistics of the United States (HSUS), Table Ba1033-1046; 2015 from Ruggles et al. (2018). See also Gordon (2016), Table 8-1.
Note: Numbers may not sum to totals because of rounding.

of the American workforce from farms to factories and offices. Table 1 illustrates the remarkable occupational evolution of the American workforce: between 1870 and 2015, the share of working people in agriculture declined from 45.9 percent to 1.0 percent. While that share was falling even before 1900, this decline was due to the relatively rapid expansion of blue- and white-collar jobs. In terms of total employment, the agricultural sector saw its peak in 1910. Thereafter, it lost jobs in every decade. The main reason why workers left the farm, as we shall see, was more lucrative job opportunities in the cities. This, in turn, sparked incentives to mechanize. Tractor use slowly expanded after World War I, but the main burst of growth began in the late 1930s, and by 1960 80.0 percent of American farms had tractors, up from 16.8 percent in 1930.

The tractor alone can account for much of the reduction in labor requirements in farming. According to estimates by the Department for Agriculture, in 1960 the tractor had saved 3.4 billion man-hours in field operations and caring for draft animals, representing the equivalent of

1.74 million farmworkers.[21] By that time, the agricultural sector had experienced a net reduction of 5.7 million farming jobs since its peak. Though labor-saving estimates for automobiles and trucks are unavailable for 1960, in 1944 they saved more than 1.5 billion man-hours in hauling and travel time and another 1.1 billion man-hours that would have been devoted to taking care of horses and mules.[22] At the same time, rural electrification relieved workers of tedious tasks such as the milking of cows by hand, and irrigation became less laborious with the arrival of electrical pumps. Like motorized vehicles, electrification virtually eliminated the need for workers in some domains: "By the middle 1920s the local Campbell Ice Cream and Milk Company had electrical processing equipment connected by pipes, so that the milk never had to be handled directly or exposed to the air, but could be pumped from one stage to the next as the machines separated the milk from the cream, heated the raw milk to pasteurize it, homogenized it, cooled it to near freezing, and then bottled it."[23]

Meanwhile, in mining, mechanization made inroads into the heavy manual work of loading coal onto electric cars in the 1920s, described at the time as "the most widespread form of drudgery existing in industry today."[24] In a mere decade, the volume of mechanized loading in coal mines expanded by a factor of twenty. And mechanized loaders were entering metal mines as well. In the Michigan copper mines in the early 1930s, it was reported that the adoption of "mechanical loaders and scrapers has replaced shovelling in large measure."[25] And as hard physical labor in the mines and the factories was gradually transferred to electrified machines and backbreaking toil on the land was taken over by motor power, workers shifted into the comforts of air-conditioned offices: as shown in table 1, people leaving agriculture after 1940 mainly took up work in offices.[26] This was largely due to the spread of office machines, not despite them. The time saved by the use of the great number of typewriters and adding and calculating machines was surely significant, but without them many tasks would have been too laborious to be economically feasible on a large scale. Much of the work done by office machines would probably not even have been done had the machines not been invented. As Harry Jerome wrote, "If letters were all written by hand, and computations all made by laborious and expensive human effort, there would be a marked shrinkage in the volume of correspondence and computation considered necessary and economical."[27]

In other words, machines were responsible for relieving workers of the most dangerous and physically demanding tasks as well as for creating new and more pleasant ones in electrified factories and air-conditioned offices. Economist Robert Gordon has calculated that the share of the workforce engaged in jobs that can be deemed physically challenging and dangerous fell from 63.1 in 1870 to 9.0 percent in 1970.[28] Of course, such estimates inevitably understate the demise of unpleasant work because the content of many unpleasant jobs changed for the better as well. As Gordon notes, "One only need contrast the 1870 farmer pushing a plow behind a horse or mule, exposed to heat, rain, and insects, with the 2009 farmer riding in the air conditioned cab of his giant John Deere tractor that finds its way across the field by GPS and uses a computer to optimally drop and space the seeds as the farmer reads farm reports and learns about crop prices on a fixed screen or portable tablet."[29]

More Jobs, Better Pay

Technology not only made jobs less hazardous and physically demanding, it also led to better-paying jobs. In the period 1870–1980, hourly compensation kept track with labor productivity (figure 9). Of course, factors other than technology also affected compensation. Though this book is about long-run trends in living standards rather than short-run fluctuations, a few arguably significant variables warrant some discussion. Part of the reason why real wage growth was rapid in the early twentieth century has been attributed to the rise of welfare capitalism in the 1910s and 1920s, which gave a boost to workers' wages as companies sought to retain their employees. With more and more capital tied up in machines, the skills required to operate them naturally became increasingly valuable. At the Ford Motor Company, the cost of

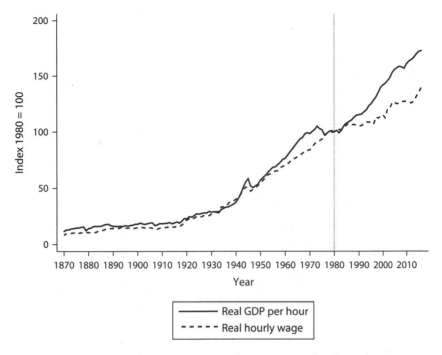

FIGURE 9: Real Gross Domestic Product per Hour and Real Hourly
Compensation of Production Workers, 1870–2015
Sources: See appendix, this volume.

constantly training new workers, as many others left the company for
better jobs elsewhere, prompted action. To keep its workers, the company introduced the five-dollar-a-day wage, effectively doubling the wages
of the employees working in its factories. Since Ford accounted for almost
half of the American production of automobiles, increasing the wage to
five dollars a day can justly be regarded as "the most dramatic event in the
history of wages."[30] In addition to raising wages, Ford also introduced a
new welfare program for its workers, and companies like Procter and
Gamble, General Electric, and Goodyear Tire soon followed suit, institu-
tionalizing similar programs that returned some of the productivity gains
to employees in terms of better wages, medical services, pension plans,
and so forth. When the BLS surveyed 431 companies in 1917, it found that
almost all of them had some form of welfare-capitalist scheme in place.[31]

However, what has been described as corporate altruism is more accurately referred to as corporate paternalism. Welfare programs were rarely unconditional. The Ford Motor Company, for example, set up a Sociological Department to advise its employees on improving their lifestyles, and department staff members visited employees' homes to inspect them for cleanliness and make sure that workers were in fact married, as the company required. It was not uncommon for Ford employees to get young amateur actresses to play the role of a loving wife during company visits.[32] But welfare capitalism nonetheless raised workers' expectations of what employers must provide. As Louis Hertz's *The Liberal Tradition in America* illustrates, the commitment to Jeffersonian individualism, small government, and strong property rights is deeply ingrained in American culture.[33]

Quite possibly, welfare capitalism helped pave the way for what the historian Jefferson Cowie has called the "great exception" in American political history—that is, the era of "collective economic rights" created by the New Deal. New Deal legislation evidently did much to shift the balance of power between capital and labor.[34] The National Labor Relations ("Wagner") Act of 1935 guaranteed the right of employees to organize into trade unions and bargain collectively for better conditions. It also provided mechanisms for solving disputes, such as the establishment of the National Labor Relations Board to mediate between employees and employers. Other legislation targeted compensation more directly. The National Industrial Recovery Acts of 1933 and 1935 authorized the president to prescribe a limited code of fair competition, including setting minimum rates of pay. And the Fair Labor Standards Act of 1938 introduced the forty-hour workweek for many Americans, while mandating employers to pay workers for any overtime hours.

Welfare capitalism and New Deal legislation surely affected compensation. But needless to say, these variables cannot account for the trajectories of real wages over a century. The strengthening of unions might, of course, also have affected wages over the long run.[35] It is true that most studies suggest that unionized workers earned higher wages than their nonunionized counterparts as late as the 2000s, when much of their power had already faded. Labor unions clearly played a role in raising pay for their members and in giving them political voice. But in the end, the bargaining power of unions depends on the value of the skills and knowledge of the workers they represent. The case of the telephone operator illustrates this point vividly. When operators went on strike in 1968, it had virtually no effect at all, as automation kept the system running. The strike was barely noticed, apart from the headline "Automation Keeps Struck Phone System Working." Citizens trying to call friends and relatives in different parts of the country were for the most part able to connect themselves. As one contemporary observer pointed out, "If a telephone strike had taken place a dozen years ago it would have been difficult and often impossible to place a long-distance call in the United States."[36] In contrast, automation allowed a few executives and supervisors to fill in for the 160,440 strikers. The simplicity of the automated system meant that the other employees had few technical difficulties taking over from the operators. In cases where the skills of a union's members became redundant, the union lacked the bargaining power to make much of a difference. The best it could do was negotiate exit deals for its members. When longshoremen saw their skills made obsolete by containerization, for example, the longshoremen's union achieved nothing more than financial compensation and retraining for the workers who lost their jobs to machines. And quite naturally, as the occupation of longshoremen diminished, so did the clout of the union.

On balance, it seems safe to conclude that the strengthening of unions following the Wagner Act can account for part of the growth in real wages relative to those before 1930. But the wages of textile workers grew steadily even during the closing decades of the nineteenth century, when textile unions were weak.[37] And as we all know, whether people unionize or not, wages can grow over the long run only if workers continue to become more productive. Figure 9 shows how hourly wages rose together with output per worker until the 1970s. In America, like in Britain, this pattern emerged in the absence of organized labor and without any significant government intervention. The rising tide of people's wages gives additional weight to the view that the first three-quarters of the twentieth century was a time when much technological change was of the enabling sort. As Gordon notes:

> Some part of the explanation of rapid real wage increases before 1940, particularly between 1920 and 1940, may be attributable to the end of mass immigration and the encouragement of labor unions by New Deal legislation. But ultimately it was technological change that drove real wages higher. Part of this was compositional—new machines that pulled, pushed, carried, and lifted shifted the composition of employment away from the common laborer to operatives doing specialized albeit repetitive tasks and to new layers of supervisors, engineers, and repairmen to plan the layout of the machines, train new workers, and tend the machines. Firms began to raise pay to reduce turnover, for the assembly line could be slowed if an experienced worker quit and was replaced by someone who could not initially keep pace. Much of this shift in the nature of employment was created by the rise of the automobile industry and its assembly line method of production and is symbolized by the contrast between the dark, satanic steel mills of the 1870s and the smoothly running Ford and General Motors assembly lines of the 1920s.[38]

As discussed, the Second Industrial Revolution spawned new tasks for labor. The general-purpose technologies of the century boosted productivity and employment, while reducing unemployment.[39] Technology also increased workers' earnings capacity by bringing steam to low-productivity sectors like farming and by allowing more citizens to shift into more productive and better-paying jobs. The growth of the electrical industry offers a remarkable illustration of the ingenuity of inventors, the entrepreneurial spirit of enterprise, and the fluidity of the American workforce. Like the automotive industry, which overtook the railroads as America's largest industry in 1940, the electrical industry became a significant operation. Together with its supplier industries, it supported millions of Americans. As electrical appliances poured into homes, relieving the American housewife of many burdens, mass production created a host of previously unimaginable occupations and industries. An early survey of new industries in 1905 takes note of the high pay in all of the electrical industries, explaining why not a single strike of any serious magnitude had been recorded. Around that time, the electrical industry was still relatively minor, providing employment for some forty-six thousand people.[40] However, in the succeeding decades the mass production of telephones, radios, washing machines, refrigerators, electric irons, and so forth required more and more operators to meet Americans' growing appetite for new consumer goods.

Old industries gave way to new ones. Employment in the automotive industry, for example, peaked nearly a century after textile employment. But even old industries continued to expand, as mass production made many consumer goods available to a growing percentage of the population.[41] Crucially, workers in these industries benefited from production technologies that allowed them to earn better wages. In several separate case studies, the BLS found that the introduction of new machines led to the creation of new tasks and better-paying jobs.[42] The growth of these industries was also the best unemployment insurance people could get. A blue-collar worker who lost his or her job had many more options at a time when semiskilled jobs were abundant. Clearly, technology did cause some occupations to vanish—like those of lamplighters, elevator operators, laundresses and so on—yet these jobs employed only a fraction of the workforce relative to the new machine-aided occupations that emerged.

Of course, as noted above, millions of farm jobs disappeared. So what happened to all the farm laborers? Following publication of an influential 1967 paper by the economist Richard Day, it was long believed that an explosion of labor-saving technologies in agricultural production forced workers to leave the countryside.[43] Indeed, Otha D. Wearin, an agricultural columnist in Iowa, wrote in 1971: "The productive capacity of power machinery has greatly reduced the farm population. Occupied farming units have become fewer and fewer, and farther and farther apart, as producers with power machinery reach out for more and more land to justify their investment. Country churches, country schools, country society and small country towns have suffered. In fact, many of them have completely disappeared."[44]

The decline of rural institutions and communities was partly offset by the automobile, which made people more mobile. But there can be no doubt that some rural communities suffered. However, mechanization of agriculture was rarely the reason. We now know that Day hugely overstated the role of mechanization in cotton harvesting, claiming that cotton in the Mississippi Delta was entirely harvested by machine in 1957, when in fact only 17 percent of it was. A recent study of the cotton harvester shows that 79 percent of the decline of hand picking was due to wages rising more rapidly elsewhere.[45] Rather than being pushed out of the farms, laborers were pulled out by better-paying jobs in the cities. In fact, the mechanization of agriculture was in large part the result of cheap labor leaving the countryside. The economists Richard Hornbeck and Suresh Naidu have demonstrated that the exodus of labor from the rural areas prompted farmers to invest in mechanization.[46]

Rural electrification, tractors, trucks, and automobiles all reduced labor requirements on the farm. As the historian Wayne D. Rasmussen writes, "With electricity farmers could run useful devices of all kinds, including not only electric lights but also milking machines, feed grinders and pumps. It took the war, however, and the accompanying shortages of farm labor, high prices for farm products and an enormous demand for those products to convince nearly all American farmers to turn to tractors and related machines."[47] He also notes that most Americans, to whom better-paid jobs were available, did not want farm jobs, leaving those jobs to immigrants or machines: "Much of the tomato crop in California had been picked by Mexican laborers who entered the U.S. under the terms of the Bracero program. When the program was ended in 1964, growers reported it was not possible to recruit U.S. citizens to do the work. Some labor leaders disputed this view, but the controversy was effectively ended by the successful mechanization of the harvest."[48]

Even when workers were forced to leave the farm, technology was rarely the reason. Natural disasters like the Great Mississippi Flood of 1927 prompted people to search for more stable employment in the cities, which induced farmers to mechanize. The Dust Bowl of the 1930s was another environmental catastrophe that eroded the livelihoods of many farmers on the Great Plains, leaving them with no viable option except to leave.[49] There can be no doubt that some farm laborers struggled as work disappeared in the countryside, especially during the Great Depression. But overall, farm laborers were drawn to the cities by the job opportunities offered by mass production. The Great Migration from the rural South to industrial cities like Chicago and Detroit was a key event in U.S. economic history. Spurred by World War I, which simultaneously increased the demand for workers in manufacturing and interrupted immigration from Europe, many people left farms for factories.[50] This, in turn, spurred mechanization on the farms. As Ivanhoe Whitted of the Iowa Department of Agriculture explained to the *New York Times* in 1919, "The Iowa farmer is turning to the tractor because it helps toward the solution of the vexatious problem of farm labor." Four decades before, he pointed out, there were few large cities, and labor in the countryside was plentiful and cheap. But thanks to manufacturing, great cities had sprung up at the expense of rural areas, leaving no farm labor to spare. The tractor, he added, "saved the day."[51] But it was enthusiastically adopted only after cheap labor had dried up.

The smokestack cities and towns of the Second Industrial Revolution, which powered American growth for almost a century, continued to churn out more stable and better-paying jobs for semiskilled workers. Perhaps the best evidence that people were drawn into the cities is the absence of resistance to the mechanization of agriculture after 1879. As noted above, before the Second Industrial Revolution, there were several episodes of unrest due to employment fears over the mechanization of agriculture, but thereafter opposition to agricultural machines practically disappeared.

In the age of mass production, despite the endless churn in the labor market, workers could for the most part expect to come out ahead. As productivity grew, so did their wages. Indeed, when Nicholas Kaldor listed six famous "stylized" facts of growth in 1957, he essentially summarized what economists had learned from the steady growth over the century, noting that the shares of national income received by labor and capital had been roughly constant over the long run.[52]

Equal Gains

As America got richer, it also became more equal. What makes the growth in the early twentieth century truly exceptional is not just its pace but how widely it was shared. The period 1900–1970 has rightly been regarded the "the greatest levelling of all time."[53] Incomes were rising for virtually everyone, and they were growing even faster at lower ranks. As Americans in the middle and at the lower end of the income distribution became the prime beneficiaries of progress, inequality went into reverse. Along with every other industrialized nation, America saw the share of income accruing to people at the top, fall. It may be telling that unlike economists of the Industrial Revolution (like Thomas Malthus, David Ricardo, and Karl Marx) who were all fond of apocalyptic economic predictions, economists living in the aftermath of the Second Industrial Revolution were largely optimistic—perhaps overly so. In any event, the idea that industrialists grew rich on the misery of workers had evidently fallen out of fashion. In the 1950s, Robert Solow advanced a model of a balanced growth path, in which progress delivered equal benefits for every social group; Kaldor put forward his stylized facts of economic growth, showing that the labor share of income had remained roughly constant, at two-thirds of national income, despite rapid mechanization; and Simon Kuznets advanced his hugely optimistic theory of economic progress in which inequality automatically decreases, regardless of economic policy choices.[54] Their optimism surely seemed warranted at the time. Schumpeterian growth did indeed make America both richer and more equal.

Like the doomsday economists of the Industrial Revolution, however, twentieth-century economists were unfortunately fond of developing iron laws of economics that could be used to explain the trajectory of capitalist development for every time and place, though it is not hard to understand their appeal. Kuznets's theory, which shaped how economists thought about inequality for half a century, was intuitive and straightforward. It predicted that the early shift from the agricultural sector to the higher-income, higher-inequality manufacturing sector would drive up inequality in the early phases of industrialization. But as manufacturing employed a growing share of the population, a larger percentage of citizens would harvest the benefits from growth, and inequality would eventually diminish. Technological progress, in other words, inevitably brings about an episode of increased inequality, but all economies have to do to achieve shared prosperity is wait for the cycle to complete itself. This was the cheerful message that Kuznets brought to the annual meeting of the American Economic Association—of which he was president—in Detroit in 1954, where he first outlined his thesis, soon to become known as the Kuznets curve.

Did American growth actually drive inequality along the lines that Kuznets suggested? Fortunately, we can explore his hypothesis empirically. Unlike the economists of the Industrial Revolution, Kuznets built his theory on a formidable statistical analysis—the first of its kind. Using newly collected data, Kuznets calculated that the share of annual national income captured by the top decile decreased by almost 10 percentage points from 1913 to 1948.[55] Thus, he demonstrated that inequality had declined in the later stages of industrialization. Subsequent analyses by economic historians also traced the patterns of inequality all the way through the nineteenth century, allowing us to examine if inequality grew in the early stages of industrialization as the Kuznets hypothesis suggests. The most detailed account of the American experience is that of Peter Lindert and Jeffrey Williamson.[56] Their findings show that both property and labor incomes became much more unequal between the dawn of the American Revolution in 1775 and the beginning of the Civil War in 1861—an unambiguous finding. Indeed, Alexis de Tocqueville, during his travels across America in the 1830s, found that the "the number of large fortunes is quite small."[57] America, at least relative to the Old World, still seemed to represent the Jeffersonian ideal of a nation consisting of independent and equal farmers. But Tocqueville also suggested that this ideal was gradually disappearing: "A manufacturing aristocracy . . . is growing up under our eyes. . . . The friends of democracy should keep their eyes anxiously fixed in this direction: for if a permanent inequality of conditions . . . penetrates into [America], it may be predicted that this is the gate by which they will enter."[58]

By the end of the Civil War, the ascent of industrial "royals" had become all the more apparent. The new industrial America had become a Gilded Age, satirized in the 1873 novel *The Gilded Age* by Mark Twain and Charles Dudley Warner.[59] The industries of the Second Industrial Revolution had not yet emerged, but steel, steam, railroads, and so on had already created unprecedented wealth. America, it seemed, was turning into a nation of the Old World, far removed from its Jeffersonian ideal and corrupted by an emerging industrial elite. Wealthy industrialists and financiers like John D. Rockefeller, Andrew Carnegie, J. P. Morgan, and Cornelius Vanderbilt were frequently labeled robber barons. In 1859, the journalist Henry J. Raymond compared Vanderbilt to a medieval nobleman, writing that he was "like those old German barons who, from their eyries along the Rhine, swooped down on commerce of the noble river and wrung tribute from every passenger that floated by."[60] To be sure, the size of the corporate giants that emerged must have been hard to comprehend. In 1893, for example, when the federal government collected $386 million in revenue, the Pennsylvania Railroad alone earned $135 million. And the combined railroad operations greatly exceeded the operations of the U.S. government, which was, of course, small by modern standards. Together, America's railroad companies earned over $1 billion in revenues, and their combined debt was close to $5 billion—almost five times the national debt.[61]

While business historians have long debated to what extent the robber baron characterization is apt, contemporaries rarely pointed to American industrialists' fortunes as the prime concern per se, instead of the methods by which they had been acquired. It goes without saying that how wealth is accumulated matters. Someone who successfully accumulates money by creating new sources of employment, comfort, and prosperity is to be regarded a public benefactor. In contrast, a person who profits from stifling competition, cheating his fellow citizens, and corrupting government is a public malefactor. Regardless of the means by which they acquired their wealth, however, the robber barons can surely account for a large share of the upsurge in the share of national income accruing to the top 1 percent. Yet to suggest that growing income inequality was entirely due to capital would be misleading.

Economists tend to rely on the Gini coefficient to measure levels of inequality across time and space. One of its virtues is that it is straightforward to interpret. If every citizen of a country had the same income, the Gini coefficient would be 0. If one person captured all the income, the income Gini coefficient would be 1. As Lindert and Williamson show, the Gini coefficient for property income was naturally higher, as property ownership is typically more concentrated, but labor income inequality rose even more rapidly: between 1774 and 1870 the property Gini grew from 0.703 to 0.808, while the earnings Gini increased from 0.370 in 1774 to 0.454 in 1860.[62] A large part of this can be explained by the displacement of the blue-collar artisan, which led to the hollowing out of middle-income jobs and, thus, greater earnings inequality.[63] Moreover, as Kuznets conjectured, inequality rose as a result of American workers shifting from low-income agricultural jobs to high-income manufacturing jobs in the city. Between 1800 and 1860, the urban-rural wage gap for a male laborer grew in both the North and the South. Because American urbanization accelerated between 1870 and World War I, while the Second Industrial Revolution spun out new industries and raised skill demand, it stands to reason that total inequality should have accelerated as well, all else being equal. It is therefore somewhat surprising that inequality (as measured by the overall Gini coefficient) plateaued around 0.5 and even fell slightly between 1870 and 1929, even though the top 1 percent share rose dramatically from 9.8 percent to 17.8 percent.[64]

Was the fall in inequality the result of industrialization's reaching its intermediate stage? While the trajectory of American inequality between the Civil War and World War I does not disprove Kuznets's hypothesis, it does suggest that there were also other factors at work. The upsurge in private wealth relative to total private incomes between 1870 and 1913 is consistent with the idea that capital played a larger role in shaping American income distribution, leading top incomes to continue to grow. In 1918, there were 318,000 corporations in America, of which the largest 5 percent captured 79.6 percent of the total net income. Thus, thirty-five years before Kuznets's address to the American

Economic Association, Irving Fisher, then the association's president, painted a very different picture at its annual meeting. Fisher devoted his presidential address in 1919 to what he held to be the most serious challenge facing America at the time—namely, inequality, which he believed threatened the very foundations of American capitalism and democracy: "Whatever we may say of theoretical socialism of various types, and however much we may and ought, in my opinion, to favor in some form an increase of socialized industry, the great fact remains that the socialist group derives its real strength from class antagonism. . . . On the face of it, we should expect that all the evils mentioned would be relieved if we had more democracy in industry, that is, if the workman and the public felt that the great industries were partly theirs, both as to ownership and as to management."[65]

In light of the concerns expressed by Fisher and others, it is not hard to understand why the idea of the Kuznets curve was received with such enthusiasm. It seemed to imply that redistribution was not necessary and would happen naturally if capitalism was allowed to take its course. Or would it? Just as Kaldor's stylized theory of growth has recently been called into question, Kuznets's assertion seems hard to reconcile with the post-1980 experience. Not only has the labor share of income consistently trended downward, but income disparities have trended upward in tandem (we shall return to this pattern in part 4). The reemergence of growing inequality seems difficult to reconcile with the Kuznets curve—a point that has been forcefully made by the economist Thomas Piketty.[66]

According to Piketty, the period observed by Kuznets was one of statistical abnormality. In the normal state of capitalism, Piketty argues, the return to wealth exceeds the overall growth rate of the economy, causing wealth-to-income ratios to rise and thus increasing income inequality, as wealth is highly unequally distributed. In Piketty's world, there are no forces within capitalism that serve to drive inequality down. From time to time, however, macroeconomic or political shocks may disrupt the normal equilibrium. Two world wars and the Great Depression served to destroy the riches of the wealthy. The great leveling was the result of violence, economic collapse, and radical political change, not the tranquil process of structural change that Kuznets described. The historian Walter Scheidel has gone so far as to suggest that by destroying the fortunes of the wealthy, mass violence and catastrophes—like war, revolution, state collapse, or plague—have been the only economic levelers since the Stone Age.[67]

Needless to say, the violent shocks of the twentieth century struck America with much less force than they did Europe. But American fortunes were also buffeted. American private wealth decreased from nearly five times the national income in 1930 to less than three and a half in 1970, albeit this was mainly due to the Great Depression. Quite naturally, as Wall Street suffered from the depression, top incomes declined: the sharp fall in incomes of those employed in finance after the Great Depression almost exactly mirrors the drop in the top 1 percent's share of income.[68] This is not to suggest that policy didn't matter. Indeed, the top income tax rate, which had been reduced to 25 percent in the administration of President Herbert Hoover, was hiked up to 63 percent under President Franklin D. Roosevelt in 1933 and continued to increase thereafter, reaching a staggering 94 percent in 1944.[69] Yet inequality declined before as well as after taxes and transfers were adjusted for. And while violence, economic collapse, and policies designed to curb the influence of capital can all explain why top incomes fell, the leveling was not concentrated at the top. How macroeconomic or political shocks should have reduced inequality among the other 99 percent—particularly in the middle and lower ranks of the income distribution—is less evident, as most Americans' incomes came from labor rather than capital.

Clearly, a number of short-run events are likely to have contributed to the leveling of American wages, but they do not provide the full story. The national minimum wage, first introduced in 1933, would have caused a compression in the lower ranks of the income distribution. Yet as Lindert and Williamson demonstrate, the great leveling was not confined to the lower ranks: compression was almost universal across occupational pay scales. Other government interventions, such as the tightening of immigration policy, provide another possible explanation, as immigrants were typically unskilled, and an influx of workers affects the growth of the labor force more broadly. It is plausible that immigration quotas caused the wages of unskilled Americans to rise, and indeed a slowdown in population growth might have caused the wage structure to become compressed. However, countries on the other side of the Atlantic similarly saw wage gaps narrow, suggesting that there were other forces at work. The fact that the wage compression was equally a European phenomenon also rules out other America-specific explanations, such as the National War Labor Board, which in 1942 was made responsible for approving all changes in American wages. Wage disparities remained quite stable after the board was dissolved in 1945, suggesting that the impact of wage-setting policies was limited.[70]

Just as the Kuznets curve was perfectly matched to its moment, Piketty's treatment of inequality seems to have touched the zeitgeist. The Kuznets curve became a powerful political weapon in showing that America could stay true to its Jeffersonian ideal under capitalism. That surely seemed to be the case. The wages of ordinary citizens were growing as mechanization propelled productivity, and America was becoming a more equal place at the same time. The popularity of Piketty's work, in contrast, reflects the post-1980 experience. Most Americans agree that inequality has reached unacceptable levels, and many quite rightly feel detached from the capitalist project: the trajectory of hourly compensation has been decoupled from productivity growth (see figure 9).

One reason why grand theories of capitalist development are unlikely to work for every place at every time has to do with the bargaining power of labor. As noted above, labor unions gained in political power after the New Deal. And as the economist Henry Farber and colleagues have shown, higher union density is associated with lower levels of wage inequality.[71] Another reason is that technology works in mysterious ways. Technological progress has at times replaced workers, putting downward pressure on wages and thereby serving to increase the share of national income that goes to capital. And at other times, it has favored labor. The relationship between technological progress and the income distribution is not monotonic: some technologies may increase inequality, while others reduce it. It depends on the degree to which technological changes are replacing or enabling. And it depends on whether the supply of workers with the right skills is able to keep pace with demand.

In the end, for most people, the main source of their income is not physical or financial capital, but human capital. The wealth of workers is in their skills. It is from their human capital that they make their living. How human capital can affect the income distribution is not hard to see: empirical studies have shown that 77 percent of the variation in workers' earnings stem from individual characteristics.[72] As we shall see in chapter 10, the lack of education and training can even exclude entire social groups from the growth engine.

Most available evidence suggests that technological change was primarily skill-biased in the period 1870–1970, serving to drive up the wages of skilled workers relative to those of the unskilled. But if technology increased the relative demand for skilled workers, why did wage inequality not grow as well? The leading explanation for the great levelling comes from pioneering work by Jan Tinbergen that conceptualized patterns of inequality as a race between technology and education.[73] Empirical work by Claudia Goldin and Lawrence Katz, two Harvard University economists, has shown that this view does a good job of explaining patterns of American wage inequality up until the 1970s.[74] Indeed, even if technological progress favors skilled workers, growing wage inequality does not have to be the result. The return to human capital depends on demand as well as supply. As long as the supply of skilled workers keeps pace with the demand for them, the wage gap between skilled and unskilled workers will not widen. While a number of short-run events and government interventions contributed to the great levelling, the most pervasive force—and certainly the best documented one—behind its long-run egalitarian impact was the upskilling of the American workforce, which depressed the skill premium.[75] Enabling technological change and the expansion of education provided the principal forces for convergence. From 1915 to 1960, the relative skill supply grew about 1 percent ahead of demand on an annual basis, causing wage differentials to become compressed. This pattern stands in contrast to the post-1980s period, when the demand for skills outpaced their supply.[76]

While the demand for skills was in part met by market forces, its supply also depended in large part on educational policies and new institutions that served to increase access to education and training. One institutional invention in particular was essential to making the gains from technological inventions widely shared: public schooling. The period 1910–40 is commonly referred to as the high school movement in American educational history. In 1910, only 9 percent of American youths obtained a high school diploma. In 1935, 40 percent did. In the period 1900–1970, over 70 percent of the increase in years of schooling was due to secondary school attendance. Some places adopted public schooling faster than others. The early adopters were characterized by greater community stability, a higher degree of ethnic and religious homogeneity, income and wealth equality, and higher levels of wealth. Thus, in short, social and financial capital helped foster the formation of human capital.[77]

The obvious reason why the high school movement happened was demand. Education was the best investment people could make for their children in the age of the Second Industrial Revolution. In the period 1890–1920, white-collar occupations, which typically required a high school diploma, paid about double the amount of occupations that didn't. High schools soon became a training camp for work—and for life more generally. As late as 1870, most Americans still worked in agriculture, and those employed in manufacturing worked in industries of the Industrial Revolution such as cotton, silk, and woolen textiles. All of these jobs required little formal training: only 10 percent of working Americans were employed in jobs that required education beyond elementary school. Child labor remained a persistent opportunity cost to education. By the turn of the century, boys as well as adult women typically earned half the hourly wage of adult men. Though school attendance was already compulsory in most states, some were reluctant to enforce it, as industry depended on access to cheap labor. Consequently, especially in the South, where child labor persisted longer, inequality was transmitted between generations as only relatively well-off families could afford to send their children to school.

A virtue of the Second Industrial Revolution was that it helped reduce the opportunity cost to education by raising skill requirements. In
1900, only four women and six boys were employed in the automotive
industry across the entire country. For Americans who wanted their
children to escape the drudgery of farming, a high school degree was
a ticket to better-paying work and gradually became essential for the
majority of jobs. Office workers, including bookkeepers, clerks, and
managers, found themselves handsomely paid for their education.
High school education was less common among blue-collar workers,
but those who did stay on in school were overwhelmingly found in
jobs associated with the Second Industrial Revolution. They were
electricians, auto mechanics, electrical engineers, machinists, and so
on. In 1902, a manager at the Decree Tractor Company made clear it
that he would not "take boys in the office unless they are at least high
school graduated. . . . In the factory we like the boys to have a high
school education if possible."[78] By 1920, a full quarter of the labor
force was working in occupations in which at least a high school degree was expected. Some leading technology firms, like General Electric, even required some years of high school for their apprentices.
And the demand for skills only continued to expand thereafter. In the
petroleum industry, for example, educational standards were continuously rising in the postwar period, for production and supervisory workers alike. At one refinery, management made a high school
education a job requirement across the board in 1948. And in 1953, a
preemployment test was introduced for applicants for production
jobs, aimed at determining "an individual's ability to memorize, concentrate, observe, and follow instructions"; it also tested mathematical knowledge, such as algebra and geometry, at the level of a high
school sophomore.[79]

Another survey, looking at the introduction of automatic reservation
systems in the airline industry, similarly shows that more sophisticated
skills were demanded as technology progressed. With airline traffic increasing to fifty million passengers in America in 1957—up by more
than 300 percent over just a decade—manual methods constituted a

key bottleneck to the capacity of airline companies to handle flight reservations. The new system significantly changed the content of jobs and the need for training:

> Classes for training instructors began while the equipment was being installed. . . . The airline lengthened this indoctrination training—which had covered from 5 to 7 days—to 8 to 10 days. After a week of subsequent on-the-job training, under his supervisor, the employee receives an additional 26–33 hours of advanced classroom instruction. . . . Seven new technician jobs were set up in connection with maintaining the new system. The technicians, who were previously employed as repairmen in the airline's radio shop, had worked directly and constantly on equipment. In contrast, the technician now works alone in an air-conditioned, noiseless control room. He works in his street clothes, and the only time he has direct contact with the automatic equipment is during preventive maintenance tests or on occasions when the equipment is out of order. The technicians were given specialized training by the manufacturer of the system, and attended classes 1 day a week for about 6 months. . . . A group of professional jobs concerned with electronic data-processing research was also created, following the advent of the new reservation system. This group is comprised of five systems engineers. These professionally trained persons perform duties which involve planning systems development and extending electronic methods to all clerical activities of the company. Their annual salaries start at $7,000. The qualifications for systems engineers include education at college level and cover a variety of airline experience. It is interesting to note that 4 of the 5 men in the group have college degrees in business administration and the social sciences. All have had considerable and varied work experience with the company.[80]

In contrast to grand theories of capitalism, the race between education and technology is a simple and robust empirical observation. It does not by any means exclude the possibility that other forces shape the trajectory of American inequality. Macroeconomic shocks, unions, tax policy, and financial-sector regulation, just to name a few, have all played a role in shaping American inequality. Even Kuznets and Piketty in their earlier work pointed to factors beyond their later theories. Before Kuznets advanced his exceedingly optimistic theory asserting that capitalist development will cause inequality to automatically fall in the long run, he explicitly mentioned the role that economic shocks might play. And in an earlier work with Emmanuel Saez, Piketty suggested that "one could indeed argue that what has been happening since the 1970s is just a remake of the previous inverse-U curve: a new industrial revolution has taken place, thereby leading to increasing inequality, and inequality will decline again at some point, as more and more workers benefit from the innovations. . . . Explanations pointing out that periods of technological revolutions such as the last part of the nineteenth century (industrial revolutions) or the end of the twentieth century (computer revolution) are more favourable to the making of fortunes than other periods might also be relevant."[81]

This is also the favored interpretation of the economist Branko Milanovic, who has recently put forward the idea of Kuznets waves accompanying every new technological revolution. His work does indeed show that the trajectory of inequality in Britain during the Industrial Revolution looks astoundingly similar to that of the computer revolution in America.[82] However, such an interpretation immediately raises the question of why American inequality during the Second Industrial Revolution seemingly followed a different pattern. The reason is that different economic models apply to different technological revolutions. As we have seen, the race between technology and education does a good job of explaining trends in the labor market over the first three-quarters of the twentieth century. But such models only apply when technological progress is of the enabling sort. This stands in stark contrast to the first seven decades of the Industrial Revolution in Britain, when the technological progress was primarily replacing and many people adjusted poorly (chapters 4 and 5). In this regard, as we shall see in chapter 9, the computer revolution more closely resembles the experience of the Industrial Revolution.

Conclusion

The Industrial Revolution didn't create the middle class, but it surely facilitated its growth. The spread of the factory system prompted the rise of industrial capitalism, and with it the expansion of the commercial and industrial bourgeoisie. Yet the history of the Industrial Revolution was not just the triumph of capital. The fact that the term "white-collar" first entered common usage during the first half of the nineteenth century suggests that labor markets were undergoing rapid change as industrialization picked up in pace. By the mid-nineteenth century, white-collar jobs supported a relatively prosperous group of families that we could also term middle class. The ascent of mechanized industry was accompanied by wage polarization as the earnings of white-collar workers pulled away from those of production workers. As discussed above, mechanization during the classic years of industrialization replaced relatively skilled artisan craftsmen with machines operated by the less skilled. Middle-income craftsmen saw their jobs disappear as mechanized factory production took over. Artisans of all kinds—cabinetmakers, watchmakers, shoemakers, and so on—closed their shops as the factories produced ever-growing quantities of goods. But as the establishments became larger and required more professional administrators, white-collar jobs expanded from 1850 onward. A more nuanced picture shows a hollowing out of the labor market, to the detriment of artisan craftsmen but to the benefit of the white-collar middle class. In America, newly collected data shows that the wages of white-collar workers were already growing steadily before the Civil War.[83]

Thus, in the nineteenth century, the ranks of the middle class were swelled by a growing number of managers and other skilled professionals, who took on an array of increasingly complex administrative and managerial tasks in the ever-growing factories. The best evidence of their societal prominence and relative affluence may be archaeological. The conditions in which the middle class lived in the late nineteenth century are most visible when one walks the streets of the old manufacturing towns in the eastern United States, from Cambridge, Massachusetts, to Hartford, Connecticut. Members of the middle class "were the first residents of the iconic nineteenth-century brownstones of New York City and the substantial Italianate and Queen Anne houses built in cities and towns throughout the northern tier of states."[84] Yet as a percentage of the American population, the middle class was still a small group. Occupational statistics provide some perspective on the percentage of middle-class households. In 1870, at the dawn of the Second Industrial Revolution, 8 percent of American workers were classified as managers, professionals, or proprietors (see table 1).

While the wages of production workers were lagging behind, the machines that replaced artisans could not run on their own. Working-class people, who gathered in the emerging factory towns, were the ones who ran the machines. The American urban working class first appeared in the early 1800s, when the economy began to industrialize, and expanded enormously over the next century as manufacturing jobs attracted millions of migrants from Europe and the American hinterland. The factories required operatives, who were "less skilled than the artisans they displaced in the sense that an artisan could fashion a product from start to finish, while the operative could perform a smaller set of tasks aided by machinery."[85] This is not to suggest that operatives did not have any skills. Factory workers had to learn how to operate the machines on the job. Though early textile machines were designed to be tended by children, the adoption of steam power eventually increased the demand for skilled operatives. In chapter 5, we saw that Engels's pause persisted as long as technological change served to replace skilled artisans with children. This was also the time when British workers regularly rioted against the mechanized factory. But things changed with the adoption of steam power, as adults regained their comparative advantage in production, where they found their skills augmented by increasingly sophisticated machines. Larger and more complex machinery also meant that a growing number of skilled engineers and mechanics were needed to design, install, and maintain the equipment. Thus, though the wages of the white-collar workforce rose significantly relative to those of the blue-collar workers below them, the wages of production workers— like machinists, furnace men, and textile weavers—increased during the second half of the nineteenth century. To be sure, the sons of manual workers rarely moved up into the ranks of the middle class. The worlds in which the white-collar middle class and production workers lived were still separated in the late nineteenth century.[86] Working-class men and women could at best aspire to a middle-class lifestyle, as could their children. But the Second Industrial Revolution meant that they finally could achieve it.

The technologies that defined the twentieth century allowed the production workers to attain a lifestyle superior to that of the upper classes in the early nineteenth century. Luxuries like hot running water and central heating systems began to be installed in large homes of the wealthy in the 1880s, and soon trickled down to working-class homes in the early twentieth century.[87] At the same time, an array of electrical inventions became available to American households during the early part of the twentieth century, relieving working-class housewives of some of their burdens. Mass production naturally targeted the mass market, and other key inventions, like the automobile, soon became available to the bulk of the population as well. As Gordon's account of the automobile revolution illustrates, there was a car for everyone:

> The Cadillac, Lincoln, and the Chrysler Imperial were for the ancien régime of inherited wealth and for the executive suite. The four-hole Buick Roadmaster connoted the vice president, while the three-hole Buick Century was for the rising midlevel executive, the owner of the local retail business or restaurant. Farther down the perceived chain of status were the Oldsmobile, the Pontiac, and the ubiquitous Chevrolet, America's best-selling car year after year, eagerly bought by the new unionized working class that, in its transition to solid middle-class status, could afford to equip its suburban subdivision house with at least one car, and often two.[88]

Working Americans did not just gain from technological change in their capacity as consumers. Perhaps more importantly, twentieth-century mechanization was primarily augmenting, and the few who lost their jobs to machines mostly faced decent alternative job options, as is reflected in the unprecedented wages taken home by blue-collar Americans: "As the wages of industrial workers increased in the thirty years after World War II, husbands were increasingly able to support a family lifestyle that included a modest home, a car, ample food and clothing, and perhaps even a vacation trip using the camper sitting in the driveway; more and more working-class families had enough income and consumption to reach the lower rungs of the broad American middle class."[89] The baby boom was in part a reflection of the growing optimism of young families, which created additional demand for goods and services and prompted the continued expansion of manufacturing and the creation of new labor-intensive services. During this period, a young male high school graduate could expect to find a secure job at a decent wage. The American economy was able to generate sufficient opportunity for blue-collar workers to attain a middle-class lifestyle on the basis of nothing more than their wages. The middle class, at its peak, was a diverse blend of white- and blue-collar workers. The result is reflected in the compression of the income distribution, prompting President John F. Kennedy to note that "a rising tide lifts all the boats."[90] Members of Marx's proletariat began to join the ranks of the middle class, which explains why worker resistance to mechanization became a distant memory.

It is indeed telling that resistance to machinery in America ended at the advent of the Second Industrial Revolution. In the nineteenth century, workers at times rebelled against mechanization. But the twentieth century did not witness such incidents. Other factors besides technology also played an important, if secondary, role. The early advent of American democracy, with universal white male suffrage achieved in the 1820s, meant that people no longer had to resort to violent protest to have their views heard. But even so, American labor history was exceedingly violent, and there was resistance to mechanization as late as 1879. The rise of the welfare state unquestionably made losing one's job less harsh, but welfare spending took off only with the Great Depression and World War II. The expansion of education and additional years of schooling made the young better equipped for the evolving labor market, but those who found their skills made redundant didn't go back to school. Perhaps more importantly, workers began to unionize and push for better pay and working conditions. But unlike the craft guilds, the unions rarely resisted new technologies. Even when unrest erupted, people didn't target machines to express their misgivings. Though workers organized and became a force of growing political power, resistance to mechanization was feeble, if not nonexistent. The benefits of progress for labor, it seems, were simply too great for the unions and their members to resist it.

PART IV

THE GREAT REVERSAL

Since the Industrial Revolution, mechanization has been controversial. Machines pushed up productivity, raising incomes per capita. But they threatened to put people out of work, to lower their wages and to divert all the gains from growth to the owners of businesses. . . . Now, it is robots that threaten work, wages and equality. . . . There have been long periods of economic history in which things did not work out well, and we must wonder whether we are in another. . . . The Luddites and other opponents of mechanization are often portrayed as irrational enemies of progress, but they were not the people set to benefit from the new machinery, so their opposition makes sense.

—ROBERT C. ALLEN, "LESSONS FROM HISTORY
FOR THE FUTURE OF WORK"

One of the greatest achievements of the twentieth century was unquestionably the creation of a diverse and prosperous middle class. It is therefore a matter of great concern that American society is now experiencing a dramatic decline in the fortunes of those people who might be described as middle class. The previous chapters have shown that technology played a key role in their rise. This part of the book will show the role it has played in their fall. As discussed above, several factors have shaped the trajectories of people's wages, but over the grand sweep of history, technology has been the predominant factor. The breathtaking rise in inequality after 1980 has without doubt been affected by other significant variables, like financial sector deregulation and superstar compensation. But these factors are primarily relevant in explaining the rise of the top 1 percent. The bigger story, however, is the decline of the middle class. The top pulling away from the rest would be much less troubling if the middle had continued to prosper. For all the talk of rising inequality per se, the greatest tragedy is that large parts of the workforce have actually seen their wages fall, adjusted for inflation. In the age of computers, the ranks of the affluent have grown—but at the cost of a withering middle class.

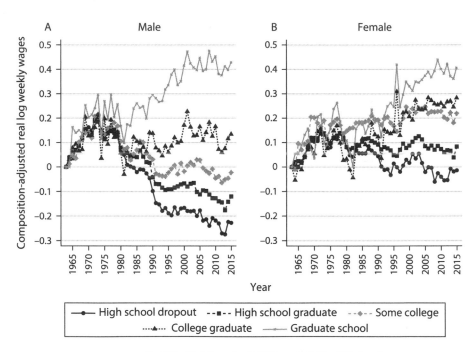

FIGURE 10: Real Weekly Wages for Full-Time, Full-Year Workers by
Educational Attainment, 1963–2015
*Sources: D. Acemoglu and D. H. Autor, 2011, "Skills, Tasks and Technologies:
Implications for Employment and Earnings," in* Handbook of Labor Economics, *ed.
David Card and Orley Ashenfelter, 4:1043–171 (Amsterdam: Elsevier).
Note: Their analysis has been extended to include the years 2009–15 using data from the
Current Population Survey and a DO file provided by David Autor.*

Since the pioneering work of Jan Tinbergen, economists tend to think about inequality as a race between technology and education. Skill-biased technological change means that new technologies increase the demand for workers with more sophisticated skills, relative to those without such skills. Thus, inequality between the skilled and the unskilled will rise unless the educational system churns out skilled workers at a greater pace than technology increases the demand for them. We saw in chapter 8 that the supply of skilled workers outpaced demand during the period of the great leveling, leading the wage differential between the skilled and others to become compressed. The post-1980 upsurge in wage inequality could simply reflect that the marketplace increasingly rewards people with more skills and the failure of the educational system to meet skill demand in the higher-tech economy. Yet if economic progress was just a race between technology and education, we would expect the wages of the skilled to pull away from those of the rest, but we would not expect the wages of the unskilled to fall. Inequality could grow, but everyone would still see their wages rise—though at different speeds. The great reversal depicted in figure 10 was first noted by Daron Acemoglu and David Autor.[1] It shows that up until the 1970s, wages rose for people at all educational levels, but after the first oil shock in 1973, wages fell and then stagnated for all Americans for about a decade. The great reversal began in the 1980s, when the wages of those with no more than a high school diploma began to fall again and continued to do so for three consecutive decades. This decline, as figure 10 shows, has primarily occurred among unskilled men who would have taken on jobs in the factories before the dawn of automation.

9

THE DESCENT OF THE MIDDLE CLASS

The computer era does not just mark a shift in labor markets. It also marks a shift in how economists think about technological progress. Daron Acemoglu and Pascal Restrepo have recently argued that the wage trends depicted in figure 10 are best understood as a race between enabling and replacing technologies. In a world of enabling technologies, the view of progress as a race between technology and education holds. New technologies augment the capabilities of some workers and enable them to perform new functions, making them more productive in a way that also increases their wages. Replacing technologies, by contrast, have the opposite effect. They render some workers' skills redundant in the tasks and jobs they perform, putting downward pressure on those people's wages.

In the 1960s, the management guru Peter F. Drucker argued that automation was no more than a more fashionable term for what once had been known as mechanization, and he took both words to mean the displacement of hand labor by machines.[1] As discussed above, replacing technologies did render the skills of some workers redundant during the first three-quarters of the twentieth century. Lamplighters, longshoremen, and elevator operators, to name just a few, saw their jobs disappear. Still, the age of automation must be distinguished from the age of mechanization. At the time when Drucker was writing, all workers saw their wages rise (figure 10). Indeed, before the spread of computers, machines could not operate on their own. They required operatives to keep production lines running. The explosive growth of semiskilled clerical and blue-collar jobs meant that even people who found themselves being replaced faced a much greater variety of job options. Factory and office machines alike were enabling technologies that made workers more productive, allowing them to take home better wages. In this regard, the computer revolution was not a continuation of twentieth-century mechanization but the reversal of it. Computer-controlled machines have eliminated precisely the jobs created for a host of machine operators during the Second Industrial Revolution. The workers that were once pulled into decent-paying jobs in mass-production industries are now being pushed out.

What Computers Do

In the *Wealth of Nations,* Adam Smith observed the division of labor in Britain's pin factories. By dividing activities into narrow tasks, he found that the first factories were able to increase efficiency enormously. While his observation concerned the division of labor between human workers, the age of automation came with a new division of labor: tasks can now be divided between humans and computers. Before the advent of the first electronic computer in 1946, the distinction between humans and computers was meaningless. Humans were computers. "Computer" was an occupation, typically performed by women who specialized in basic arithmetic.[2]

The division of labor between human and machine crucially depends on the tasks computers can do more effectively. Prior to the age of artificial intelligence (AI)—to which we shall return in chapter 12—computerization was largely confined to routine work. The simple reason is that computer-controlled machines have a comparative advantage over people in activities that can be described by a programmer using rule-based logic. Until very recently, automation was technically feasible only when an activity could be broken down into a sequence of steps, with an action specified at every contingency. A mortgage underwriter, for example, decides whether a mortgage application should be approved on the basis of explicit criteria. Because we know the "rules" for obtaining a mortgage, we can use computers instead of underwriters.[3]

But in other cases, we know the rules for only some of the tasks involved in an occupation. As is evident by the existence of ATMs, we can easily write a set of rules that allows computers to substitute for bank tellers in accepting deposits and paying out withdrawals. Yet we struggle to define the rules for dealing with an unsatisfied customer. Naturally, banks have taken advantage of this by reorganizing work so that tellers are no longer checkout clerks but relationship managers, advising customers on loans and other investment products. Consequently, as the handling of money has been automated, tellers have taken on nonroutine functions.

On the eve of the computer revolution, many occupations—like mortgage underwriter—were essentially rule-based. The majority of Americans still worked in what economists call routine occupations. In 1974, Harry Braverman, an American Marxist, drew attention to the dehumanizing nature of routine work, which he observed had persisted since the birth of the factory system. "The earliest innovative principle of the capitalist mode of production," he argued, "was the manufacturing division of labor, and in one form or another the division of labor has remained the fundamental principle of industrial organization."[4] In this regard, Braverman merely revived an old concern. As discussed above, in the 1830s, non-Marxist writers like Peter Gaskell and Sir James Kay-Shuttleworth argued that the repetitive motions of machines absorbed workers' attention to an extent that adversely affected their moral and intellectual capabilities. Braverman, who lived through the age of mass production, found that the Fordization of America had accelerated routinization. Machine operations had become even more subdivided. Workers' jobs were turned into mechanical motions, in which conveyors brought the task to the worker. Such specialization greatly increased productivity in American factories but brought greater monotony for the worker. From this point of view, factory automation can be regarded as a blessing because it meant that industrial robots, controlled by computers, could eliminate the need for direct human intervention in operating machines. Instead of having workers specializing in machine tending, many routine tasks could suddenly be performed by robots with a higher degree of accuracy. As automation progressed, more complex and creative functions became more plenti-ful. Computers, as Norbert Wiener declared, made possible "more human use of human beings."[5]

On the downside, these allegedly mindless, degrading, machine-tending, routine jobs were the ones that employed a large share of the American middle class. Numerous studies have shown that routine jobs were overwhelmingly clustered at the middle of both the skill and the income distribution.[6] As computer-controlled machines reduced the need for routinized chores, middle-class Americans saw their jobs dis-appear. As recently as 1970, more than half of working Americans were employed in blue-collar or clerical jobs. While few of them got rich, these jobs supported a broad and relatively prosperous middle class. And perhaps more important, most of these jobs were open to people with no more than a high school degree.[7] What Braverman was chal-lenging, however, was the notion that mechanization had increased the demand for skilled workers. He had little data to prove his point, but the idea that jobs had become more routine and required more skills does seem like a contradiction in terms. Many of the routine jobs that emerged over the course of the twentieth century were surely not too intellectually demanding, yet as we saw in chapter 8, the growing com-plexity of heavy industrial machinery and the expanding array of office machines did require more skilled operators.

The great reversal, depicted in figure 10, is in large part a consequence of computers making the skills of machine-tending workers obsolete. As the scope of automation has expanded from one routine task to another, those workers have faced worsening options in the labor mar-ket. But like electrification and the adoption of steam power, comput-erization did not happen overnight. Its impact on the labor market came decades after the birth of the electronic computer. William Nordhaus's heroic study of computer performance over the centuries shows that the first major discontinuity occurred around World War II.[8] The real cost of computing fell by a factor of 1.7 trillion over the 1900s, with the greatest leap occurring in the second half of the century. The timing is no mystery: the first programmable and fully electronic computer—the Electronic Numerical Integrator and Calculator (ENIAC)—arrived in

1946 and was accompanied by the invention of the transistor a year later. But for all its virtues, ENIAC was hardly fit for office use. It contained 18,000 vacuum tubes, 70,000 resistors, and weighed thirty tons. And while it was a general purpose computer, it was primarily built to calculate artillery firing tables. As discussed above, computers were a key source of automation anxiety in the 1950s and 1960s. But like the hype surrounding autonomous vehicles and AI today, concerns over computers' taking people's jobs merely reflected a few first use cases (chapter 7). At the 1958 annual convention of the National Retail Merchants Association, for example, there was much excitement over the new computers and merchandise handling systems, but few attendees opened their wallets. Computers were still too bulky and expensive for widespread adoption.[9]

Though ENIAC can justly be regarded as the symbolic inception of the computer revolution, it was the personal computer (PC) that heralded the dawn of the age of automation.[10] When *Time* displaced humans on its front cover and declared the PC the "Machine of the Year" in 1982, America had just begun to computerize. According to *Time*, "Now, thanks to the transistor and the silicon chip, the computer has been reduced so dramatically in both bulk and price that it is accessible to millions. . . . In contrast to the $487,000 paid for ENIAC, a top IBM PC today costs about $4,000, and some discounters offer a basic Timex-Sinclair 1000 for $77.95. One computer expert illustrates the trend by estimating that if the automobile business had developed like the computer business, a Rolls-Royce would now cost $2.75 and run 3 million miles on a gallon of gas."[11]

In America's largest five hundred industrial companies, only 10 percent of typewriters had given way to the word processor. Robots, to which computers provided the mechanical brain, had taken over some of the nation's dull and dirty jobs, but few industries had robotized. Of the 6,300 robots operating in America's factories in 1982, 57 percent of them were at four companies: General Motors, Ford, Chrysler, and IBM.[12] Yet from the 1980s onward, a growing share of routine tasks were transferred to computer-controlled machines. As computers became smaller, cheaper, and more powerful, routine employment began to

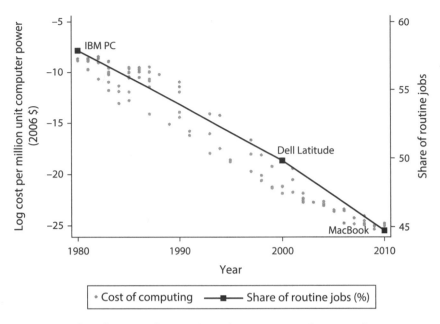

FIGURE 11: The Falling Cost of Computing and Disappearance of Routine Jobs, 1980–2010
Sources: C. B. Frey, T. Berger, and C. Chen, 2018, "Political Machinery: Did Robots Swing the 2016
U.S. Presidential Election?," Oxford Review of Economic Policy 34 (3): 418–42;
W. D. Nordhaus, 2007, "Two Centuries of Productivity Growth in Computing," Journal of
Economic History 67 (1): 128–59; N. Jaimovich and H. E. Siu, 2012, "Job Polarization and Jobless
Recoveries" (Working Paper 18334, National Bureau of Economic Research, Cambridge, MA).
Note: The figure shows how routine employment has contracted as the cost of computing has fallen.
All dots denote the year in which a new computer technology was introduced and its cost.

shrink (figure 11). But we now know that the consequence was not wide-spread technological unemployment, as many had predicted in the 1950s and 1960s. While automation replaced workers in some jobs, it also created new ones. Robots replaced workers in repetitive assembly work, but the machines also required skilled personnel capable of programming, reprogramming, and occasionally repairing them. Job titles like robot engineer and computer-software programmer are a direct consequence of automation. Thus, the erosion of old jobs gave rise to new ones. When automatic flight reservation systems arrived, for example, "the strictly routine tasks of posting each sale on a sales control chart and the cumbersome method of using a visual display board to denote availability of flight space were both eliminated."[13] But another outcome was the enlargement of the sales function: "The job title of *clerk* was replaced by *sales* or *service* agent. An upgrading took place for two employees who perform the functions of Specialist (Reservisor Information) and Assistant to the Specialist."[14]

Focusing only on the rise and fall of individual occupations, however, inevitably glosses over much of the transformation of the workplace. Many of these changes have taken place within occupations. For instance, while the job of the secretary has not disappeared, it no longer has much in common with the jobs held by secretaries in the 1970s. Before the computer revolution, the Bureau of Labor Statistics described the role of the secretary as follows: "Secretaries relieve their employers of routine duties so they can work on more important matters. Although most secretaries type, take shorthand, and deal with callers, the time spent on these duties varies in different types of organizations."[15] The impact of the computer age becomes evident when we look at the description of the same job from the same source in the 2000s: "As technology continues to expand in offices across the Nation, the role of the secretary has greatly evolved. Office automation and organizational restructuring have led secretaries to assume a wide range of new responsibilities once reserved for managerial and professional staff. Many secretaries now provide training and orientation to new staff, conduct research on the internet, and learn to operate new office technologies. In the midst of these changes, however, their core responsibilities have remained much the same—performing and coordinating an office's administrative activities and ensuring that information is disseminated to staff and clients."[16] What is true of secretarial positions is also true of many other jobs. For example, in the 1970s, American men and women could make a good living as bank tellers by accepting deposits and paying out withdrawals. As noted, the job has not disappeared, but the skill requirements have changed so dramatically that it requires a different breed of worker.

All the same, computerization has clearly not been as dismal for labor as some people predicted when ENIAC arrived. Although computers have taken over an ever-growing share of routine work, labor has retained its comparative advantage in other domains. One reason is because of what the economist David Autor has called "Polanyi's paradox."[17] A key bottleneck to automation that engineers have found hard to overcome is well summarized by Michael Polanyi's famous phrase: "We know more than we can tell."[18] (We shall take a closer look at the AI-enabled inroads on Polanyi's paradox in chapter 12.) Humans constantly draw upon large reservoirs of tacit knowledge that we struggle to articulate and define even to ourselves, making it exceedingly hard to specify it in computer code. To illustrate Polanyi's point, it is helpful to contrast the task of repetitive assembly with that of designing a new car, writing a piece of music, or giving a galvanizing speech. The rules for what makes a good song or a great speech are hard to define because there are none. Artists and other creative professionals constantly break and redefine them. From an automation point of view, Polanyi's insight is critical, because it means that there are many tasks humans are able to perform intuitively but that are hard to automate because we struggle to define rules that describe them. For activities that demand creative thinking, problem solving, judgment, and common sense, we understand the skills only tacitly. But more importantly, from an economic standpoint, Polanyi's observation also means that some human skills are complemented by computers. Things done by computer technology, such as the storing and processing of information, make humans more productive problem solvers, decision makers, and analysts. As computerization reduced the costs of critical inputs to these tasks, humans have become more productive in computer-using jobs.[19] In 1970, a lawyer in Grinnell, Iowa (a town without a law library), would have had to drive to the next city to do legal research. In the computer era, she can leave her car in the garage and connect her word processor to Westlaw, which provides digitized records on case law, state and federal statutes, administrative codes, and so on. Indeed, most of the professions have found their practitioners' skills augmented by computers. Consider, for example, the evolving office of Stephen Saltz, a Boston cardiologist:

> In September 2001, Dr. Saltz took an echocardiogram of an elderly male patient we will call Harold. Harold had suffered a small heart

attack. His condition was complicated by diabetes, a disease that creates "silent" heart blockages not detected in standard tests. When Saltz had trained in Boston's Brigham Hospital in the early 1970s, an echocardiograph was an oscilloscope-like device that provided limited information on the heart's blood flow and valve flaps. Over time, advances in computerization allowed the instrument to create a full two-dimensional image of virtually all aspects of the heart's functioning, including blood flows, blockages, and valve leakages. Using this image, Saltz saw that the entire front wall of Harold's heart was malfunctioning. The information led Saltz to refer Harold to a surgeon who would perform a bypass or insert a stent. Either operation would improve the length and quality of Harold's life. The computerized imaging had made Saltz a better diagnostician.[20]

As engineers have expanded what computers can do, technological progress has continuously moved in the direction of favoring skills that require higher education, such as complex problem solving and creative thinking, because computers have taken over the more mundane tasks. When Robert Reich surveyed the transformation of the labor market in his classic 1991 book, *The Work of Nations*, he found that work could be divided into three broad categories. A new class of what he called "symbolic analysts" had emerged, who were reaping the benefits of the new economy.[21] Among these analysts, we find managers, engineers, attorneys, scientists, journalists, consultants, and other knowledge workers. In the age of computers, they had all become more productive analysts. Besides symbol-analytic services, Reich reckoned, there are also routine jobs and in-person services. As noted above, routine jobs have gradually been taken over by computers. But in-person service jobs have become more plentiful. Indeed, most Americans do not work in technology industries or professional services. Few are employed directly by software companies, law firms, or biotechnology start-ups. But these occupations still support the livelihoods of many citizens. While today's technology companies provide less opportunity for the unskilled relative to the smokestack industries of the Second Industrial Revolution, many people are employed indirectly by those firms, whose employees demand many services that the unskilled provide. When America's symbolic analysts shop locally, they support the incomes of hairstylists, barkeepers, waiters, taxi drivers, and store clerks. These jobs may not have seen the technological miracles of biotechnology and software production, "but that is where most Americans work and their fates are tied to their ongoing ability to sell their time to those workers who sell exportable goods and services."[22]

If Polanyi's paradox were the only hurdle to automation, most remaining jobs would be for symbolic analysts. A second reason why there are still so many jobs is explained by Moravec's paradox, named after the computer scientist Hans Moravec. The paradox he noted was the fact that it is hard for computers to do many tasks that are easy for humans, and conversely, computers can do many things that we find exceedingly difficult: "It is comparatively easy to make computers exhibit adult level performance in solving problems on intelligence tests or playing checkers, and difficult or impossible to give them the skills of a one-year-old when it comes to perception and mobility."[23] A computer would have an easy time beating the world chess champion Magnus Carlsen, but it would be unable to clean the chess pieces after the game and put them back in the right place. Any human cleaner still outperforms computer-controlled machines in perception, dexterity, and mobility. Today's computers far exceed human capabilities in storing and processing information, yet they cannot climb a tree, open a door, clear coffee cups off a table, or play football. A powerful explanation is that our unconscious sensorimotor abilities have evolved in the human brain over millions of years, making them exceedingly difficult to imitate. From an early age, humans can walk, identify and manipulate objects, and understand complex language. Giving a computer these basic abilities, mastered by any four-year-old child, has turned out to be among the hardest engineering problems.

Relative to Polanyi's paradox, the critical difference is that many of the skills that are hard to automate because of Moravec's paradox have not been made more valuable by computers. Together, the persistence of these engineering bottlenecks explains why the labor market has evolved the way it has. As computers have made symbolic analysts more affluent, the analysts have spent a larger percentage of their income on personal services that are hard to automate. But the automation of routine jobs has meant fewer employment opportunities for high school graduates, so there has been a flow of workers from productive and automated sectors to low-productivity service jobs—like those of janitors and gardeners, child-care workers, receptionists, and so on.[24] Unfortunately, this has meant that millions of workers have migrated into jobs where the productivity ceiling is low, and consequently, their wages have fallen behind those of symbolic analysts. Even so, economists would expect the wages of people in technologically stagnant jobs to rise, as employers need to pay them more to prevent workers from leaving for more productive jobs of higher pay. As noted, the fact that the wages of men without college degrees have fallen over the course of three decades, in other words, suggest that they are faced with fewer alternative job options for which their skills are suitable. Together with Autor, in their pioneering 2004 book, *The New Division of Labor*, Frank Levy and Richard Murnane, two economists at the Massachusetts Institute of Technology, were among the first to note this pattern:

> As computers have helped channel economic growth, two quite different types of jobs have increased in number, jobs that pay very different wages. Jobs held by the working poor—janitors, cafeteria workers, security guards—have grown in relative importance. But the greater job growth has taken place in the upper part of the pay distribution—managers, doctors, lawyers, engineers, teachers, technicians. Three facts about these latter jobs stand out: they pay well, they require extensive skills, and most people in these jobs rely on computers to increase their productivity. This hollowing-out of the occupational structure—more janitors and more managers—is heavily influenced by the computerization of work.[25]

While their work focused on the United States, the polarization they observed was not just an American phenomenon. As figure 12 shows, the hollowing out of the middle is a feature of labor markets across the

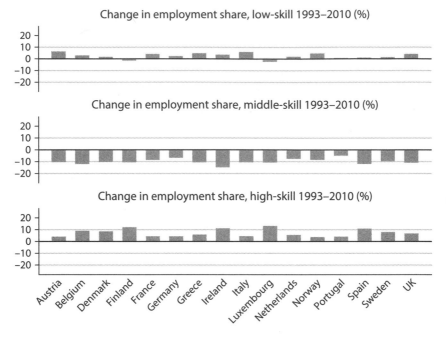

FIGURE 12: Job Polarization in 16 European Countries, 1993–2010
Source: M. Goos, A. Manning, and A. Salomons, 2014, "Explaining Job Polarization:
Routine-Biased Technological Change and Offshoring,"
American Economic Review 104 (8): 2509–26.

industrial world. This dynamic of job growth at the top and the bottom of the skill and income distribution has contributed to the growing divide between college and high school graduates.

The Cognitive Divide

Terms like "middle class" and "working class" were invented to describe the profound changes that accompanied the Industrial Revolution. Yet as of late, they have become increasingly problematic. President Barack Obama mentioned the "middle class" ten times in his first State of the Union address. He spoke about the "working class" only once, when he referred to Vice President Joseph Biden as "a working-class kid from Scranton."[26] The disappearance of industrial jobs has meant that fewer citizens can be regarded as working class. Most adults with a high school education no longer work in factories. There is no stable working class for young men and women to break into. Thus, the term has become a shunned pejorative, while the "middle class" is now used to refer to almost everyone except the very wealthy and the very poor.[27]

What we regard as "middle class" has, of course, always been elastic. During the classic years of industrialization, it was largely reserved for the commercial and industrial bourgeoisie. But in the succeeding centuries, it expanded to the point that it converged with the working class. Welding-machine operators and other blue-collar workers became associated with relatively stable, decent-paying jobs by the mid-twentieth century. The conditions of the working classes in the golden postwar years were strikingly different to those described by Karl Marx and Friedrich Engels a century earlier, and they were able to attain a middle-class lifestyle. But in a world of robots, there are fewer jobs for the so-called blue collar aristocracy. With some rare exceptions, only college educated workers qualify as middle class. What distinguishes Reich's symbolic analysts from the rest is that nearly all the jobs they hold require a college degree. As clerical and blue-collar jobs have disappeared from the lower and middle parts of the income distribution, the employment prospects for young people with no more than a high school education have become more similar to those of high school dropouts than to people with a college education. Thus, sociologists mostly use college education, rather than occupation, as an indicator of a citizen's class in the post-1980 period.[28]

As has been widely documented, education has reinforced the divide between those who thrive in the new economy and their less-educated peers. This pattern becomes all the more evident when we look at how workers have adjusted to automation. Those with analytical skills have moved up into the expanding sets of high-wage jobs, while people who lack valuable skills have dropped down and are competing for unskilled service jobs at declining wages. In the postwar era, workmen on the assembly lines who experienced displacement could still find work in other routine jobs requiring similar skills. But since the computer revolution, unemployed Americans who used to work in a routine occupation have become much less likely to find new employment in routine jobs.[29] Fewer job options, especially for non-college-educated production workers, has led to cascading competition for low-skilled jobs.

To be sure, as the jobs of machine operators dried up, new highly skilled jobs were created, as computer programmers were needed to design numerically controlled machine tools. In 1985, when the first wave of automation swept through America's car factories, the *Wall Street Journal* published a story about Lawrence Maczuga, a thirty-seven-year-old machine operator at Ford Motor Company's transmission plant in Livonia, Michigan. While Maczuga was working the machines at the factory, he had been going to college at night to get a degree in computer science. As automation took over, he gave up his semiskilled job on the assembly line to take one of the plant's "super-jobs" as a manufacturing technician. Maczuga had for some time considered giving up his job at Ford to become a computer programmer. But as Ford revealed its new plans for computerizing production, he accepted the new role.[30]

The problem is that Maczuga was the exception, not the pattern. Few manual workers got degrees in computer science or any other college concentration. Thus, as automation eroded the demand for routine skills and physical strength, blue-collar people found themselves in an ever-weaker position. Indeed, the economists Matias Cortes, Nir Jaimovich, and Henry Siu found that prime-aged men without college degrees were the main victims of the contraction of routine employment. Many adjusted by taking on low-paying service jobs, like those of food-preparation workers, gardeners, and security guards. Unskilled men were more likely to find themselves pushed down or even out of the workforce as routine jobs dried up than they were to move up.[31]

The adverse consequences of automation have manifested themselves not only in falling wages but also in rising joblessness among groups in the labor market. For several decades now, the percentage of prime-aged men ages 25–54 who do not go to work in the morning has steadily risen (figure 13). While economists still debate the relative importance of supply and demand factors in explaining men's detachment from the labor market, there is an emerging consensus that demand factors should be given more weight for recent years. Welfare programs,

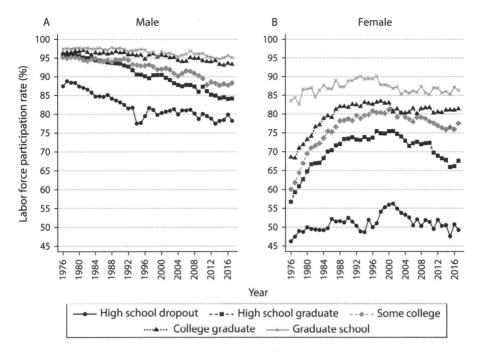

FIGURE 13: Labor Force Participation Rate (Ages 25–54) by Educational Attainment, 1976–2016

Source: 1963–91: Current Population Survey data from D. Acemoglu and D. H. Autor, 2011, "Skills, Tasks and Technologies: Implications for Employment and Earnings," in Handbook of Labor Economics, ed. David Card and Orley Ashenfelter, 4:1043–171 (Amsterdam: Elsevier). 1992–2017: author's analysis using data from S. Ruggles et al., 2018, IPUMS USA, version 8.0 (dataset), https://usa.ipums.org/usa/.

spousal employment, and changing social norms have all played a role in shaping some men's decision not to work over the course of the twentieth century (see also figure 7). But since 2000, most of the rise in joblessness is seemingly involuntary. When the economists Katharine Abraham and Melissa Kearney recently set out to review what we know about the causes of male joblessness, they found trade and robots to be the prime reasons why fewer men ages 25–54 have been working since the new millennia.[32]

However, the experience of women has been rather different. As is well known, the great leap forward of the "pink-collar" workforce came to an end in the 2000s, when computers began to take over more clerical work (figure 13). Just a few decades ago, people who called Amtrak to make a train reservation would have heard a woman answering the phone on the other end. Today, they hear a recorded voice saying, "Hi, I'm Julie, Amtrak's automated agent." But consistent with what we know from studies in neuroscience pointing out that women perform better in interactive and social settings, women have adjusted much better than men to an increasingly interactive world of work.[33] Instead of being pushed back into low-wage service jobs, where women had traditionally been dominant, many have moved up into professional and managerial jobs. Women also are more likely to graduate from college than men, and consequently their skills are more suitable for the computer age. Indeed, while men have found themselves increasingly likely to be replaced by computer-controlled machines, women are more likely to use a computer at work.[34] The rising share of women in the professions and the decline of male-dominated blue-collar sectors have allowed many women to overtake their male counterparts in terms of career advancement. Of course, women still have some way to go before they surpass men in terms of earnings, but a shift is under way. American women age thirty or younger have higher earnings power than their male counterparts—with the exception of the three largest metropolitan areas, where skilled men have clustered.

Though this process has been going on for decades, it has intensified in recent years, as new technologies, like multipurpose robots, have come into play. These robots are automatically controlled so that they do not require human operators. And they can be reprogrammed to perform various manufacturing tasks, like welding, assembling, or packaging. Thus, they must be distinguished from single-purpose robots and other computer-controlled machine tools, which are designed for one specific purpose.

All analyses of the role of technology in rising male joblessness are regrettably limited to multipurpose robots, as systematic data on single-purpose robots remain sparse. But even if this means that we underestimate the pervasiveness of robots in the economy, such data are still informative. Daron Acemoglu and Pascual Restrepo estimate that each multipurpose robot has replaced about 3.3 jobs in the U.S. economy. Blue-collar people in heavily robotized industries—like automobile

manufacturing, electronics, metal products, chemicals, and so on—
have naturally felt the force of automation most keenly. But where robots have been adopted, people in nearly all occupations suffered both
wage and employment losses. Job losses were unsurprisingly more significant among workers without college degrees. And men have been
twice as likely as women to find themselves replaced by robots.[35]

Those estimates focus on the period 1993–2007. However, statistics
on the adoption of multipurpose robots from the International Robotics Federation show that American industry has continued to use robots
more extensively in the postrecession period: the number of robots in
use grew by almost 50 percent between 2008 and 2016. But needless to
say, the worker-replacing effects of robots may have been counterbalanced by other technologies. Computers, as noted above, have also
spun off new tasks for labor, in which they augment workers' skills. All
the same, there is compelling evidence that technological change in general has become more worker replacing in recent years. As we saw in
figure 10, real wages for non-college-educated men have fallen since the
1980s. Of course, this might also reflect some other permanent factor,
like jobs being sent abroad. But there is more direct evidence to suggest
that technology has become increasingly labor replacing. In a major
study of eighteen Organisation for Economic Co-operation and Development (OECD) countries, published in 2018, the economists David
Autor and Anna Salomons found that whether automation is measured
by productivity gains, patenting flows, or the implementation of multipurpose robots, it has reduced the share of national income captured
by labor. Technological progress, they show, only turned replacing as
computers became more pervasive in the 1980s, and its negative effects
on the labor share became more pronounced throughout the 2000s.[36]

The Return of Engels's Pause

The age of automation is not without parallel, however. We saw in chapter 5 how the Industrial Revolution caused a hollowing out of middle-
income jobs, put downward pressure on workers' wages, and prompted
an upsurge in inequality. The classic years of industrialization are known

as Engels's pause, when the mechanized factory displaced domestic industry, worsening the economic prospects of many citizens even as the British economy took off. During the classic years, output experienced an unprecedented expansion, yet the gains from growth didn't trickle down to most people. Output per worker grew more than three times faster than average weekly wages. As middle-income artisan jobs were trimmed, the gains of the Industrial Revolution went to industrialists who saw the rate of profit double. Engels's assertion that industrialists grew rich on the misery of workers was broadly right for the period he observed. The pause came to an end only around 1840.

If Friedrich Engels were living today, what would he have written about the computer era? Working conditions in the industrial West clearly do not have much in common with the "dark, satanic mills." But the trajectories of per capita output and people's wages look exceedingly similar. In America, labor productivity has grown eight times faster than hourly compensation since 1979.[37] Even as the American economy has become much more productive, real wages have been stagnant, and more people are out of work; consequently the labor share of income has fallen. Corporate profits have swept up an ever-greater share of national income while the share going to workers has rarely been smaller. And official measures of labor compensation include the paychecks of CEOs and superstars in music, sports, and media, meaning that the percentage of income going to the average worker has declined even further. As was the case in the classic years of the Industrial Revolution, the gains from growth have shifted from the bottom to the top of the income distribution and from labor to owners of capital. During the postwar years, the labor share hovered around 64 percent, but since the 1980s it has steadily declined to its lowest postwar level after the Great Recession, averaging around 58 percent in recent years.[38] This is in accord with the trends depicted in figure 9, where we saw a widening gap between labor productivity and worker compensation emerging in the 1980s. And it is not just a U.S. phenomenon. The economists Loukas Karabarbounis and Brent Neiman, for example, have documented that the share of national income that goes to labor has declined dramatically in most countries since the 1980s, which, they argue, is thanks to cheaper computers.[39]

There are good reasons to think that rising profits and the falling labor share is linked to the automation of routine middle-income jobs (such as those of machine operators, bookkeepers, and mortgage underwriters) and the shift of labor into low-income service jobs (for example, those of janitors, waiters, and receptionists). In 2017, the International Monetary Fund (IMF) published a report showing that "technological advancement, measured by the long-term change in the relative price of investment goods, together with the initial exposure to routinization, have been the largest contributors to the decline in labor income shares in advanced economies."[40] Consistent with the hollowing out of labor markets, as computer-controlled machines have taken over the jobs of the middle class, the IMF found that the decline in labor shares has been particularly sharp for middle-skilled workers.

The changing face of technology is also reflected in long-run trends in the Gini coefficient (figure 14). As Branko Milanovic has noted, "This revolution [the computer revolution], like the Industrial Revolution of the early nineteenth century, widened income disparities."[41] These periods were not only times when the profit share of income reached historical heights and the wages of ordinary citizens were stagnant. As noted above, both were episodes when technology replaced middle-income workers. In the computer era, the increase in inequality happened in large part because new technologies strongly rewarded more highly skilled symbolic analysts while driving up the capital share of national income. At the same time, as middle-income routine jobs were shredded, unskilled labor moved toward low-paying service occupations, causing wage disparities to rise. In similar fashion, technological change during the Industrial Revolution pushed people out of middle-income jobs in the domestic industry, to the detriment of many craftsmen, while creating low-income production jobs in the factories and high-paying skilled jobs for white-collar workers to manage and administer production. Indeed, the economists Lawrence Katz and Robert Margo have pointed out that the effects of computers on the labor

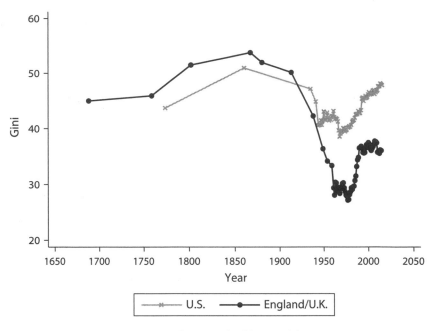

FIGURE 14: Income Inequality in England/U.K. and the U.S., 1688–2015
Sources: See appendix, this volume.

market today have been similar to those that accompanied the spread of the mechanized factory in the nineteenth century.[42]

So far, new computer technologies have not caused widespread unemployment as has been so widely feared. Though industries and occupations have lost jobs due to automation, job losses have been offset by the creation of new tasks, customers and suppliers benefiting from cheaper goods, and increases in overall consumer spending.[43] But computer technologies have shrunk the size of the middle class, put downward pressure on unskilled workers' wages, and reduced labor's share of income. And, as the experience of the Industrial Revolution illustrates, even when new jobs are being added, it can take a long time for workers to acquire the necessary skills to successfully move into the newly emerging jobs. In many cases, new or changing job roles require a different breed of worker. Case studies of office automation have shown that computers reduced clerical staff in routine activities, while opening up only a "relatively small number of better paid positions for programming and operating the new systems."[44] Aptitude tests provided a means of selecting staff for the new and substantially more skilled jobs: "Those selected were chiefly men in their late twenties with some college education and some company experience in . . . related work."[45] Few older workers or employees displaced from the affected units were chosen for the newly created positions. The adverse impacts of automation were much greater for middle-age and older employees.

When replacing technologies make the skills of existing workers redundant, they reduce the earning capacity of significant parts of the population. Though new tasks may be spun off in the process, new skills take time to learn and are often seen in workers' wages only years later. As discussed in the context of the Industrial Revolution, the wages of power-loom weavers took off long after the jobs of hand-loom weavers were shredded. A modern equivalent is the case of typographers, which also illustrates this point vividly. One advantage of the computer was that a file could be saved to its memory, eliminating tedious rekeyboarding of typewritten text to correct errors. The effect on typographers' jobs and wages was significant. James Bessen has calculated that between 1979 and 1989, employment among typesetters and compositors fell from 170,000 to about 74,000, while their median wages declined by 16 percent, adjusted for inflation. "Membership in the International Typographical Union fell sharply, and in 1986, much weakened, it merged with another union."[46]

As computer publishing eliminated tedious rekeyboarding of text and made typesetting less costly and more versatile, desktop publishers and graphic designers took on much of the typographical work, and employment opportunities in these occupations soared. But making this transition required workers to learn graphic design software, such as page layout programs. While we lack the data to determine the extent to which typographers successfully took on jobs as graphic designers, these jobs required very different skills from those used in typography, and few typographers probably made this transition. And even those who managed to become graphic designers didn't see their wages rise, as average pay for designers remained stagnant. Bessen explains:

After accounting for inflation, the average hourly pay of graphic designers has been stagnant in recent years; the average pay of all types of designers has actually fallen since the 1970s. While designers are paid, on average, a bit more than typographers, the median designer of 2007 earns only about a dollar more per hour than the median typographer of 1976. Designers seem to have shared little in the benefits of this technology. Why don't average designers earn more, now that they have acquired substantial new skills and job responsibilities? Because the technology and organization of work for designers seem to be in constant flux. The print designers who replaced the typographers have been partly replaced by web designers, who are partly being replaced by mobile designers. Technology is continually redefining what publishing is and how it is done. Each of these changes requires new, specialized skills—skills learned largely through experience or by sharing knowledge, rather than in school. Each year, designers have to learn new software and new standards in order to keep up. A few years ago they learned Flash; now it is HTML5. Next year, perhaps something else.[47]

When occupational skills are replaced by machines, the investment workers have made in building up the human capital associated with that occupation has gone industrially bankrupt. A worker displaced from a steel mill will not be able to begin a new career as a barber the next morning, and he or she is rarely equipped to switch into a professional, managerial, or engineering job. The higher the cost of accumulating new human capital, the longer the transition will take. Even low-skilled service jobs in restaurant, hotels, and gasoline stations require some skills. Experience is valuable in just about every occupation. But unquestionably, the cost of acquiring new human capital to move into well-paying jobs has become much greater in the ever-higher tech economy, leading to a hardening division between those who went to college and whose who did not. What's more, as we shall see, location matters about as much as education. The most serious adjustment problems have occurred among unskilled workers in declining towns and cities.

10

FORGING AHEAD, DRIFTING APART

As discussed above, Engels's pause was a time when citizens saw rapid change, not just in the workplace but also in the communities in which they lived. Indeed, in the early nineteenth century, social critics like Peter Gaskell made the impact of mechanized industry on society the center of public debate. Gaskell's *The Manufacturing Population of England: Its Moral, Social, and Physical Conditions*, published in 1833, was the first of its kind and a source of inspiration for the essay Friedrich Engels later wrote on the conditions of the English working classes. Gaskell believed that the struggle between human power and steam-powered machinery was approaching a crisis, and he argued that mechanization was changing the "very framework of the social confederacy."[1] In addition to depriving craftsmen of jobs in rural industry, he declared, the factory system was creating a new socially deprived class in Britain's emerging industrial cities: "The universal application of steam power as an agent for producing motion in machinery, has closely assimilated the condition of all branches, both in their moral and physical relations. In all, it destroys domestic labour; in all, it congregates its victims into towns, or densely peopled neighbourhoods; in all, it separates families."[2]

The computer revolution, as we shall see, has caused to the demise of many of the factory cities that industrialization once gave rise to. And as was the case with the Industrial Revolution, its social consequences for individuals, families, and their communities have been profoundly negative. Since its peak in 1979, more than seven million American manufacturing jobs have disappeared. And America's industrial towns and smokestack powerhouses, where blue-collar jobs were clustered, have felt the consequences most keenly. Where middle-class jobs withered, whether due to automation or globalization, a range of societal problems have emerged. In the 1990s, the sociologist William Julius Wilson attracted enormous attention with his study of urban ghettos where work had vanished. Using large-scale surveys and ethnographic interviews, he concluded that "the consequences of high neighborhood joblessness are more devastating than those of high neighborhood poverty. A neighborhood in which people are poor but employed is different from a neighborhood in which people are poor and jobless. Many of today's problems in the inner-city ghettos—crime, family dissolution, welfare, low levels of social organization, and so on—are fundamentally a consequence of the disappearance of work."[3] While the work of Wilson focused on black neighborhoods in inner cities suffering from economic restructuring and suburbanization, many of their troubles are now shared by communities of the white working class.

When Jobs Disappear

No town or city could represent all of America, but the closest we can get is probably Port Clinton, New York, on the shores of Lake Erie. In *Our Kids*, sociologist and political scientist Robert Putnam looks back at his Port Clinton high school class of 1959, in what was then a blue-collar middle-class town, and in almost every respect a remarkable microcosm of America. He recalls that few of his classmates' parents were educated. A full third had not even graduated from high school. Yet like just about everyone in town, they were reaping the benefits of postwar prosperity: "Some dads worked the assembly lines at the local auto part factories, or in the nearby gypsum mines, or at the local Army base, or on small family farms."[4] But in an era when technological change was of the enabling sort, few families had experienced joblessness or felt economic insecurity. Though few families in Port Clinton were affluent, very few were poverty-stricken. And their children were like everybody else's. Regardless of social background, they participated in many extracurricular activities, including sports, music, and drama: "Friday night football games attracted much of the town's population."[5] Fast-forwarding half a century, Putnam's classmates had also made a great leap forward relative to their parents. Three-quarters of them were better educated, and the overwhelming majority had moved up the economic ladder. Perhaps most strikingly, many children with less well-off parents had climbed farther up the rungs than those from more privileged, better-educated backgrounds. Upward mobility among the kids from relatively disadvantaged backgrounds was almost as great as among the affluent ones.

Yet today, the American dream in Port Clinton is a split-screen nightmare. The children of the next generation are facing a very different reality. Those whose parents were symbolic analysts or the like have stayed on track, while those whose modest family fortune depended on factory jobs have founded themselves stuck on the wrong side of the tracks. As Putnam points out, the causes of faltering upward mobility in America are many. But one is surely that the manufacturing foundation upon which Port Clinton's prosperity rested has dwindled. As well-paying blue-collar jobs were trimmed, births out of wedlock rose sharply, child poverty skyrocketed, and upward mobility went into reverse. Because blue-collar workers tend to shop close to where they live, they also support many other jobs in the service sector. Thus, the disappearance of blue-collar jobs meant a blow to the local service economy, with shops closing down and people leaving town. As work vanished, many people left in search of a better future elsewhere, leaving Port Clinton in despair: "The Port Clinton population, which had jumped 53 percent in the three decades prior to 1970, suddenly stagnated in the 1970s and 1980s, and then fell by 17 percent in the two decades after 1990. Commutes to jobs got longer and longer, as desperate local workers sought employment elsewhere. Most of the downtown shops of my youth stand empty and derelict, driven out of business partly by the Family Dollar and the Walmart on the outskirts of town, and partly by the gradually shrinking paychecks of Port Clinton consumers."[6]

The Port Clinton story is regrettably a typically American one. In the postwar years there were many places like Port Clinton, whose modest prosperity was built on the foundation of manufacturing that provided stable, thriving communities and opportunities for those starting at the bottom. But the employment prospects of blue-collar Americans are no longer what they were in the 1950s and 1960s, and neither are the communities in which they live. In *Coming Apart*, political scientist Charles Murray shows that ordinary middle-class Americans who could have prospered in Port Clinton in the postwar years have become increasingly detached from the rest of American society more broadly. To illustrate this, he creates a statistical construct he calls Fishtown—after the predominantly white, blue-collar community of the same name in Philadelphia, Pennsylvania—drawing upon demographic data from the Current Population Survey. Those assigned to Fishtown are all white citizens who did not go to college. If they have a job, they work in blue-collar occupations, provide in-person services, or are employed in low-income clerical jobs. What these people have in common is that their skills are insufficient for them to compete successfully in the new economy. As Murray rightly points out, "The higher-tech the economy, the more it relies on people who can improve and exploit the technology, which creates many openings for people whose main asset is their exceptional cognitive ability."[7] Thus, to paint the growing divide between the fortunes of America's cognitive elite and the misfortunes of the white working class, he creates another statistical construct that he calls Belmont, after the upper-middle-class suburb near Boston, Massachusetts. Those assigned to Belmont have much in common with Robert Reich's symbolic analysts, who have been reaping the benefits of computerization. They hold at least a bachelor's degree, work in tech jobs or skilled professions, and enjoy a secure and affluent lifestyle.

Murray then examines in more detail what has happened to the citizens of Fishtown since the postwar boom years. One worrying albeit unsurprising trend is that fewer people work. In 1960, there was not much difference between Belmont and Fishtown in terms of working habits: in 90 percent of Belmont households, there was at least one adult who worked forty hours per week or more, and the same was true of 81 percent of Fishtown households. But by 2010, this gap had increased dramatically. In 87 percent of Belmont households, at least one adult still put in a minimum of a forty-hour work week. But in Fishtown dramatic changes in the likelihood of an adult being in the workforce had occurred, with only 53 percent of households still containing at least one working person.

As joblessness in Fishtown grew, so did crime. Inmate surveys carried out by the federal government in the period 1974–2004 show that about 80 percent of whites in state and federal prisons came from Fishtown and less than 2 percent were from Belmont.[8] Hence, the upsurge in crime and imprisonment was concentrated in one segment of the white population—the working class. And just as Wilson found that joblessness was a cause of the growing share of single-parent households among blacks, there was a fall in marriage rates among blue-collar whites.[9] As late as 1970, only 6 percent of births among non-college-educated white women were out of wedlock. Four decades on, 44 percent of births were. Only a third of Fishtown's children now grow up in families that include both biological parents. This matters, because children in single-parent households have much worse chances later in life. Indeed, the economist Raj Chetty and collaborators found that the strongest predictors of upward mobility are measures of family structure. The greater the percentage of single parents in an area, the lower the chances of their children moving up.[10]

Murray does not see the computer revolution as the cause of Fishtown's misfortunes—instead, he sees joblessness as a consequence of a deteriorating work ethic and welfare dependency. Scholars still debate the relative importance of economic and cultural factors in shaping the patterns of joblessness, crime, and marriage rates, but it seems most reasonable to emphasize the importance of both. However, there can be no doubt that economic dislocation can explain a large part of such patterns. Those assigned to Fishtown clearly belong to groups that have seen diminishing opportunity due to trade and technological change. And many of Fishtown's social ills can be directly linked to outcomes in the labor market. Crime rates, for example, are related to the expected costs and benefits of illegal activity.[11] If workers' expected earnings in the labor market fall, so does the opportunity cost of spending time in prison. Thus, unsurprisingly, people have historically been more likely to become involved in illegal pursuits when the relative payoff of criminal activity has increased. For example, as Britain industrialized and workers saw their skills made redundant by the mechanized factory in the nineteenth century, older workers—especially those in artisan jobs—began to commit more crimes for economic gain.[12]

While unemployment is often cyclical and short-lived, the decline in the wages of the unskilled due to trade and technological change has been going on for several decades and will likely have more long-lasting effects on criminal activity than short-term unemployment does. Indeed, when the economists Eric Gould, Bruce Weinberg, and David Mustard studied the relationship between vanishing opportunity in the labor market and crime, they found the former to have caused the latter. Both unemployment and falling wages affected crime rates among unskilled men. Over the period the authors analyzed (1979–97), the wages of unskilled men fell by 20 percent, while the property crimes rose by 21 percent.[13]

Other misfortunes have also afflicted the people of Fishtown, which are less obviously but nonetheless directly linked to the state of the labor market. While marriage has become less common across all spectra of society, the reason it has become so much more uncommon in Fishtown is that work has disappeared. Among the white population, the skilled have consistently seen the highest percentage of married men in America since the 1880s.[14] The difference in marriage rates between people in skilled occupations and blue-collar Americans narrowed until the end of the manufacturing boom years, when it experienced a reversal. The narrowing happened as the position of blue-collar workers in the labor market strengthened. But in today's higher-tech economy, blue-collar workers are becoming less likely to marry as their position in the labor market has gotten weaker. The economists David Autor, David Dorn, and Gordon Hanson have indeed found that the disappearance of blue-collar jobs disrupted marriage markets and reduced marriage rates. Deteriorating labor market prospects made men less marriageable. One reason had to do with joblessness and the declining economic stature of blue-collar men. But more alarmingly, vanishing factory jobs also increased the mortality gap between young men and women. As work disappeared, young men became more likely to experience an early death.[15]

Their findings are consistent with those of other studies. Workers who were displaced due to plant downsizings in Pennsylvania during the recession of the early 1980s suffered annual earnings losses averaging 25 percent even six years after they lost their jobs, and immediate losses of more than 40 percent.[16] But layoffs did not just affect workers' earnings. They also brought a significantly higher risk of death. When the economists Daniel Sullivan and Till von Wachter went back and traced the fates of the workers who had been displaced by the downsizings, they found that the ones who saw their jobs disappear experienced a 50–100 percent increase in short-term mortality rates after being laid off.[17] Even if the effect tailed off over time, a middle-aged man saw his life expectancy drop by one and a half years, an effect comparable to that of being forty pounds overweight at the age of forty. So in 2015, when two Princeton economists, Anne Case and Angus Deaton, winner of the Nobel Prize in Economics, shockingly found that annual death rates among middle-aged whites had risen since the turn of a century after decades of improvements, they naturally suggested that the reversal might reflect the long-standing process of diminishing opportunity in the labor market for working-class whites, whose departure from the middle of society's spectra has come with so much distress.[18] Rising mortality, they found, was not caused by typical killers like heart disease and diabetes but by suicide and substance abuse.[19]

Reports on subjective well-being, it is true, have consistently shown that people who experience unemployment are significantly less happy, even when a wide range of factors (including income and education) are controlled for.[20] Men fare the worst mentally from unemployment, especially if it occurs in their prime years.[21] One widely cited study even found that "joblessness depressed well-being more than any other single characteristic, including important negative ones such as divorce and separation."[22] But while there is compelling evidence to suggest that health and well-being are closely related to labor market outcomes, to what extent the loss of jobs due to technology and trade can account for the recent upsurge in the "deaths of despair" documented by Case and Deaton remains an open question. The growing misuse of and addiction to opioids has turned into a serious national crisis that affects public health and social welfare. America's opioid crisis is certainly part of the story, but part of it might also be the consequence of rising joblessness. What is beyond question is that disappearing middle-income jobs have caused much material and emotional suffering, which has had a devastating impact on a broad swath of the middle classes.

The Geography of New Jobs

The drifting apart of American society is about more than unequal gains. Much more worrying than rising income inequality is the fact that large groups in the labor market have been left worse off economically and in terms of subjective well-being. Their reality has also become harder to comprehend because they have become increasingly segregated from the rest of society. Though Fishtown and Belmont are statistical constructs, they speak to a well-documented rise in geographic polarization. When the eminent sociologist Douglas Massey, together with Jonathan Rothwell and Thurston Domina, recently examined patterns of segregation in America, they found that a kind of incipient cognitive class apartheid had worked itself through the country in the last quarter of the twentieth century. College-educated parents and their children had become increasingly detached from the realities facing those families who have seen work ebb. The growing split between the college educated and the rest is not just one of neighborhood segregation.[23] A broader shift has also taken place between cities, which is linked to the geography of new jobs.

In the 1980s and 1990s, it was believed that the exact opposite would happen. With the advent of the World Wide Web, email, and cell phones, pundits proclaimed that location would soon become irrelevant and the curse of geography a distant memory.[24] Futurists like Alvin Toffler even predicted that the death of distance would eventually render the city obsolete.[25] And in 2005, the cover of the first edition of Thomas Friedman's best-selling *The World Is Flat* pictured a world in which geographic divisions were history.[26] Information technology, these authors declared, was making face-to-face interactions unnecessary, so that the time when companies and workers had to cluster in expensive places like Manhattan or Silicon Valley would soon be over. But the truth is that even with modern computers, many complex interactions remain too subtle to be performed via technology. As Harvard University's distinguished economist Edward Glaeser pointed out at the time, digital and in-person communication are best seen as complements rather than substitutes for each other.[27] More efficient information technology has made it possible to maintain a greater number of relationships, which increases the number of in-person contacts. Though the computer revolution rendered New York's advantages as a manufacturing city obsolete, it amplified its competitive edge in innovation. Cities specializing in knowledge work (that is, developing and exporting ideas) became more productive.

Place still matters because of "agglomeration economies," which derive from the value of proximity. Workers want to be close to jobs. Companies want to have access to talent and be close to customers. Parents want to live near good schools. And the elderly may prefer places with good health care and a pleasant climate. Agglomeration, in short, comes down to the desire to reduce the costs of moving goods, people, and ideas.[28] Of course, the cost of transporting goods has become a much less important factor in where companies chose to locate, simply because shipping has become so much cheaper. One reason why industrial cities like Detroit began to decline before the age of automation, when manufacturing employment was still expanding, is that production began to move away from the Great Lakes to right-to-work states in the Sunbelt, where union security agreements between companies and labor unions are prohibited.[29] Such locational freedom, however, may have reduced the desire to transport the smart people and ideas that have become so valuable in the higher-tech economy. And indeed, that is what has happened.

In *The New Geography of Jobs*, the economist Enrico Moretti tells an intriguing story of two places in California: Menlo Park and Visalia. The story begins in 1969, with a young engineer turning down a job offer at Hewlett-Packard in Menlo Park (in the heart of Silicon Valley) to move to the midsize town of Visalia, three hours' drive away. At the time, many professionals were leaving cities for smaller communities, which were considered better places for family life. At the time, both places in California had prospering middle classes, similar rates of crime, and comparable quality of schools. And while incomes in Menlo Park were higher on average, America was on an equalizing path.

Yet today, Menlo Park and Visalia are in different universes. As Silicon Valley has grown to become the world's hub for innovation, Visalia has become a backwater. It has the second lowest share of college-educated workers in America, and its crime rates are high and trending upward while its relative earnings are in decline.[30] And these are not isolated examples:

> [They reflect] a broader national trend. America's new economic map shows growing differences, not just between people but between communities. A handful of cities with the "right" industries and a solid base of human capital keep attracting good employers and offering high wages, while those at the other extreme, cities with the "wrong" industries and a limited human capital base, are stuck with dead-end jobs and low average wages. This divide—I will call it the Great Divergence—has its origins in the 1980s when American cities started to be increasingly defined by their residents' levels of education. . . . At the same time that American communities are desegregating racially, they are becoming more segregated in terms of schooling and earnings.[31]

This trend began with the age of computers. It is true that some professional services, like accounting, can now be delivered electronically from a distance. But new jobs, spawned by computer technologies, are highly concentrated, suggesting that place has become more significant as production has become more skill intensive. This is underlined by the dramatic shift in the geography of new job creation beginning in the

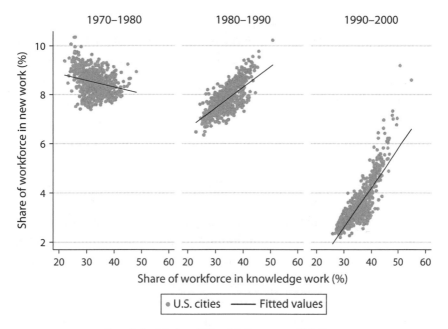

FIGURE 15: Knowledge Work and New Job Creation in U.S. Cities, 1970–2000
Source: T. Berger and C. B. Frey, 2016, "Did the Computer Revolution Shift the Fortunes of U.S.
Cities? Technology Shocks and the Geography of New Jobs," Regional Science and Urban
Economics 57 (March): 38–45; J. Lin, 2011, "Technological Adaptation, Cities, and New Work,"
Review of Economics and Statistics 93 (2): 554–74.
Note: These figures show the percentage share of each city's workers that were employed in jobs that
did not exist by the beginning of each respective decade against the initial share of "knowledge
workers" in occupations that involve abstract tasks across 321 American cities.

1980s. Occupational classifications, which are updated every decade,
allow us to identify new jobs that did not exist a decade earlier. My work
with Thor Berger, an economic historian, shows that before the computer revolution, when some emerging occupations were still routine,
new jobs didn't just emerge in skilled cities. But as a wide range of
computer-related occupations—like those of computer programmers,
software engineers, and database administrators—became more plentiful in the 1980s, the comparative advantage in new job creation firmly
shifted toward cities initially specializing in knowledge work (figure 15).[32]
We found in a follow-up study that data on the location of new industries,
as opposed to new occupations, reveal a similar pattern. New industries

that appeared in official statistical classifications for the first time in the 2000s primarily relate to digital technologies, such as online auctions, web design, and video and audio streaming. Ironically, it is precisely the technologies that futurists once believed would flatten the world that have made it more uneven: digital industries have overwhelmingly clustered in cities with skilled populations.[33]

The location decisions of technology companies, which are at the forefront of digital technology, provide the best evidence of the value of in-person communication: "The fact that Silicon Valley is now the quintessential example of industrial agglomeration suggests that the most cutting-edge technology encourages, rather than eliminates, the need for geographic proximity."[34] The past two decades have supported that view, as geographic clusters like Silicon Valley remain strong, despite the abundance of long-range electronic communication tools. In fact, geography has become more important as new jobs have become more skill intensive. Indeed, for most of the 1900s places with lower average incomes were catching up with richer cities and regions. One of the most widely cited studies in economics is a paper by Robert Barro and Xavier Sala-i-Martin that shows persistent and speedy income convergence across American regions over the century preceding the computer revolution.[35] Not just in America but also across the Atlantic, the great convergence within countries was pervasive and persistent throughout the postwar decades. But income convergence came to a halt in the 1980s, as cognitive segregation became more widespread. In a paper titled "Why Has Regional Income Convergence in the U.S. Declined?," the Harvard University economists Peter Ganong and Daniel Shoag argue that historically, income convergence was driven by people migrating from poorer regions to wealthier ones. The steady influx of workers held down wage growth in rich places and helped incomes rise in poor regions as people left. But the clustering of new jobs in skilled cities in combination with stricter land-use regulations disrupted this trend: the rising costs of living in booming cities meant that migration ceased to be an option for the unskilled, while highly skilled workers continued to migrate.[36]

Why, then, are not firms moving to the hinterland, where labor and housing are cheap? The answer is that innovative companies want to stay close to other such companies and skilled people. Giles Duranton and Diego Puga are two economists who have aptly conceptualized cities as nurseries for innovation, where firms in their early stages need to experiment to grow.[37] During this innovative phase, firms benefit from the exchange of ideas that is facilitated by the density of cities and their proximity there to related businesses, which increases the demand for the relatively skilled. These nursery cities serve as incubators for new job creation. When a prototype has been developed and operations have become more standardized, it makes economic sense to relocate to places where real estate is cheaper and the costs of production lower. New jobs, in other words, will eventually spread to other locations. As long as nursery cities do not churn out new jobs faster than they diffuse geographically, convergence in employment will follow. But new jobs spread only as they become standardized, and since the beginning of the computer age, jobs that have become standardized have not diffused across the country: they have either been automated away or sent abroad. The flourishing cities of America have become nursery cities for innovation. But the rest is done abroad or by machines.

The places where work has been replaced, rather than complemented, by machines, are the ones that are in decline. The uneven map of multipurpose robots depicted in figure 16 makes a simple point about technological progress. Automation—like most other economic trends—will not occur everywhere in the same way or at the same pace. More than half of America's robots are in just ten states, most of which are in the eastern heartland—where male joblessness and life dissatisfaction is the highest.[38] As a matter of fact, Michigan alone has almost as many robots as the entire American West. And like the presence of robots, joblessness has not risen uniformly across America. The share of middle-aged men who do not go to work is 51 percent in Flint, Michigan, but only 5 percent in Alexandria, Virginia. The economists Benjamin Austin, Edward Glaeser, and Lawrence Summers recently conducted a detailed study of the geography of joblessness among men in their

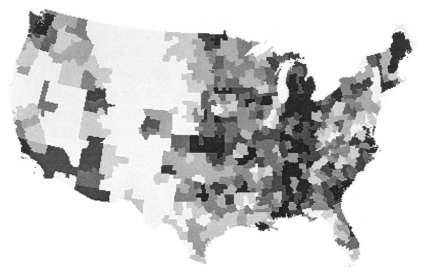

FIGURE 16: The U.S. Geography of Industrial Robots, 2016
Source: International Federation of Robotics (database), 2016, World Robotics: Industrial Robots, Frankfurt am Main, https://ifr.org/worldrobotics/; S. Ruggles et al., 2018, IPUMS USA, version 8.0 (dataset). https://usa.ipums.org/usa/.
Note: This figure shows the number of industrial robots per thousand workers across America in 2016. Darker shades correspond to more robots per thousand workers. County boundaries are based on maps from IPUMS NHGIS (www.nhgis.org).

prime. Some of the authors' findings may not come as a surprise. One is that on average, men are more likely to go to work in places where educational attainment is high.[39] Another is that men are less likely to go to work if durable manufacturing was historically a key source of employment in the area where they live. Indeed, as noted above, since the dawn of the computer revolution, new jobs have overwhelmingly appeared in cities with skilled populations. Meanwhile, in places specializing in routine manufacturing work, automation has had the opposite effect: it has replaced workers rather than creating new functions for them.

Because jobs have come and gone so unevenly across America, the computer revolution has made the country less flat. Where new tech jobs have appeared, the local service economy has been given a boost. Moretti estimates that each new tech job creates sufficient demand to support another five jobs in a given city. Where blue-collar jobs disappeared, in contrast, the local economy has taken a hit. Each manufacturing job lost, Moretti finds, costs another 1.6 jobs in the local service sector.[40] This process has meant that while skilled cities have prospered, smokestack cities have become obsolete. As late as 1970, average incomes were higher in Cleveland, Ohio, and Detroit, Michigan, than in skilled cities like Boston, Massachusetts, and Minneapolis, Minesota. But in recent decades, skilled cities have pulled away while manufacturing cities and industrial towns have faded. In the words of Glaeser, "If these patterns continue, we may come to see an ever more uneven America with rich, successful, skilled regions that compete well worldwide and poor, unskilled regions that are repositories of despair."[41]

11

THE POLITICS OF POLARIZATION

What happens to liberal democracy when the social fabric fractures and the middle class starts to shrink? Societies with extreme gaps between rich and poor have historically been more prone to oligarchy and populist revolutions. A broad middle class, as many political scientists have pointed out, is an essential pillar of a stable democracy. Indeed, the long persistence of extreme inequalities helps explain why liberal democracy did not arrive earlier. In preindustrial societies, the landowning elites had little interest in extending the franchise, while avoiding starvation was the main preoccupation of the poor. Without a middle class with a different set of expectations, there was little demand for democracy. Barrington Moore's classic *Social Origins of Dictatorship and Democracy* is perhaps best known for the blunt remark "no bourgeoisie, no democracy."[1] Though this assertion has been subject to much criticism, Moore's point was not that the bourgeoisie must always produce democracy. What he argued was that the replacement of the landed elite was essential in bringing democracy about, as was the case in England—where industrialization set the stage for democratic transitions.

While social scientists have long noted a close correlation between economic development and democracy, it is not immediately clear what is driving that relationship. However, one compelling explanation is that industrialization spins off new social groups that begin to demand more political power as they get richer. In *Political Order and Political Decay*, Francis Fukuyama provides a vivid account of how the Industrial Revolution changed the underlying nature of society in ways that challenged the old authoritarian order.[2] The rise of democracy surely had much to do with the spread of values favoring equality, but these ideas didn't come out of a vacuum. The profound changes set in motion by the Industrial Revolution not only generated sustained economic growth but also dramatically changed the composition of societies by creating and mobilizing new groups of people—notably, the bourgeoisie and the factory working class. Thus, Fukuyama's road to liberal democracy starts from Karl Marx's theory of social class. According to Marx, the first new social class to emerge out of the old feudal order was the bourgeoisie. This class included merchant townsmen who had grown wealthy through trade and invested heavily in the factory system and thus the Industrial Revolution. Industrialization in turn mobilized Marx's second new class, the proletariat, whose members left rural areas for the emerging factory towns. These groups were excluded from political participation in the feudal order, but as they grew richer and became more organized, they demanded more political power—which in turn created pressures for democracy.[3] Fukuyama writes:

> Increasing industrialization induced peasants to leave the countryside and enter the working class, and by the beginning of the twentieth century they were the largest social group. Under the impact of expanding trade, the number of middle-class individuals began to swell, first in Britain and the United States, then in France and Belgium, and by the late nineteenth century in Germany, Japan, and other "late developers." This then set the stage for the major social and political confrontations of the early twentieth century.... The key insight is that democracy is desired most strongly by one specific social group in society: the middle class.[4]

Democracy and the Middle Class: A Very Brief History

Marx predicted that the proletariat's struggle against the bourgeoisie would prompt a socialist revolution, but in practice, industrialization meant the incorporation of working people into a broader middle class. However, as the Luddite uprisings and other machinery riots suggest, the objectives of the bourgeoisie and the average worker were hardly aligned from the onset. One obvious reason was Engels's pause, during which labor did not see the benefits of industrialization. In fact, Marx's belief that capitalism was set for a crisis of overproduction, in which mechanized industry extracted surpluses from working people, which led to greater wealth in the hands of the few and the impoverishment of the proletariat, seemed quite plausible at the time. Cascading levels of inequality, Marx predicted, would eventually create a shortfall in aggregate demand and lead to the collapse of the capitalist system. Such a crisis, he argued, could be avoided only though a revolution in which the proletariat would take ownership of the means of production and redistribute the benefits of mechanization. Had Engels's pause persisted longer, some version of this outcome might well have occurred. But mercifully it didn't.

We saw in chapter 5 that things changed around the mid-nineteenth century. As Marx was writing, the modern growth pattern emerged. By the closing decades of the Industrial Revolution in Britain, productivity growth had accelerated with the adoption of steam, and real wages had begun to rise in tandem, as the nature of technological progress shifted and people acquired new skills. As noted above, augmenting technical change has a symbiotic relationship with the bargaining power of labor, while worker-replacing technical change works in the opposite direction. During Engels's pause, early textile machinery replaced skilled craftsmen in production. As the domestic system gradually vanished, new jobs emerged in the factories, but the textile machines were designed to be tended by children.

However, the arrival of more complex machines, following the more widespread adoption of steam power, broke the deskilling pattern according to which skilled craftsmen were replaced by child labor. Heavy machinery required more skilled workers of greater physical strength in the factories, and technical change turned from worker replacing to augmenting. This served to increase the bargaining power of labor, as workers' skills became more valuable over time. And consequently, workers began to see the benefits of mechanization in their pockets. While a causal link is hard to establish, it is even harder to explain the end of widespread resistance to machinery in isolation from the arrival of the modern growth pattern. As we have seen, workers and the trade unions representing them rarely questioned the desirability of mechanization thereafter.

The modern growth pattern came about in the absence of organized labor or any significant government intervention, but that should not be taken as downplaying the importance of the labor movement. Another reason why Marx's socialist revolution did not happen is that throughout the industrializing West, workers joined forces in trade unions and began to push for democracy, higher wages, safer working conditions, and more redistribution of wealth and income. The political branch of the labor movement de facto accepted a laissez-faire regime with respect to mechanization, but they insisted on establishing a tax-financed welfare system. From the mid-nineteenth century onward, industrialization put labor on a virtuous cycle, where mechanization increased workers' earning power and trade unions increased their bargaining power. These two processes were mutually reinforcing. The augmenting nature of technological change made workers' skills more valuable, thus increasing labor's bargaining power. The labor movement successfully took advantage of this by agitating for workers' rights to organize formally and vote.

To be sure, the two components of liberal democracy—the rule of law and universal suffrage—are distinguishable political goals. And quite naturally, they tended to be favored by different groups.[5] In Britain, the commercial and industrial bourgeoisie, which provided the base support for the British Liberal Party in the nineteenth century, was more interested in legal protection for private property and free trade policies, while the working class was more interested in democracy. The Great Reform Act of 1832 was the first of a series of reforms on Britain's long road to universal suffrage. It is justly regarded as "a major turning point in English history," as it helped trigger democratization and key economic reforms like the implementation of the personal income tax in 1842 and the repeal of the Corn Laws four years later.[6] Because Parliamentary reform had become a party-political question, supported by Whigs and Radicals (who would go on to form the Liberal Party) and opposed by most Tories, a Whig majority in the House of Commons was necessary for the Great Reform Act to make it through Parliament. The chief motive of the Liberals was to strengthen their own position by increasing the power of Parliament and reducing that of the crown, though over time some Whigs came to support expanding the franchise for its own sake. What the coalition of Liberals had in common was that they were in favor of more personal freedom, checks on the powers of the crown and the church, and above all free trade. However, their electoral success depended on working-class agitation. The economists Toke Aidt and Raphael Franck have shown that "the reform-friendly Whigs would not have obtained a majority of seats in the House of Commons in the 1831 election had it not been for the violence of the Swing riots."[7] The social unrest brought about by the rioters' fear that threshing machines would take their jobs set in motion a process of democratization. The perceived threat of social upheaval, as political scientists have noted, prompted the ruling classes to extend the franchise.[8] Voters and patrons who didn't support Whigs and Radicals in earlier elections, the authors show, voted for the candidates promoting Parliamentary reform, but only after experiencing the violence caused by machinery uprisings, giving the reformers a majority in Parliament.

The Great Reform Act of 1832 can hardly be described as a working-class victory: people who didn't own any property remained politically disenfranchised. But workers did help set the wheels of democratization in motion. From then on, electoral competition between the Liberals and the Tories created pressures for the continued extension of the franchise, meaning that labor gradually came to represent a growing percentage of the electorate. This was spurred on by the struggles to expand trade unions, which prompted the rise of the Labor Party and eventually its displacement of the Liberals as the main opposition to the Tories. However, in the mid-nineteenth century, long before the rise of the Labor Party, the Tories had already shifted their position from representing wealthy landowners to gathering support from the new middle class. As the historian Gertrude Himmelfarb has argued, Benjamin Disraeli's decision to pursue the Reform Act of 1867, which extended the franchise to about 40 percent of the male population, reflected the belief that the Tories were a national party, capable of appealing to a wider share of the populace.[9]

As more and more people won the franchise, workers in the industrial West used their newfound political power to vote for welfare state policies and social legislation. In *Growing Public*, Peter Lindert shows that between 1880 and 1930, heavily voting democracies taxed more and spent more on social transfers—such as aid to the poor, elderly, sick, and unemployed.[10] One reason why welfare spending was so restricted before the twentieth century, Lindert convincingly demonstrates, is that workers lacked political voice. In the classic years of the Industrial Revolution, when Engels's pause persisted, the British government even rolled back welfare spending. Before the Great Reform Act, Britain was the one place where poor relief had become much more generous. In the light of workers' lack of political clout, this might seem puzzling. Why did landowners tax themselves 2 percent of gross domestic product (GDP) to help the poor? At the time, no government had ever done so.

Following in the footsteps of George Boyer, Lindert argues that rural landlords, who depended on seasonal hires, had an interest in keeping farmers in the area to secure access to cheap labor. Poor relief provided a way of keeping workers from emigrating to Britain's emerging industrial centers. But things changed as the new merchant class gained political clout. The Great Reform Act extended the franchise to merchant manufacturers in urban centers. Industrialists saw little benefit in supporting a system that kept workers in stagnant rural areas. The rise and fall of the Poor Laws, in other words, were shaped by the self-interest of the groups that controlled the levers of political power.

While some people have always worried about the wealthy using their economic power to bend democracy to their own ends, there has also been concern over voting majorities using their democratic power to tax away the fortunes of the rich. In *Democracy in America*, published in 1835, Alexis de Tocqueville noted that universal suffrage, which "gives

the government of society to the poor," will lead to the poor redistribut-
ing the wealth of the rich because in virtually every nation, the vast
majority of citizens do not have any property.[11] "Tocqueville's observation has received modern expression, with an optimistic twist, in leading theories of income redistribution offered by political scientists," in
the words of Jacob Hacker and Paul Pierson.[12] Median voter theories,
so popular among political scientists and economists alike, build on the
premise that in a majority-rule voting system, the decisive swing
voter—or the median voter—ultimately drives politics and the supply
of redistributive policies. Because the median voter almost invariably
earns less the than the national average, he or she will seek more income
redistribution through government. This theory then leads to the
simple prediction that more redistribution will follow from greater
inequality.[13]

We all know that political development has taken very different turns
in different places. While a link between political voice and welfare
spending exists, voting rights for more people does not necessarily
mean more redistribution. The relationship between the expansion of
the franchise and social spending is a complex one. Even in liberal democracies, citizens do not have the right to vote separately on every
single issue. They must rely on their elected representatives, who trade
influence on many different issues. The United States was born out of a
revolution against the concentration of power of the British monarchy
and founded on the principle of equality and self-rule by the commoner.
In this spirit, America from the very beginning had a much larger franchise than European nations. Universal white male suffrage came in the
1820s, but that did not lead to much redistributive taxing and spending,
as Tocqueville feared would be the case. Before the Great Depression,
U.S. government poor relief, for example, never exceeded 0.6 percent of
the national income. And there were very few private funds directed to
supporting the poor. In 1896, even the founder of the New York Charity
Organization Society, Josephine Shaw Lowell, made clear that she
didn't believe in poor relief: "Their distress is due to inherent faults,
either physical, mental, or moral . . . [R]elief is an evil—always. Even
when necessary, I believe it is still an evil. One reason that it is an evil is

because energy, independence, industry, and self-reliance are undermined by it."[14]

Why didn't the working poor demand more redistribution? One reason is that the early democratization of America came with its own set of problems. In the succeeding decades, the party system that emerged had to gather support from a new group of poor and uneducated voters. The promise of a job or other personal favors turned out to be the most effective way of mobilizing them. Clientelism soon became widespread at virtually every level of government. The lucrative short-term benefits offered to the working poor meant that their long-term interests suffered. Because ordinary citizens got individual favors in return for political participation, it proved much harder to recruit them into the kind of working-class or socialist parties that popped up in Europe, where people demanded more redistribution, universal health care, and so on.[15] Both the Republican and Democratic Parties gained support from working-class Americans by offering short-term benefits rather than long-term policy involvement. As the historian Richard Oestreicher has argued, the rise of clientelism is one reason why socialism never arrived in America.[16]

The reform agenda of the Progressive Era, which ended the problem of clientelism, is not the focus here. Rather, nineteenth-century clientelism illustrates a broader point about the role of the middle class in stable democracy: poor and uneducated voters cannot achieve much politically on their own. Redistributive taxing and spending depend on whether the middle-income voters feel an affinity with people with lower incomes. Long-term divisions between middle- and lower-income voters can thus undermine the pursuit of common political goals. When inequality is rampant, as it was in the nineteenth century, there is little loyalty between the middle class and the working poor. In other words, its easiest to assemble a political coalition to compensate the losers to technological progress—or other sources of dislocation—when there is a broad middle class, and thus when there's the least need for it. As Lindert writes, "The more a middle-income voter looks at the likely recipients of public aid and says 'that could be me' (or my daughter, or my whole family), the greater that voter's willingness to vote for taxes to fund such aid."[17] A broader middle class with a different set of expectations was required to open the door for a new kind of middle-class politics.

The Great Depression ended a period of extraordinary wage growth for ordinary people. In the period 1900–28, the yearly income of full-time manufacturing workers grew by more than 50 percent. Workers in transportation and construction also experienced similar pay rises. They were reaping the gains from growth, just like the white-collar workforce above them. The Depression, as we all know, spawned the New Deal and the rise of the welfare state. But both depended on loyalty between the white-collar middle class and the working class. In the decades that followed, such loyalty only grew stronger, as the working class joined the ranks of the middle class. As Robert Putnam's colorful descriptions of life in Port Clinton in the 1950s illustrate, "The children of manual workers and of professionals came from similar homes and mixed unselfconsciously in schools and neighborhoods, in scout troops and church groups. . . . Everyone knew everyone else's first name."[18] Port Clinton was not an exception in this regard. At that time, manual workmen and their families could live on the same street as a white-collar family. Such middle-class living provided the foundation for middle-class politics. Robert Gordon explains:

> This rough economic equality was a political fact of the first importance. It meant that, in a break with the drift of things in pre-war America, postwar America had no working class and no working-class politics. It instead had a middle-class politics for an expanding middle class bigger in aspiration and self-identification than it was in fact—more people wanted to be seen as middle-class than had yet arrived at that state of felicity. Socialism in America, the German political economist Werner Sombart wrote in 1906, foundered upon "roast beef and apple pie," a metaphor for American plenty. The expanding middle class of the postwar era—property owning, bourgeois in outlook, centrist in politics—hardly proved him wrong. The clear overlap between blue-collar and white-collar ambitions and success from the 1940s until the 1970s symbolized the egalitarian experience of a diverse and stable middle class.[19]

Thus, when Americans came out of two tumultuous world wars and the Great Depression, they found themselves in a position where their politics were no longer sharply polarized. Workers had seen their incomes rise steadily as mechanization made their skills more valuable, and they had won greater privileges for themselves. The growing pie was divided equally between labor and capital. The postwar years were a time of rapid wage growth, steady profit growth, job security and stability, and fewer instances of labor unrest. The living standards of ordinary citizens kept rising, to the point where many workers or their children were able to afford a middle-class lifestyle. To paraphrase Fukuyama, Marx's communist utopia failed to materialize in the industrial world because his proletariat turned into a growing middle class. The ascent of the middle class put America on a virtuous cycle, with economic and political convergence going hand in hand. More and more workers shifted into middle-income jobs, where they became more middle class in political outlook.

Writing in 1961, Robert Dahl famously began his landmark work on modern political science with the question, "In a political system where nearly every adult may vote but where knowledge, wealth, social position, access to officials, and other resources are unequally distributed, who actually governs?"[20] His work examined the experience of New Haven, Connecticut, in the late 1950s, and the answer was that political power was highly dispersed. New Haven, and indeed America, was run by the middle-class median voter. And just as important, as Americans had grown closer together economically, they had also grown closer politically. In the period 1900–75, the percentage of moderate Democrats and Republicans increased in the House of Representatives and the Senate, while the number of extremists fell in both parties: "In the middle of the twentieth century, the Democrats and the Republicans did dance almost cheek to cheek in a courtship of the political middle."[21]

When the political scientist Larry Bartels recently revisited Dahl's question, he found that its "significance . . . has been magnified, and the pertinence of his answer has been cast in doubt, by dramatic economic and political changes in the United States over the past half-century."[22] As economic inequality has skyrocketed in recent decades, why has not a massive upsurge in redistributive taxing and spending followed, as median voter theories predict? If there is a link between political voice and redistribution, should we not expect more of the latter? Yet since 1980, American social expenditure on unemployment, housing, family allowances and cash benefits, and labor market programs has been stagnant as a percentage of gross domestic product.[23]

One possible reason could be that low-income workers might want to keep taxes low in the expectation that they will earn more later in life. However, this explanation will not do it: as we have seen, Americans are now much more pessimistic about their own prospects and those of their children than they were a generation ago. Thus, quite naturally, political scientists have begun to wonder if the pluralistic democracy that Dahl observed has not been undone by a small and increasingly wealthy elite whose members deploy their economic power to their own political advantage. A key concern is that growing economic inequality has made the political system less responsive to the needs of ordinary citizens, which in turn has solidified economic inequality. As the middle class has shrunk, the number of moderate members of Congress has fallen sharply, and politics has become polarized: "Conservative and liberal have become almost perfect synonyms for Republican and Democrat."[24] The relationship between economic and political polarization has aptly been characterized as a "dance" with much back and forth. Economic inequality feeds political polarization and vice versa, making it harder to redress inequality that may arise from nonpolitical changes in technology, trade, compensation practices, and so on.[25]

Another related concern is that the growing concentration of wealth is undermining the legitimacy of democracy. Expensive political campaigns, for example, have increased the reliance of elected officials on people with economic power. Yet the diminishing political voice of average working Americans has been driven by a much broader set of interests than those of a few wealthy individuals. More worrying is the fact that corporate spending on lobbying has greatly increased while membership of labor unions has fallen, "eroding the primary mechanism for organized representation of working people in the governmental process."[26] As an illustrative example, consider the erosion of the minimum wage. We know from a wealth of survey evidence that there is long-standing and broad support to increase it, among both Democrats and Republicans. During the 2006 and 2008 election campaigns, the Cooperative Congressional Election Study asked almost 70,000 Americans whether they favored or opposed raising the minimum wage. Among Democrats, 95 percent were in favor, almost regardless of their earnings. Among low-income Republicans, about 75 percent favored raising the minimum wage, while only 45 percent of those earning more than $150,000 per year were in favor. This is just one of many such surveys, and "the breadth and consistency of public support for raising the minimum wage make it all the more surprising that the real value of the minimum wage has declined so substantially since the 1960s."[27] However, as the journalist Marilyn Geewax has noted, despite favorable opinion polls, few voters ever contact their elected representatives about the issue, while "restaurateurs and small-business owners were organized, energized and informed by top-notch lobbyists who never stopped telling Congress that higher wages would cut profits and limit the ability to create jobs."[28] In addition, the decline of organized labor has hardly made the case of labor easier. Analyzing the correlates of year-to-year fluctuations in the real minimum wage in the period 1949–2013, Bartels shows that the real value of the minimum wage has been 40–55 cents higher under Democratic presidents than Republican ones. Yet the fates of the minimum wage supporters, he finds, have been even more dependent on organized labor.[29]

The point is not that raising the minimum wage would be the best way of addressing workers' concerns. Higher minimum wages also spur efforts to automate, suggesting that its benefits for labor might be short-lived.[30] However, Bartels's analysis illustrates a broader point, which is that the failure to raise the minimum wage, despite widespread support, suggests that workers are losing political influence. Unionization, as we all know, was at a high point in the mid-1950s, when factory and office machines made the skills of non-college-educated workers more valuable and thereby increased their earning power. This also made the unions stronger, though there were cases (for example, with lamplighters and longshoremen) when technology made workers' skills–and thus the unions representing them—obsolete. Yet their members were a mere fraction of the working population. The single largest industry in the mid-twentieth century was automotive, and the United Auto Workers (UAW) union achieved significant benefits for its members, including higher wages, generous pensions, and health insurance, while conceding to management decisions regarding mechanization and other key capital investments. In 1950, UAW's president, Walter Reuther, negotiated what *Fortune* magazine would call "the Treaty of Detroit" with General Motors and made similar deals with Ford and Chrysler, protecting car companies from strikes while gaining higher pay and more holidays for its members in return, among other things. These deals also influenced collective bargaining in many other mass-production industries of the Second Industrial Revolution. But as the sociologist Andrew Cherlin writes:

> Whereas the 1950s workers, backed by their powerful unions, trusted management enough to pledge uninterrupted labor on the assembly line in return for good wages and the promise of retirement pensions down the road, today's workers and employers do not trust each other to pledge much of anything. . . . The power of unions has faded: the overwhelming majority of less-educated young adults do not work in a place where a union has successfully organized. Absent an agreement between labor and management, young adults have neither the right to decent wages and benefits nor the obligation to be loyal workers. . . . It is as though the 1950s Treaty of Detroit, so lauded by Walter Reuther and other leaders of organized labor at the time, has been replaced by pervasive distrust.[31]

In the mid-twentieth century, trade unions were an institution that gave workers a coherent political voice and created social ties among the unskilled. For example, Putnam has persuasively argued that workers' social capital has declined as union membership has dropped.[32]

What's more, the type of workers represented by unions has changed as well. When union density was at its peak in the 1950s and 1960s, union members were relatively unskilled. Now union members are just as skilled as nonmembers.[33]

A similar shift away from representing the unskilled can also be observed in party politics. In the 1950s and 1960s, the non-college-educated members of the middle class constituted the base support for left-wing political parties. Indeed, drawing upon postelectoral surveys, Thomas Piketty has shown that in those decades, left-wing parties in France, Britain, and America favored greater redistribution and were elected by voters with limited education. But afterward traditionally labor-supporting social-democratic parties have become associated with voters who have more education. In the 2000s and 2010s, Piketty argues, this shift gave rise to a multiple-elite party system, in which highly educated elites now vote for the new left, whereas the wealthy vote for the right.[34]

Thus, the unskilled have become increasingly detached from the main political parties. Labor unions, which gave workers additional bargaining power and political voice while facilitating social ties among the unskilled, are in decline. At the same time, the impact of cognitive segregation is that symbolic analysts are less likely to have firsthand knowledge of the lives of the working class because they do not see each other in the communities in which they live. Increasing economic segregation has meant that the unskilled have become more and more detached from those who have prospered, which explains why political preferences have also become polarized along geographic lines.[35]

Globalization, Automation, and Populism

Elected officials have become unresponsive to the concerns of millions of unskilled citizens, leaving their political interests unserved or ignored. A year before General Motors closed its plant in Janesville, Wisconsin, in 2008, President Barack Obama gave a spirited speech at the factory, suggesting that "this plant will be here for another hundred years."[36] After its closure, the chairman of Obama's White House Council on Automotive Communities and Workers paid one visit to

Janesville but failed to offer any significant help or relief. When Obama told a cheering crowd that "the auto industry is back on top" at a campaign rally in Madison in 2012, the citizens of Janesville who happened to see him on the news must have wondered what he was talking about. To paraphrase Amy Goldstein, those words would have been hard to repeat in Janesville.[37]

Populism and identity politics have been fueled by diminishing economic opportunity for the unskilled and the lack of a political response to their concerns. During the 2016 presidential election, Donald Trump infuriated just about every conceivable group except perhaps one: the white working class. It has been argued that the election outcome was a result of anxiety about the future status of white Americans as the dominant group rather than about the consequence of economic hardships. As the political scientist Diana C. Mutz puts it, "In many ways, a sense of group threat is a much tougher opponent than an economic downturn, because it is a psychological mindset rather than an actual event or misfortune."[38] However, this explanation ignores the fact that white Americans are seeing themselves, and their identity, threatened because of fading opportunity in the labor market. The working class was always more than an economic category—it was a cultural phenomenon, too. In the manufacturing era, industrial male workers had to find ways of taking pride in monotonous toil on a factory's assembly line. Their solution, the sociologist Michèle Lamont has convincingly argued, was to construct an identity as "the disciplined self."[39] It took discipline to get up early each morning, go to a factory, and perform the same routine job hour after hour, day after day. And it took discipline to be a family breadwinner, bringing home a paycheck every week of the year. When Lamont interviewed blue-collar men in the 1990s, she also found that they sharply contrasted their own type of discipline with that of other groups of Americans. The college-educated elites or symbolic analysts were perceived as untrustworthy. Blue-collar Americans believed that those people lacked integrity and would do anything to move up in the ranks. Blue-collar whites also distanced themselves from the black population, whose members they believed lacked discipline and too often lived on welfare.

The "whiteness" associated with the working class has historical roots. In the words of Cherlin:

> Many unions did not recruit black members, and even among unions that did, the local chapters were often segregated. When the American Federation of Labor (AFL) became the most powerful union organization in the 1890s, its leader, Samuel Gompers, urged its member unions to admit blacks so that employers could not use low-paid black workers to weaken the position of white workers. But the Federation did little to back up its rhetoric, and a number of important unions, such as the National Association of Machinists, were allowed to join the Federation even though they refused to admit black members. It was a fateful choice. In this way, "working-class" became a term that had the connotation of whiteness, which it retained throughout the nineteenth century and much of the twentieth century.[40]

In Rust Belt cities and townships, where joblessness is now widespread, the "disciplined self" identity has become harder to maintain, making dormant grievances come alive. We know from a wide range of studies that relative income matters in shaping people's aspirations and subjective well-being.[41] White blue-collar workers felt that they had moved up in the world, but now they feel that they have been left behind. As shown by the General Social Survey, there are considerable racial differences in how people perceive the recent past as well as in their optimism or pessimism about the future. Since 1994, the survey has asked Americans questions like, "Compared to your parents when they were the age you are now, do you think your own standard of living now is much better, somewhat better, about the same, somewhat worse, or much worse than theirs was?" Among non-college-educated citizens, the percentage of negative responses among blacks has decreased since 1994, while negative responses among whites has risen dramatically.[42] This shift in attitudes can go a long way toward explaining Trump's appeal among white working-class voters:

> Among less-educated workers, racial tensions remain. White workers, without realizing it, are drawing upon a long history of animosity

toward black workers. To be sure, the overt racism of late-nineteenth- and early-twentieth-century industrialization—a time when white unions largely resisted the incorporation of African Americans—has greatly diminished. Civil rights legislation, changing attitudes, and increased education are among the factors that have improved the relative position of African American workers. No one could imagine a union leader today issuing the kind of blanket denunciation of the "Negro race" that was common in 1900. Nevertheless, there is still a connection between whiteness and the working class. . . . The white men we interviewed saw the deterioration in their labor market prospects compared to the previous generation, and they were right. In an environment in which overall opportunities for blue-collar labor are constricting, white workers perceive black progress as an unfair usurpation of opportunities rather than as a weakening of the privileged racial position they held.[43]

We all know that Trump's campaign involved many racial provocations as well as attacks on America's elites. His appalling rhetoric was surely appealing to some, and there can be no doubt that he was speaking to the working-class identity that Lamont described in her study. Of course, much of the Trump campaign centered on issues like immigration. But would his tactics have been so successful if there had been well-paying jobs in abundance for the unskilled and their wages were rising? By any account, technology and globalization have played a quantitatively larger role in putting downward pressure on unskilled people's wages than immigration has: on the contrary, the evidence shows that immigration has boosted employment, innovation, and productivity without having any significant adverse impacts on the wages of the unskilled.[44] And the slogan "Make America Great Again" clearly targeted people in the smokestack cities and towns of the Second Industrial Revolution that once flourished but are now in despair. Take, for example, social mobility. Virtually every citizen cares deeply about his or her chances of realizing the American dream. But people's prospects of moving up the income ladder greatly depends on where they happen to grow up. Among America's largest cities, the probability that a child born into a family in the bottom quintile of the national income distribution will reach the top quintile varies between 4.4 percent in Charlotte, North Carolina, and 12.9 percent in San Jose, California.[45] (While 12.9 percent may seem low, intergenerational mobility by quintile cannot exceed 20 percent: if there is a 20 percent chance that a child born into the bottom quintile will move into the top quintile, its chances are just as good as those of any other child.) The cities of the American South, with their long history of racial segregation, still have the lowest rates of social mobility in the country. But in manufacturing cities like Cleveland, Ohio, and Detroit and Grand Rapids, Michigan, workers' prospects of climbing the economic ladder have become almost as dismal. What drives such inequalities of opportunity? The places where the American dream has turned into a nightmare have a few things in common: many children grow up in single-parent households, crime is widespread in the community, income disparities are large, and the middle class has withered. In short, they suffer from many of the social ills that we saw characterized in Charles Murray's Fishtown.

Blue-collar Americans have much to be unhappy about. As noted above, they have felt a blow to family finances, some have experienced divorce, and others are in deteriorating health. It is true that Trump voters were a mixed bag, with an overrepresentation of high-income earners, but many economists believe that economic distress across working-class white Americans—those who have seen their jobs taken by machines or the Chinese—swung the election for him. This explanation is not just appealing because it is intuitive; it is also empirically grounded. Globalization made American politics more polarized even before the days of Trumpism. Ideological extremists in Congress have gained more votes where the labor market prospects for the unskilled have deteriorated. Since China's admission to the World Trade Organization in 2001, Congressional districts exposed to the forces of globalization have been more likely to replace moderate incumbents with more ideologically extreme successors in both parties.[46] Of course, Trump made globalization a key plank of his campaign, and unsurprisingly he also enjoyed the greatest gains—relative to George W. Bush's 2000 results—in areas most exposed to Chinese imports.[47] But while globalization is the most frequently cited villain, automation has also helped shatter the communities of the so-called blue-collar's aristocracy. Even if production is brought back to the United States, it will not replace the vast numbers of jobs for non-college-educated members of the middle class that have been lost to the process of deindustrialization. The computer revolution, which has also been an underlying facilitator of globalization, has meant diminishing opportunity for the unskilled across the board: routine work is now disappearing in parts of the developing world as well.[48]

In America, this process has been going on for decades, yet it was hidden by other factors. Though many blue-collar men have seen their incomes decline in real terms, family incomes were still rising for some, as more and more women joined the workforce. Women helped offset the work deficit among men up until 2000, when the growth in female labor force participation was reversed. But there was still another source of relief: the everyday consequences of technological change for the middle class were counterbalanced by subsidized mortgages for low-income households, which meant that consumption was broadly unaffected even as incomes fell. This was made possible by the flow of liquidity from China, leaving even unskilled Americans under the illusion that their standards of living were rising until the housing bubble burst in 2007.[49] In addition, the housing boom masked the vanishing of industrial jobs for the unskilled as a growing abundance of construction jobs simultaneously pushed in the other direction. The Great Recession, in other words, revealed the long-term disappearance of routine blue-collar jobs that had been hidden by excessive cheap credit and the consequent housing bubble.[50]

To be sure, the recession itself also directly led to job loss across the country. But in areas where factories closed, unemployment has fallen since. The problem is that while many jobs have returned, well-paying jobs have not. Amy Goldstein's brilliant account of citizens' postrecession experience in Janesville, where General Motors closed its factory in 2008, describes the state of the town as follows:

So, seven and a half years after the Great Recession technically ended, how is Janesville faring? Surprisingly well, or not, depending on how you measure. By the most recent count, unemployment in Rock County has slid remarkably to just under 4 percent, the lowest level since the start of the century. As many people are working now as just before the Great Recession; distribution centers have arrived, Beloit plants such as ones for Frito-Lay and Hormel Foods have been hiring, and some people are working further away. Good news. But not everyone who now has a job is earning enough for the comfortable life they expected. Real wages in the county have fallen since the assembly plant shut down. . . . For the moment, the big job news is that Dollar General has decided to put a distribution center on the south side of town. The city government is providing an $11.5 million package of economic incentives—a new Janesville record. The workforce that Dollar General says it will need—about 300 at first and perhaps 550 eventually—will bring the biggest hiring spurt in years. Most of the jobs will pay $15 or $16 per hour—far below the $28 wage GM'ers were being paid when the plant closed, but decent enough money in town these days. In a sign of a lingering hunger for work or better pay, when Dollar General held a recent job fair, three thousand people showed up.[51]

Janesville is not an isolated example. Across the Rust Belt, middle-income jobs have not come back, and with recent advances in automation technology, it seems increasingly unlikely that they will. In 2017, the *Washington Post* ran a story on Wilmington, Ohio—a predominantly white town that in 1995 had been featured in Norman Crampton's *The 100 Best Small Towns in America*.[52] Trump had visited Wilmington twice during his campaign, and that had paid off. Wilmington had turned into a place where making America great again had become an expectation rather than just a slogan. When the German freight company DHL packed up and left Wilmington in 2008, 7,000 jobs disappeared from a town with a population of 12,500. Michael O'Machearley, who makes custom knives in his backyard, now earns half of what he did before he was laid off from his DHL shipping job in 2008. But considering the circumstances, he thinks that he has been doing well. As he explains, "There were people in this town that went through divorces because of it [the DHL closure], that lost their homes because of it. You couldn't sell a house in this town for any kind of money. . . . Our downtown used to be a precious place. It died."[53] The big hope in town is that Wilmington might become an Amazon shipping and distribution hub. The problem, O'Machearley tells us, is that they "won't hire the same amount of people because they use a lot of robots."[54]

O'Machearley did not lose his job to automation. But he was right in thinking that robots have meant fewer and fewer available jobs for unskilled workers who suffer bad luck. Had he lost his job in the peak postwar years, when well-paying jobs for the unskilled were abundant, things would not have been as harsh. Before the age of automation, workers accepted the churn in the labor market with the expectation that they would eventually come out ahead. Yet it has become less likely that they will today. In Ohio, a state that Trump won in a landslide, 350,000 factory jobs have disappeared since 2000, and the middle class has shrunk more than in perhaps any other state. Health care has become the largest employer, but jobs in that sector often pay less than the production jobs that disappeared. The median yearly income has dropped from $57,748 in 2000 to $49,308 in 2015, adjusting for inflation. One reason is that Ohio is second only to Michigan in terms of the number of robots operating in its factories.

As noted above, since the Great Recession, the number of robots in America's factories has grown by 50 percent. The robot revolution is largely a Rust Belt phenomenon, and this is also where Trump made the greatest gains for the Republican Party. The Rust Belt, which swung the election for him, used to be part of what pundits and political analysts call the Blue Wall—that is, safe Democratic states. Not every manufacturing town there voted for Trump. But electoral districts specializing in industries that have invested heavily in automation overwhelmingly did. Whether voters had found their jobs taken over by robots or simply faced diminishing outside opportunities as a consequence, they were more likely to support Trump. My research with Thor Berger and

Chinchih Chen shows that Michigan, Pennsylvania, and Wisconsin would have swung in favor of Hillary Clinton, leaving the Democratic Party with a majority, had the number of robots in America's factories not increased since the 2012 election. We accounted for a variety of alternative explanations, including globalization and immigration. While counterfactuals must always be taken with a grain of salt, there is clearly a relationship between levels of automation and voting patterns, which provides a powerful explanation for why the three states—which went Democratic in every presidential election since 1992—ended up being won by Trump.[55]

The role of technological progress in triggering protests is thus as salient today as it was in the early nineteenth century. As was the case then, protest recently has been driven by fundamental concerns over fading opportunity in the labor market. As Toke Aidt, Gabriel Leon, and Max Satchell argue in a recent study, "The current situation of ex-miners in the north of England and factory workers in the American Midwest is not unlike that of rural farm workers made redundant by the adoption of threshing machines in the early 1830s."[56] But the authors also show that while the Captain Swing riots were driven by concerns that threshing machines would take people's jobs, there was also significant contagion when potential rioters in one parish learned about riots in neighboring parishes—which suggests that information flows exacerbated workers' concerns.[57] Such contagion is much greater today. The Swing riots predated the construction of the railroads and the telegraph network (meaning that information had to travel by foot, horseback, or carriage), so contagion primarily occurred through meetings at markets and fairs. In the age of social media, by contrast, the information diffusion process has dramatically speeded up. And as is well known, the artificial intelligence used by Facebook and other companies learns about its users' preferences and thus reinforces their political beliefs and prejudices. Social media undoubtedly became an important channel that allowed the Trump campaign to tap into people's discontents, as the Cambridge Analytica scandal bears witness, but it was not in itself the cause of people's concerns.

The New Luddites

Globalization has moved to center stage of the political debate. During the 2016 U.S. presidential election, Bernie Sanders and Donald Trump both made blistering assaults on trade agreements a main theme of their campaigns. Trump's win was in part attributable to the adverse impacts of trade on parts of the labor market, and it stands to reason that his campaign promises to renegotiate trade deals, which he claimed had benefited other countries at the expense of American workers, appealed to those who felt that they had lost out to globalization. His assaults on trade were surely excessive, but there is no question that many manual workers and their families have felt the consequences of low-cost competition intensely—especially since China's admission to the World Trade Organization (WTO). Globalization did not lift all of the boats, but neither did automation. As the economist Dani Rodrik writes, "Globalization was hardly the only shock which gutted established social contracts. By all accounts, automation and new digital technologies played a quantitatively greater role in de-industrialization and in spatial and income inequalities."[58]

Rodrik also offers a powerful explanation for why globalization has become the political target that automation has not. "What gives trade particular political salience," he argues, "is that it often raises fairness concerns in ways that the other major contributor to inequality—technology—does not."[59] Inequality is more problematic when it occurs due to unfair competition. When a better technology makes an old one obsolete, nobody has reason to complain: "Banning the light bulb because candle makers would lose their jobs strikes almost everyone as a silly idea."[60] But when a firm competes by outsourcing production to countries where firms compete according to different ground rules—where labor's bargaining rights are repressed and child labor is prevalent, undercutting the social contract and institutional arrangements of Western nations—opposition is more likely. Though globalization and automation have hit the same people, they feel less sanguine about trade, because firms in countries like China and Vietnam violate trade rules and compete on terms that would be illegal under American laws and regulations. Trade, in other words, is not problematic just because it redistributes income. Almost every policy intervention or market transaction does that in some way. Technological progress has been a source of ceaseless churn in the labor market for more than two centuries. But, as Rodrik writes, "when we expect the redistributive effects to even out in the long run, so that everyone eventually comes out ahead, we are more likely to overlook reshufflings of income. That is a key reason why we believe that technological progress should run its course, despite its short-run destructive effects on some."[61]

Still, we must distinguish between different types of technologies. If people believe that they will eventually be made better off by technological progress, they are more likely to accept the churn. But if citizens do not see their incomes improve over several decades as their alternative job options gradually fade, they are more likely to resist the force of technology. As we have seen, during the early days of industrialization not everyone came out ahead, and quite rationally those who were negatively affected vehemently opposed the introduction of new technologies. While Engels's pause came to an end eventually, and ordinary people were much better off in the very long run, many of those who lost their jobs to machines never saw the gains from growth. We are now living through another period of worker-replacing technical change. As Rodrik points out, "The potential benefits of ongoing discoveries and applications in robotics, biotechnology, digital technologies, and other areas are all around us and easy to see. . . . Many believe that the world economy may be at the cusp of another explosion in new technologies. The trouble is that the bulk of these new technologies are labor-saving."[62]

In the end, there is nothing that ensures that citizens will accept the verdict of the market, regardless of whether outcomes have been shaped by automation, globalization, or some other factor. Many citizens quite understandably do not feel sanguine about recent advances in technology either. On May 23, 2018, almost all of the Las Vegas Culinary Workers Union's twenty-five thousand members voted to go on strike. In addition to higher pay, they demanded greater job security against robots. Chad Neanover, a cook at a Las Vegas hotel, said: "I voted yes to go on strike to ensure my job isn't outsourced to a robot. . . . We know technology is coming, but workers shouldn't be pushed out and left behind." The secretary treasurer of the Culinary Workers Union added: "We support innovations that improve jobs, but we oppose automation when it only destroys jobs. Our industry must innovate without losing the human touch."[63]

More broadly, a 2017 survey by the Pew Research Center of 4,135 American adults shows that 85 percent would support policies to restrict workforce automation to hazardous jobs, with 47 percent favoring the idea "strongly." Another 58 percent responded that "there should be limits on the number of jobs that businesses can replace with machines, even if those machines are better and cheaper." Respondents were more divided on the question of who should bear the responsibility for workers who lose their jobs, with about half saying it should be the government, "even if it means raising taxes substantially." Respondents identifying as Democrats typically favored a stronger role for government, while Republicans were more likely to think it is the responsibility of individuals. Yet 85 percent of Democrats and 86 percent of Republicans thought that automation should be limited to dangerous or unhealthy jobs. And perhaps unsurprisingly, groups in the labor market that have seen jobs ebb were more likely to favor policies to restrict automation: among the respondents with only a high school diploma or less, seven in ten said that there should be limits on the number of jobs firms can allow machines to perform, with four in ten of those with a college degree sharing that view.[64]

History tells us that political elites may block technological progress if they fear it may cause political unrest.[65] As discussed above, preindustrial monarchs, who held most of the political power, were worried that they might have to share it with the increasingly wealthy merchant class. And they were also alarmed by the threat from below, fearing that machines' taking workers jobs would lead to social and political upheaval. But while such concerns persisted well into the nineteenth century, growing competition among nation-states changed the calculations of the ruling classes. As noted in chapter 3, the weakening of the guilds and growing international competition meant that the external threat of replacement became greater than the internal threat from below. Cascading competition reduced incentives to block progress, in large part because "falling behind technologically makes countries vulnerable to foreign invasion."[66]

The main reason why British governments began to side with the innovators, even if progress came at the expense of middle-income workers, was concern over Britain's competitive position in trade. They were also aware that a strong war machine depended on having a strong economy. The external threat, in other words, was perceived to be greater than the internal threat posed by machinery rioters. As the Luddite riots swept across Britain, yet another Parliamentary committee heard petitions from cotton workers suffering as a result of mechanization. The 1812 report that came out of the hearings attests to the government's determination to keep the technological genie out of the bottle, even though it acknowledged the distress inflicted upon the workforce. As noted, Lord Liverpool, who became prime minister in 1812, was convinced that any measures to relieve workers' hardship would only impede their redeployment, to the detriment of British competitiveness.[67]

Governments in the nineteenth century did not see technology as an unstoppable force. Rather, they had to use considerable force to make sure that the Luddites and other groups were unable to block mechanization. And the working class didn't view mechanization as inevitable either. One attempt to bring the spread of machines to a halt was followed by another. If the Luddites had been successful, the mechanized factory would have failed to replace domestic industry, and it is quite likely that the Industrial Revolution would not have begun in Britain.

The past few decades have not seen any slackening of competition. Just as the Industrial Revolution was partly caused by competition in trade, the drive to computerize has in part been driven by high labor costs in the West and intensifying global competition. The ascent of Japan, Korea, and more recently China has left many American companies with the choice of either moving production offshore or automating. Donald Bennett, a union leader at General Electric's Louisville, Kentucky, factory, told the *New York Times* in 1984: "The automation had to be done, otherwise we would have lost the plant altogether. Some jobs have been lost for the moment, but we had to accept some changes to keep the factory here. We sure as hell didn't want those jobs to go somewhere else."[68] Even China has recently made a drive to automate to keep low-cost competition at bay. It is now the largest market in the world for industrial robots.

The race for world technological leadership has if anything intensified in recent years. Supercomputers are the elephants of the computer kingdom, and some readers may be surprised to learn that the fastest supercomputer is no longer on American soil. This matters because the country with the fastest supercomputers will also be at the forefront of a variety of other domains—which is why the White House Office of Science and Technology Policy has singled out supercomputers as "essential to economic competitiveness, scientific discovery, and national security."[69] In 2015, the administration of President Barack Obama created the National Strategic Computing Initiative to ensure that the American lead in supercomputing is maintained. Despite these efforts, the world's fastest supercomputer is now in China. America is feeling the pressure, because as every American government knows, shifts in technological supremacy come with shifts in political power.

Yet in liberal democracies with shrinking middle classes, the internal threat of political unrest is becoming ever greater. And with populism on the march, the concerns of the unskilled have become harder to ignore. Even if governments are concerned with international competition, populists may choose to promote policies to restrict automation, the way they are now clamping down on globalization. Automation does not need be seen as an inexorable fact of life. Instead, it provides opportunities and challenges that governments can seek to control politically. Restraining technological innovation, for example, is not the same as restricting some of its uses. If there are strong political preferences for job conservation, policies that favor jobs at the expense of productivity might still be implemented. One reason that you are likely to pass dozens of bookstores while taking a walk in Paris is that France recently passed a so-called anti-Amazon law, which says that online sellers cannot offer free shipping on discounted books. The law is part of a drive in France to promote "biblio-diversity" by helping independent bookstores compete.[70] France decided to forgo productivity and consumer benefits in the interest of keeping jobs.

This example is not offered as an endorsement of anti-automation policies. As history shows, labor-saving technology and rising productivity are a prerequisite for rising living standards over the long run. One reason why growth was so slow before the Industrial Revolution was precisely the resistance to technologies that threatened to render the skills of the workforce obsolete. The point is that there is nothing to ensure that technology will always be allowed to progress uninterrupted. It is perfectly possible for automation to become a political target. The twentieth century was an extraordinary period in human history in that it saw very little resistance to machines. Though political parties often try to represent the interests of particular groups over time, it is also true that as the composition of the electorate shifts and new economic and societal issues arise, as do new political agendas. Politicians are autonomous actors who seek power by mobilizing voters through continuously shifting their agendas to reflect voters' concerns, and one such concern these days is clearly automation. Consequently, in Britain, the leader of the Labor Party, Jeremy Corbyn, has pledged to tax robots to slow down the pace of automation, which he thinks threatens workers' jobs.[71] And in South Korea, President Moon Jae-in has already downsized tax breaks on investments into robotics and automation due to employment concerns.[72]

In America, Andrew Yang, who will make automation the key theme for his 2020 run for the White House, thinks it is hard to tax robots directly. Instead, he proposes a special value-added tax on companies that use automation.[73] Though few candidates have made automation a central political theme, we know from recent events that populist ideas can spread quickly if they gain some traction. Trump's promise to slap U.S. tariffs on steel and aluminum imports has earned him praise from Democrats in the Rust Belt states who seek to appeal to the electorate. For example, Senator Sherrod Brown recently told Reuters that "this welcome action is long overdue for shuttered steel plants across Ohio and steelworkers who live in fear that their jobs will be the next victims of Chinese cheating."[74]

Looking forward, as labor markets become more shielded from the impacts of trade and the replacing effects of technology become greater, the populist target could shift in the coming years. The rise of China as a major industrial power has already happened, and the jobs that can be sent abroad have already left American and European soil. They are not going to come back in larger numbers, as voters will find out eventually. Those who think that tariffs on steel will bring American jobs back would do well to visit European steel mills. In Austria, fourteen employees are needed to produce 500,000 tons of steel wire per year. As a visitor at an Austrian plant notes, "There's barely anybody there. At most, there are three technicians monitoring the output on flatscreens."[75]

Because most Americans now work in nontradable sectors of the economy, they are also increasingly shielded from the direct impacts of trade. Michael Spence, winner of the Nobel Prize in Economics, and Sandie Hlatshwayo have shown that nontradable services might have accounted for as much as 98 percent of total U.S. employment growth in the period 1990–2008.[76] But as we shall see, the rise of artificial intelligence and autonomous driving means that a large percentage of nontradable jobs are now susceptible to automation. As Obama noted when leaving office, "The next wave of economic dislocations won't come from overseas. It will come from the relentless pace of automation that makes a lot of good, middle-class jobs obsolete."[77]

Conclusion

The rise of the middle class was in large part a consequence of two industrial revolutions. From the mid-nineteenth century until the age of computers, technological change helped a steadily growing share of workers join the ranks of the middle class. In this regard, the computer revolution was not the continuation of a century of mechanization but the complete reversal of it. Recently, automation has cut out the jobs that were created by the spread of office and factory machines over the course of the twentieth century. The restructuring of the American economy has not worked in favor of the middle class. The experience of the decades succeeding the 1980s in many ways resembles that of the early nineteenth century, when the arrival of the mechanized factory caused a similar hollowing out of the labor market, put downward pressure on workers' wages, and caused the labor share of income to fall—to the detriment of average people. The recent populist backlash is less puzzling when considering the reversal of fortunes experienced by the non-college-educated middle class. Blue-collar families (in particular, those who felt that they had moved up in the world) now feel that they have fallen behind. It is hard to believe that populism would be as appealing if the wages of the lower middle class were rising and well-paying employment opportunities were plentiful.

The point is obviously not to suggest that America would have been better off by stopping the technological clock at the onset of the computer revolution. We can be thankful that the Industrial Revolution was not brought to a halt by the Luddites, and like the Industrial Revolution, the age of automation has delivered enormous benefits, especially for consumers. But—again like the Industrial Revolution—it has also fundamentally restructured the economic and social fabric—to the detriment of large groups in the labor market. Parallels to the early nineteenth century can surely be exaggerated. It is hard to believe that contemporary Americans would trade their jobs for the "dark satanic mills." And the hardships suffered by those we regard as poor today look less harsh when compared to the material conditions of the Luddites. In 2011, the Heritage Foundation published a provocative report entitled, "Air Conditioning, Cable TV, and an Xbox: What Is Poverty in the United States Today?" The authors rightly noted that the material standards of poor Americans have greatly improved over the past century. Innovative products that once were luxuries had become common in all households: "In 2005, the typical household defined as poor by the government had a car and air conditioning. For entertainment, the household had two color televisions, cable or satellite TV, a DVD player, and a VCR. If there were children, especially boys, in the home, the family had a game system, such as an Xbox or a PlayStation. In the kitchen, the household had a refrigerator, an oven and stove, and a microwave. Other household conveniences included a clothes washer, clothes dryer, ceiling fans, a cordless phone, and a coffee maker."[78] However, even the Luddites had access to a wide range of consumer items that had been unavailable to their great grandparents:

> The quantitative study of probate records and other sources has prompted historians to conclude that the increase in consumer durable goods such as clocks, furniture, toys, books, rugs, carriages, jewelry, flatware, coffee and tea paraphernalia, paintings and other domestic decorations, peaked between 1680 and 1720. Most of these goods remained primarily within the confines of the middle class—indeed they may have been the signs that defined the middle class. Yet in the eighteenth century they kept trickling down to working-class people, if not, perhaps, to the unskilled poor, cottagers and paupers, who constituted the bottom 20 percent of the income distribution.[79]

The fact that low-income households today have access to many things that weren't available to Renaissance monarchs is indeed evidence of enormous material progress over the centuries. The capitalist achievement, as Joseph Schumpeter noted, was not to provide more silk stockings for monarchs but "bringing them within the reach of factory girls in return for steadily decreasing amounts of effort."[80] But that does not make concerns over the well-being of the shrinking middle class moot. Nor does it make moot concerns over the fact that there are many necessities that are not getting any cheaper. With inflation growing faster than some workers' wages, health care, education, and housing are becoming less affordable to many Americans, regardless of how many television sets, microwaves, smartphones, and computers they have. Many households that are not poor but were once firmly middle class are feeling squeezed. And one reason is precisely the cheapening of many consumer products, whose cost has been brought down by automation and moving production offshore. Most Americans are both consumers and producers. In times of worker-replacing technological change, the flip side of the cheapening of goods is that significant parts of the workforce may suffer in the labor market. That is what happened during the early stages of the Industrial Revolution and again in the age of automation. Even if we assume that the redistributive effects of auto-mation will even out in the long run, as was the case with the mecha-nized factory in the late nineteenth century, and technological change ends up lifting all boats, the short run can be a lifetime for some.

PART V

THE FUTURE

What of the future? . . . The history of computing to date shows
no slackening of innovation in the fundamental computational
processes or in applications of computation throughout the
economy. Perhaps, aside from humans, computers and software
are the ultimate general purpose technology. They are a
technology that has the potential for penetrating and
fundamentally changing virtually every corner of economic life.
At current rates of improvement, computers are approaching the
complexity and computational capacity of the human brain.
Perhaps computers will prove to be the ultimate outsourcer.

—WILLIAM NORDHAUS, "TWO CENTURIES OF
PRODUCTIVITY GROWTH IN COMPUTING"

The Luddites of the early 19th century surely had their voice heard,
as did their likeminded emulators over the following decades.
However, they could hardly expect to make a dent on their fate:
democracy was still highly limited and living standards still very low
for the vast majority, so that most people were just consumed by the
need to provide for their basic needs. Much has changed since, and
nowadays virtually every individual in advanced western countries
has come to expect to be entitled, at least in principle, to full
participation in every realm of society: the political, the economic,
the cultural. The expectation is not just to vote in periodic elections
but to have an influence via "participatory democracy"; not just to
hold a job, but to partake in the benefits of economic growth—this
is what constitutes "the democratization of expectations."

—MANUEL TRAJTENBERG, "AI AS THE NEXT GPT: A
POLITICAL-ECONOMY PERSPECTIVE"

The Danish physicist Niels Bohr supposedly once quipped that "God gave the easy problems to the physicists." Since the scientific revolution, the steady accumulation of scientific knowledge has given the physical sciences much improved means of predicting outcomes. In economics, the opposite is true. While the laws of physics apply across time and space, in economics and other social sciences, boundary conditions are not timeless. Arguably, the predictability of economic outcomes peaked before the Industrial Revolution, when growth was slow or stagnant.

It is true that technological progress follows an evolutionary process, meaning that invariant statements cannot be made over the long run. As we have seen in the preceding chapters, the potential scope of automation has steadily expanded over time. But we can establish some near-term engineering bottlenecks that currently set the boundaries for the type of tasks computers can perform. As we saw in chapter 9, routine jobs were eliminated in large numbers beginning in the 1980s. But already in the 1960s, the Bureau of Labor Statistics made the following observation: "Mechanization may indeed have created many dull and routine jobs; automation, however, is not an extension but a reversal of this trend: it promises to cut out just that kind of job and to create others of higher skill."[1] They predicted the Great Reversal two decades before it happened by observing what computers can do. Because it takes time before technologies are adopted and put into widespread use, we can infer the exposure of current jobs to future automation by examining technologies that are still imperfect prototypes.

There is no economic law that postulates that the next three decades must mirror the last three. Much depends on what happens in technology and how people adjust. It is possible that we are on the cusp of a series of enabling technological breakthroughs that will create an abundance of new jobs for middle-class people. However, the empirical reality of the last decades points in the opposite direction, and there are good reasons to think that current trends will continue at least for some time, unless policies are implemented to counteract them. The employment prospects for the middle class crucially hinge upon what computers can and cannot do. And the division of labor between man and machine is constantly evolving. Recent breakthroughs in artificial intelligence (AI) mean that for the first time in history, machines are now able to learn. To better understand the next wave of automation, let's begin by looking at exactly what computers can do in the age of AI.

12

ARTIFICIAL INTELLIGENCE

A perfect storm of advances, including larger databases, Moore's Law, and clever algorithms, has paved the way for much of the recent progress in artificial intelligence (AI). Most significantly over the past decade, this has led to automation extending beyond routine jobs and into new and unexpected areas. In the past rule-based era of computing, automation was limited to deductive instructions that had to be specified by a computer programmer. By discovering ways of automating things that we struggle to articulate or explain, like how to drive a car or to translate a news story, AI allows us to unravel Polanyi's paradox, at least in part (see chapter 9).[1] The fundamental difference is that instead of automating tasks by programming a set of instructions, we can now program computers to "learn" from samples of data or "experience." When the rules of a task are unknown, we can apply statistics and inductive reasoning to let the machine learn by itself.

Outside of the technology sector, AI is still in the experimental stage. Yet the frontiers of AI research are steadily advancing, which in turn has expanded the potential set of tasks that computers can perform. The victory of Deep Mind's AlphaGo over the world's best professional Go player, Lee Sedol, in 2016 is probably the best-known example. With the defeat of Sedol, humans lost their competitive edge in the last of the classical board games, two decades after being superseded in chess. As we all know, in a six-game match played in 1996, the chess master Garry Kasparov prevailed against IBM's Deep Blue by three wins but lost in a historic rematch a year later.

Relative to chess, the complexity of Go is striking. Go is played on a board that is nineteen by nineteen squares, whereas chess uses a board that is eight by eight squares. As the mathematician Claude Shannon demonstrated in 1950, in his seminal paper on how to program a machine to play chess, a lower-bound estimate of the number of possible moves in chess is greater than the number of atoms in the observable universe, and the number of possible moves in Go is more than twice that number.[2] Indeed, even if every atom in the universe was its own universe and had inside it the number of atoms in our universe, there would still be fewer atoms than the number of possible legal moves in Go. The illimitable complexity of the game means that not even the best players are capable of breaking it down into meaningful rules. Instead professionals play by recognizing patterns that emerge "when clutches of stones surround empty spaces."[3] As discussed above, humans still held the comparative advantage in pattern recognition when Frank Levy and Richard Murnane published their brilliant book *The New Division of Labor* in 2004.[4] At the time, computers were nowhere near capable of challenging the human brain in identifying patterns. But now they are.

Much more important than the fact that AlphaGo won is how it did so. While Deep Blue was a product of the rule-based age of computing, whose success rested upon the ability of a programmer to write explicit if-then-do rules for various board positions, AlphaGo's evaluation engine was not explicitly programmed. Instead of following prespecified rules of the programmer, the machine was able to mimic tacit human knowledge, circumventing Polanyi's paradox. Deep Blue was built on top-down programming. AlphaGo, in contrast, was the product of bottom-up machine learning. The computer inferred its own rules from a series of trials using a large data set. To learn, AlphaGo first watched previously played professional Go games, and then it played millions of games against itself, steadily improving its performance. Its training data set, consisting of thirty million board positions reached by 160,000 professional players, was far greater than the experience any professional player could accumulate in a lifetime. The event marks what Erik

Brynjolfsson and Andrew McAfee have called the "second half of the chessboard."[5] As *Scientific American* marveled, "An era is over and a new one is beginning. The methods underlying AlphaGo, and its recent victory, have huge implications for the future of machine intelligence."[6]

Deep Blue may have beaten Kasparov at chess. But ironically, at any other task, Kasparov would have won. The only thing Deep Blue could do was evaluate two hundred million board positions per second. It was designed for one specific purpose. AlphaGo, on the other hand, relies on neural networks, which can be used to perform a seemingly endless number of tasks. Using neural networks, DeepMind has already achieved superhuman performance at some fifty Atari video games, including Video Pinball, Space Invaders, and Ms. Pac-Man.[7] Of course, a programmer provided the instruction to maximize the game score, but an algorithm learned the best game strategies by itself over thousands of trials. Unsurprisingly, AlphaGo (or AlphaZero, as the generalized version is called), also outperforms preprogrammed computers at chess. It took AlphaZero four hours to learn the game well enough to beat the best computers.

Much recent progress, like AlphaGo's triumph, has been aided by exponentially growing data sets, collectively known as big data. When things are digitized, they can be stored and transferred at virtually no cost. The digitization of just about everything generates billions of gigabytes on a daily basis through web browsers, sensors, and other networked devices. Digital books, music, pictures, maps, texts, sensor readings, and so on constitute massive bodies of data, providing the raw material of our age. As an ever-growing percentage of the world's population becomes digitally connected, more and more people gain access to a significant share of the world's accumulated knowledge. This also means that more and more people are able to add to this knowledge base, creating a virtuous cycle. As billions of people interact online, they leave digital trails that allow algorithms to tap into their experience. According to Cisco, worldwide internet traffic will increase nearly three-fold over the next five years, reaching 3.3 zettabytes per year by 2021.[8] To put this number in perspective, researchers at the University of

California, Berkeley estimate that the information contained in all books worldwide is around 480 terabytes, while a text transcript of all the words ever spoken by humans would amount to some five exabytes.[9]

Data can justly be regarded as the new oil. As big data gets bigger, algorithms get better. When we expose them to more examples, they improve their performance in translation, speech recognition, image classification, and many other tasks. For example, an ever-larger corpus of digitalized human-translated text means that we are able to better judge the accuracy of algorithmic translators in reproducing observed human translations. Every United Nations report, which is always translated by humans into six languages, gives machine translators more examples to learn from.[10] And as the supply of data expands, computers do better.

Google Translate draws on a plethora of algorithms, but it would be far less pervasive without the great leap in computer hardware powered by Moore's Law. Many of the building blocks of computing—processing speed, microchip density, storage capacity, and so on—have seen decades of exponential improvements. For example, the idea of artificial neural networks (that is, layers of computational units that mimic how neurons connect in the brain) has been around since the 1980s, but the networks performed poorly due to constraints imposed by computational resources. So up until recently, machine translations relied on algorithms that analyzed phrases word by word from millions of human translations. However, phrase-based machine translations suffered from some serious shortcomings. In particular, the narrow focus meant that the algorithm often lost the broader context. A solution to this problem has been found in so-called deep learning, which uses artificial neural networks with more layers. These advances allow machine translators to better capture the structure of complex sentences. Neural Machine Translation (NMT), as it is called, used to be computationally expensive both in training and in translation inference. But due to the progression of Moore's Law and the availability of larger data sets, NMT has now become viable.

In machine translation, deep learning is not without its own drawbacks. One major challenge relates to the translation of rare words. For example, if you type the Japanese word for "once-in a lifetime encounter" into an NMT-based system, your output is likely to be "Forrest Gump." While this might seem strange at first, this happened to be the subtitle of the Japanese version of the film. And because the word is rare, it did not show up in many other contexts. However, machine learning researchers have found some creative ways of circumventing this problem, at least in part, by dividing words into subunits. As a team of Google researchers demonstrated in a 2016 *Nature* article, the use of "word-units" and neural networks collectively reduced error rates by 60 percent, compared to the old phrase-based system.[11] Though Google's NMT system still lags behind human performance, it is catching up.

Like steam, electricity, and computers, AI is a general purpose technology (GPT), which has a wide range of applications. As the economists Iain Cockburn, Rebecca Henderson, and Scott Stern have shown, there has been a dramatic shift in AI-related publications, from computer science journals to application-oriented outlets. In 2015, the authors estimate, nearly two-thirds of all AI publications were outside the field of computer science.[12] Their finding is consistent with the general observation that AI is being applied to a cascading variety of tasks. The same technology that has shown promising results in machine translation is also performing visual tasks, such as image recognition. Starting from the individual pixels in an image, these algorithms work up through increasingly complex features, like geometric patterns.

Image recognition has seen exponential progress in recent years. Error rates in the labeling of images have fallen from 30 percent in 2010 to 2 percent in 2017.[13] While in many cases the technology is still at an experimental stage, it is already showing promising results. In Germany, for example, trials of automatic face recognition technology to identify people passing through Berlin's Suedkreuz railway station have proven successful, aiding the work of security officials. Interior Minister Thomas de Maiziere reported that the right person had been recognized 70 percent of the time, while the algorithm had flagged the wrong person in less than 1 percent of cases, despite poor image quality.[14] The same type of AI that identifies faces has also proven adept at diagnosing disease. New research published in *Nature Medicine* shows that AI is already capable of distinguishing between different types of lung cancers, using pathology images. And it does so with 97 percent accuracy.[15] Another *Nature* article, published in 2017, used neural networks and a data set of 129,450 clinical images to test AI's performance against twenty-one board-certified dermatologists and found that AI has already reached human level performance: "The [algorithm] achieves performance on par with all tested experts across both tasks, demonstrating an artificial intelligence capable of classifying skin cancer with a level of competence comparable to dermatologists. Outfitted with deep neural networks, mobile devices can potentially extend the reach of dermatologists outside of the clinic. It is projected that 6.3 billion smartphone subscriptions will exist by the year 2021 and can therefore potentially provide low-cost universal access to vital diagnostic care."[16]

Machines are not just turning into better translators and diagnosticians. They are becoming better listeners, too. Speech recognition technology is improving at staggering speed. In 2016, Microsoft announced a milestone in reaching human parity in transcribing conversations. And in August 2017, a research paper published by Microsoft's AI team revealed additional improvements, reducing the error rate from 6 percent to 5 percent.[17] And like image recognition technology promises to replace doctors in diagnostic tasks, advances in speech recognition and user interfaces promise to replace workers in some interactive tasks. As we all know, Apple's Siri, Google Assistant, and Amazon's Alexa rely on natural user interfaces to recognize spoken words, interpret their meanings, and respond to them accordingly. Using speech recognition technology and natural language processing, a company called Clinc is now developing a new AI voice assistant to be used in drive-through windows of fast-food restaurants like McDonald's and Taco Bell.[18] And in 2018, Google announced that it is building AI technology to replace workers in call centers. Virtual agents will answer the phone when a customer calls. If a customer request involves something the algorithm cannot yet do, he or she will automatically be rerouted to a human agent. Another algorithm then analyzes these conversations to identify patterns in the data, which in turn helps improve the capabilities of the virtual agent.[19] As the technology evolves, its effects on the labor market could be significant. Despite decades of companies' moving jobs offshore, roughly 2.2 million Americans still work in 6,800 call centers across the country, and several hundred thousand do similar jobs in smaller sites.[20]

One of the greatest leaps forward has taken place in autonomous driving. In 2004, the Defense Advanced Research Projects Agency (DARPA)—set up by President Dwight Eisenhower in 1958, in response to the Soviet Union's launch of the first artificial earth satellite, Sputnik 1—held its first "grand challenge" for driverless cars. The goal was to drive 142.0 miles through the Mojave Desert within ten hours without any human assistance. The farthest any of the vehicles got was 7.1 miles, and several cars did not even get off the starting line. The $1 million prize went unclaimed. Yet in 2016, the world's first self-driving taxis were picking up passengers in Singapore.

Recent progress in autonomous driving is thanks to big data and clever algorithms. It is now possible to store representations of a complete road network in a car, which simplifies the navigation problem. The changing of seasons, which brings challenges like snow, was long a key bottleneck to algorithmic navigation. But by storing records from the last time that snow fell, AI can now handle this problem.[21] AI researchers have shown that algorithmic drivers now are able to identify major changes in the environment in which they operate, such as roadwork.[22] In a major study, my Oxford engineering colleagues Bonolo Mathibela, Paul Newman, and Ingmar Posner concluded: "A vehicle can therefore prepare for the possibility of encountering humans on the road, or areas where [the vehicle] may not be stationary—thus gaining a dynamic sense of situational awareness, like a human."[23]

While it is still early days, autonomous vehicles are being deployed in a number of settings. Some agricultural vehicles, forklifts, and cargo-handling vehicles are already autonomous; and in recent years hospitals have begun to use autonomous robots to transport food, prescriptions, and samples.[24] In 2017, Rio Tinto, an Anglo-Australian metals and mining giant, announced that it will expand its fleet of autonomous hauling trucks in its Pilbara mine by 50 percent by 2019, making operations fully autonomous.[25] But so far, the adoption of autonomous vehicles has mostly been limited to relatively structured environments like warehouses, hospitals, factories, and mines. When computer programs can better anticipate the range of objects and scenarios a vehicle may encounter, automation is relatively straightforward. Using explicit if-then-do rules, the programs can just tell the vehicle to stop or slow down if another object approaches it. But in unstructured environments, like the streets of major cities, there are so many possible scenarios that this approach would require an almost infinite number of such rules.

AI combined with cheap and powerful digital sensors has recently raised the prospects of having fully autonomous vehicles also in unstructured environments. By equipping vehicles with a host of sensors, car companies have now collected millions of miles of human driving data for algorithms to learn from. As Ajay Agrawal, Joshua Gans, and Avi Goldfarb write, "By linking the incoming environmental data from sensors on the outside of the car to the driving decisions made by the human inside the car (steering, braking, accelerating), the AI learned to predict how humans would react to each second of incoming data about their environment."[26] Still, one obvious limiting factor for all AI models is that they struggle to predict outcomes when new situations arise that are not included in their training data. And in city traffic, vehicles constantly encounter new situations. One way forward has been to reduce the complexity of the environment. In Frisco, Texas, the company Drive.ai deploys autonomous minivans to transport people, but they are used only within specific office and retail areas. Instead of trying to mimic a human driver, engineers try to simplify things. All pickups and drop-offs take place at designated stops: "Riders hail the vans using an app and go to the nearest stop; a vehicle then appears to pick them up."[27]

We all know that the path to autonomous driving has been one of impressive progress but also one of setbacks. In 2018, one of Uber's self-driving vehicles tragically killed a woman who was crossing a street on her bicycle in Tempe, Arizona, sparking concerns over safety and, more fundamentally, over the future of autonomous driving. Yet similar and equally tragic setbacks were just as prevalent with earlier transportation technologies. As noted in chapter 4, the first public railroad demonstration in 1830 ended with a member of Parliament being fatally injured because the brakes on the train were slow to respond. The incident was reported in nearly every British media outlet, but that did not hinder the adoption of railroad technology. And in 1931—just before tractor adoption accelerated—the *New York Times* reported that, in Somerville, New Jersey, a tractor had crushed a four-year old boy to death, and one tractor was reported to have exploded, killing several people.[28] It is also worth recalling that as engineers are pushing autonomous driving forward, accidents involving human drivers are happening every minute. A survey prepared for the National Highway Transportation Safety Administration of car crashes found that human error was responsible for 92.6 percent of them.[29] And the number of casualties are many: just in 2013, 1.25 million people died in car accidents globally and 32,000 in the United States alone.[30] Thus, autonomous cars do not need to be perfect to be justifiable. Human drivers are certainly not.

There are still situations that autonomous vehicles struggle to handle, especially in crowded cities where pedestrians and cyclists provide additional complicating elements. In Singapore, autonomous taxis have a safety driver in them who takes over in emergencies, to minimize the possibility of accidents. But while self-driving cars are still at an experimental stage, successful trips in city traffic have already been accomplished. In Tokyo a self-driving taxi—also with a safety driver—has already driven paying passengers, "raising the prospect that the service will be ready in time to ferry athletes and tourists between sports venues and the city centre during the 2020 Summer Olympics."[31] These events are important, because the underlying AI systems require the collection of millions of miles of real-world data from vehicles' sensors. And the quantity of data is not all that matters. Driving on the interstate highway or through some quiet Midwestern town is hardly the same as driving in Manhattan. This is just as true for algorithms as it is for human drivers.

Allowing algorithms to practice in city traffic is therefore an important step toward the age of driverless transportation.

Progress is likely to be more rapid outside of cities, however, where there are fewer complicating elements. In May 2015, Daimler-Benz put the first autonomous big rig on the road. Approved by the state of Nevada, the autonomous system will take hauls only on highways, to keep things simple for now. And in Colorado in October 2016, an autonomous semitrailer successfully drove fifty thousand cans of Budweiser beer from Fort Collins to Colorado Springs. The truck drove itself 100 miles on the interstate, but when it reached the city limits, a human driver took over.

These achievements produce mixed responses. There are 1.9 million Americans working as heavy and tractor-trailer truck drivers today. Worries that autonomous trucks will cause a "tsunami of displacement" are widespread, though this is unlikely to happen in the next few years.[32] In light of these concerns, it is also important to remember that the barriers to technology adoption are not just technological. As we have seen in the preceding chapters, replacing technologies are likely to be resisted if workers face poor alternative options—an issue to which we shall return.

All human performers of transportation and delivery tasks are not at immediate risk from the rise of autonomous vehicles, of course. As AI skeptics like Robert Gordon have pointed out, even if "the car drives up in front of my house, how does the package get from the Amazon car to my front porch? Who carries it up when I'm away from home?"[33] At the same time, we have been able to overcome seemingly more complicated engineering problems in the past through clever task redesign. As Hans Moravec has noted, it is hard for computers to do many tasks that are easy for humans, and vice versa. But while this remains true, engineers have also been able to take steps toward resolving Moravec's paradox (see chapter 9) by making simple tasks even simpler.

Indeed, a common misconception is that for a task to be automated, a machine must replicate the exact procedures of the worker it is intended to replace. Simplification is mostly how automation happens. Even state-of-the-art robotics would not be able to replicate the motions and procedures carried out by medieval craftsmen. Production became automatable only because previously unstructured tasks were subdivided and simplified in the factory setting. The factory assembly line turned the nonroutine tasks of the artisan shop into repetitive tasks that were automatable once robots arrived. In similar fashion, we did not automate the jobs of laundresses by inventing multipurpose robots capable of chopping down trees; carrying water, wood, or coal from the outside to the stove; and performing the motions that washing clothes by hand entails. And we did not automate the jobs of lamplighters by inventing robots capable of climbing lampposts.

A contemporary example of task simplification is prefabrication:[34] "On-site construction tasks typically demand a high degree of adaptability, so as to accommodate work environments that are typically irregularly laid out, and vary according to weather. Prefabrication, in which the construction object is partially assembled in a factory before being transported to the construction site, provides a way of largely removing the requirement for adaptability. It allows many construction tasks to be performed by robots under controlled conditions that eliminate task variability—a method that is becoming increasingly widespread, particularly in Japan."[35] Not just in construction but also in retailing, clever task redesign has yielded promising results. For example, Kiva Systems, acquired by Amazon, solved the problem of warehouse navigation simply by placing bar-code stickers on the floor that inform robots of their precise location. With clever task redesign, engineers are already breaking the rules about what robots can do.

In the late 1990s, computers lent steam to retailing operations. But productivity growth could not be sustained, as companies soon ran into bottlenecks. Goods still needed to be moved from the factory to the warehouse, then to the retail store, and finally to the ultimate buyer. Freight trucking was an "inherently unproductive activity, as delivery drivers navigate congested and potholed streets, search for parking spaces, ring doorbells, and wait for an answer."[36] To work around this, Amazon is now experimenting with using drones (which can bypass congested streets) for delivery. To return to Gordon's question of "How does the package get from the Amazon car to my front porch?," it looks increasingly likely that many packages will not arrive by car. In London, for example, a company called Skyports is already acquiring rooftop spaces that it plans to convert into vertiports, where drones can take off and land. And in March 2018, Amazon was granted a patent for a delivery drone that responds to human gestures. The technology should help address the issue of how "flying robots might interact with human bystanders and customers waiting on their doorsteps. Depending on a person's gestures—a welcoming thumbs up, shouting or frantic arm waving—the drone can adjust its behavior, according to the patent. The machine could release the package it's carrying, alter its flight path to avoid crashing, [and] ask humans a question or abort the delivery, the patent says."[37]

Aided by AI, engineers have also come up with clever ways of reducing labor requirements within stores, without offloading the tasks done by cashiers onto consumers through complicated self-service checkout procedures. One example is Amazon Go, an archetypical example of a replacing technology. Today, some 3.5 million Americans work as cashiers across the country. But if you go to an Amazon Go store, you will not see a single cashier or even a self-service checkout stand. Customers walk in, scan their phones, and walk out with what they need. To achieve this, Amazon is leveraging recent advances in computer vision, deep learning, and sensors that track customers, the items they reach for, and take with them. Amazon then bills the credit card passed through the turnstile when the customer leaves the store and sends the receipt to the Go app. While the rollout of the first Seattle, Washington, prototype store was delayed because of issues with tracking multiple users and objects, Amazon now runs three Go stores in Seattle and another in Chicago, Illinois, and plans to launch another three thousand by 2021. Globally, companies like Tencent, Alibaba, and JD.com are also investing in AI to achieve the same goal.

Chinese companies like JD.com have also started to invest more heavily in unmanned warehouses. Inside JD.com's Shanghai warehouse, machines are guided by image scanners. They handle all the goods, most of which are consumer electronic products: "Packages travel along a highway of belts. Mechanical arms stationed throughout the network place the items on the right tracks, wrap them in plastic or cardboard and set them onto motorized pucks [that] carry the parcels across a floor that resembles a giant checkerboard and plunk them down chutes to sacks. Computerized shelves on wheels retrieve the loads and transport them to trucks, which deliver most orders within 24 hours of a shopper's click."[38] While JD.com employs some one hundred sixty thousand workers throughout Asia today, it has made its intent clear to trim that number to fewer than eight thousand over the next decade. And those jobs, it expects, will require a very different set of skills.[39]

The main reason why warehouses still employ large swaths of the population is that order picking remains a largely manual process. Humans still hold the comparative advantage in complex perception and manipulation tasks. But here, too, AI has made many recent breakthroughs possible. At the OpenAI lab in San Francisco, California, set up by Elon Musk, a robotic five-fingered hand called Dactyl bears witness to impressive progress in recent years: "If you give Dactyl an alphabet block and ask it to show you particular letters—let's say the red O, the orange P and the blue I—it will show them to you and spin, twist and flip the toy in nimble ways."[40] Though this is an easy task for any human, the achievement lies in the fact that AI allows Dactyl to learn new tasks, largely on its own through trial and error.

For robots to become effective manipulators, however, they must also learn to identify and distinguish between various items. In this domain, the state of the art only a few years ago is probably best exemplified by the Gripper—a machine equipped with a two-fingered gripper, which is much easier to control than a five-fingered hand. The Gripper is able to identify, manipulate, and sort familiar objects like a screwdriver or a ketchup bottle. But when faced with an object it has not seen before, all bets are off.[41] This might not be a problem in a warehouse with a limited set of items, but in warehouses that store thousands of objects and receive a steady flow of new items, robots are needed that can pick up just about anything. Researchers at Autolab, a robotics lab inside the University of California, Berkeley, are now building such systems using AI:

> The Berkeley researchers modeled the physics of more than 10,000 objects, identifying the best way to pick up each one. Then, using an algorithm called a neural network, the system analyzed all this data, learning to recognize the best way to pick up any item. In the past, researchers had to program a robot to perform each task. Now it can learn these tasks on its own. When confronted with, say, a plastic Yoda toy, the system recognizes it should use the gripper to pick the toy up. But when it faces the ketchup bottle, it opts for the suction cup. The picker can do this with a bin full of random stuff. It is not perfect, but because the system can learn on its own, it is improving at a far faster rate than machines of the past.[42]

Thus, while robots are still far from having human-level capabilities when it comes to perception and manipulation tasks, they are becoming sufficiently sophisticated to handle gripping tasks in a structured warehouse setting, like picking items and parcels off a pallet and placing them into cartons or boxes. Just as robots entered the factories, they are gradually making an appearance outside manufacturing. Warehouse automation today is probably where factory automation was in the 1980s.

It is true that many of the AI technologies discussed above are still imperfect prototypes. But it is important to remember that just about every technology was imperfect in its early days. To most observers, for example, the first telephones seemed ridiculous. Getting used to hearing a disembodied voice through an earpiece was an experience entirely different from any previous form of communication. An early article in *Scientific American* argued that it was a silly invention, for which people would find little use: "The dignity of talking consists of having a listener, and it seems absurd to be addressing a piece of iron."[43] Retrospectively, this might seem like a silly thing to think. But early telephony was made with a single-wire system that suffered great losses in clarity: "In 1878 the recently invented telephone was hardly more than a scientific toy.

In order to use it a person was required to briskly turn a crank and to scream into a crude mouthpiece. One could faintly hear the return message only if the satanic screechings and groanings of static permitted."[44] But just a decade later the technology looked much more promising. In 1890, a reporter for *Time* was invited by the American Telephone and Telegraph Company (AT&T) to inspect the status of long-distance telephony. General Superintendent A. S. Hibbard made a test call to showcase the technology: "Boston, 300 miles away was rung up, and then ensued a pleasant conversation. The operator at the other end was a young woman, who at once struck up a spirited discussion of the latest development in Theosophic Buddhism. Her voice was not so much raised as in ordinary talking, and its expressiveness was perfect."[45]

The Next Wave

More and more jobs lend themselves to automation. But anecdotes alone cannot tell us much about the extent to which jobs will be replaced in the future, or the types of job that will be affected. So in a 2013 paper titled "The Future of Employment: How Susceptible Are Jobs to Computerisation?," my Oxford University colleague Michael Osborne and I set out to identify the near-term engineering bottlenecks to automation as a way of estimating the exposure of current jobs to recent advances in AI. As noted above, until recently, computers had a comparative advantage in tasks involving routine rule-based activities, while humans did better at everything else.

Routine jobs began to disappear in large numbers in the 1980s, but some economists made accurate predictions about the domains in which humans would be replaced much earlier simply by observing what computers do. One Bureau of Labor Statistics case study, conducted in 1960, found that "a little over 80 percent of the employees affected by the change were in routine jobs involving posting, checking, and maintaining records; filing; computing; or tabulating, keypunch, and related machine operations. The rest were mainly in administrative, supervisory, and accounting work."[46] But if there was a Nobel Prize for predicting the future of work, it should have gone to Herbert Simon for his essay titled "The

Corporation: Will It Be Managed by Machines?," first published in 1960.[47] (Of course, Simon did win one in economics for his work on the decision-making process within economic organizations.) While Simon did not lay out an explicit framework, he got things spectacularly right by looking at trends in technology. He was right to think that computers would take over many routine factory and office jobs. He correctly predicted that there would still be many jobs responsible for the design of products, processes, and general management. And he saw that a growing share of the population would be employed in personal service jobs. In other words, he basically predicted the hollowing out of middle-class jobs decades before it happened.

The question now is: What can and can't computers do in the age of AI?

Identifying engineering bottlenecks to automation is obviously not an economics question, so I was fortunate that Michael had been researching this subject for some time. The trouble economists face in researching technological change is that we are bound to be behind the curve (Simon was not just an economist but also a highly regarded computer scientist). I was hardly up to date on everything going on in the labs. But as I was writing economics papers, Michael had been developing the algorithms that have expanded the set of tasks computers are now able to perform.

In the spirit of Simon, we set out to determine in which domains humans still retain the comparative advantage. Rather than asking unanswerable questions about the prospects of superintelligence or trying to predict the great inventions of the future, we looked at technologies on the horizon. In the words of Thomas Malthus, writing at the onset of the Industrial Revolution, "many discoveries have already taken place in the world that were totally unforeseen and unexpected. . . . But if a person had predicted these discoveries without being guided by any analogies or indications from past facts, he would deserve the name of seer or prophet, but not of philosopher."[48] Many of the technologies discussed here are still prototypes, but their arrival in the marketplace is not unforeseeable, and while they are still imperfect, every technological revolution began with imperfect technology. The early steam engines were used only to drain mines, and they did not even do that particularly well. Yet Thomas Savory, Thomas Newcomen, and James Watt, all realized that the steam engine was a GPT, and they conceived many applications for it. As noted above, AI is another GPT, and it is already being used to perform both mental and manual tasks.

Because its potential applications are so vast, Michael and I began by looking at tasks that computers still perform poorly and where technological leaps have been limited in recent years. For a glimpse of the state of the art in machine social intelligence, for example, consider the Turing test, which captures the ability of an AI algorithm to communicate in a way that is indistinguishable from an actual human. The Loebner Prize is an annual Turing test competition that awards prizes to chat bots that are considered to be the most humanlike. These competitions are straightforward. A human judge holds computer-based textual interactions with both an algorithm and a human at the same time. Based on these conversations, the judge must then try to distinguish between the two. In a paper written in 2013, Michael and I noted: "Sophisticated algorithms have so far failed to convince judges about their human resemblance."[49] Yet a year later the computer program Eugene Goostman managed to convince 33 percent of the judges that it was a person. Some people subsequently argued that we had underestimated the accelerating pace of change. However, such claims exaggerate the capabilities of Eugene Goostman, which simulated a thirteen-year-old boy speaking English as his second language. Even if we assume that algorithms at some point will be able to effectively reproduce human social intelligence in basic texts, many jobs center on personal relationships and complex interpersonal communication. Computer programmers consult with managers or clients to clarify intent, identify problems, and suggest changes. Nurses work with patients, families, or communities to design and implement programs to improve overall health. Fund-raisers identify potential donors and build relationships with them. Family therapists counsel clients on unsatisfactory relationships. Astronomers build research collaborations and present their findings on conferences. These tasks are all way beyond the competence of computers.

Many jobs also require creativity, like the ability to come up with new, unusual, and clever ideas. Survey data show that the work of physicists, art directors, comedians, CEOs, video game designers, and robotics engineers, to name a few, all involve such activities.[50] The challenge here, from an automation point of view, is not one of generating novelty but generating novelty that makes sense. For a computer to produce an original piece of music, write a novel, develop a new theory or product, or make a subtle joke, in principle the only things that are required is a database with a richness of experience that is comparable to that of humans and solid methods that allow us to benchmark the algorithm's subtlety. It is also entirely possible to give an algorithm access to a database of symphonies, label some as bad and others as good, and allow it to generate an original recombination. Algorithms already exist that generate music in many different styles, reminiscent of specific human composers. But people do not just generate ideas on the basis of related existing works. They draw upon experiences from all aspects of life.

As discussed above, many challenges remain when algorithms have to interact with a variety of irregular objects in uncontrolled environments. Some perception tasks, like identifying objects and their properties in a cluttered field of view, have proven hard to overcome. Robots remain unable to match the depth and breadth of human perception, which translates into further difficulties in manipulation. Distinguishing a pot that is dirty and needs to be cleaned from a pot holding a plant is straightforward to any human. Yet robots still struggle to mimic human capabilities in such tasks, making many jobs—like those of janitors and cleaners—exceedingly hard to automate. Though single-purpose robots exist, capable of doing individual tasks like cleaning floors, there is no multipurpose robot that could find and remove rubbish. In controlled environments like factories and warehouses, it is possible to circumvent some engineering bottlenecks through clever task redesign. But a home is a different matter. In addition to the hard perception tasks, like identifying rubbish, there are "yet further problems in designing manipulators that, like human limbs, are soft, have compliant dynamics and provide useful tactile feedback."[51] While there has been recent progress in very simple tasks like spinning an alphabet block, picking up familiar objects with a gripper, and even teaching robots to recognize the best way of picking up any item using AI, the ability of advanced robotics to manipulate various objects is still very limited. Most industrial manipulation makes use of work-arounds to address these challenges.

With these engineering bottlenecks in mind, Michael and I proceeded to explore the automatability of jobs based on twenty thousand unique task descriptions.[52] This sort of detailed information comes with one problem: it is a lot of data to process. So instead of examining each individual task, we took a sample of seventy occupations that a group of AI experts deemed either automatable or nonautomatable on the basis of the tasks the occupations entail. This gave us what machine learning researchers call a training data set. While the task descriptions for each occupation were unique, our database also provided a set of common features. Based on these features, our algorithm was able to learn about the characteristics of automatable occupations, allowing it to predict the exposure to automation of another 632 occupations. Thus, our final sample covered 702 occupations, in which 97 percent of the American workforce is employed.

Using AI for our analysis did not just have the benefit of saving time and labor. Our analysis also underlined the fact that algorithms are now infinitely superior to humans in pattern recognition. We were quite convinced that the work of waiters and waitresses would not lend itself to automation, but our algorithm told us that we were wrong. Analyzing the similarities between the jobs of waiters and other jobs in a more comprehensive manner than we possibly could have done, it predicted that waiters are susceptible to automation. Indeed, in the months after our original analysis, we learned that McDonald's had plans to install self-ordering kiosks. We learned about Chili's Grill and Bar's plans to complete the rollout of its tablet ordering system. We learned that Applebee would introduce tablets to eighteen hundred restaurants. And in 2016, a new and almost fully automated restaurant chain called Eatsa opened. Customers order their food at an iPad kiosk. They then wait a few minutes in front of a giant vending machine that churns out freshly prepared quinoa bowls. At the other side of the machine, kitchen staff members cook the food, but Eatsa does not employ any waiters.

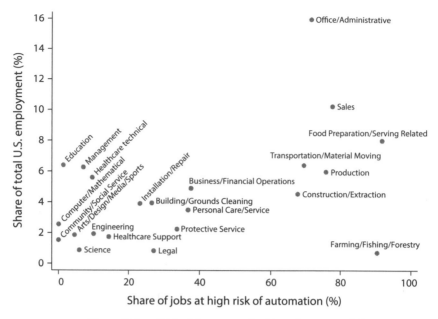

FIGURE 17: Share of Jobs at Risk of Automation by Major Occupational Categories
Source: C. B. Frey and M. A. Osborne, 2017, "The Future of Employment: How Susceptible Are Jobs to Computerisation?," Technological Forecasting and Social Change 114 (January): 254–80.

Of course, this does not mean that all waiting jobs will be replaced. In many settings, consumers are likely to prefer human waiters as part of the service experience. What it does tell us is that their jobs are automatable in principle. We shall return to the question of the determinants of technology adoption shortly.

Figure 17 plots the exposure of major occupational categories to automation by their employment share. Office and administrative support, production, transportation and logistics, food preparation, and retail jobs loom large in terms of both their exposure to automation and the percentage of Americans they support. Overall, our algorithm predicted that 47 percent of American jobs are susceptible to automation, meaning that they are potentially automatable from a technological point of view, given the latest computer-controlled equipment and sufficient relevant data for the algorithm to draw upon. What most of these jobs have in common is that they are low-income jobs that do not require high levels of education (figure 18).

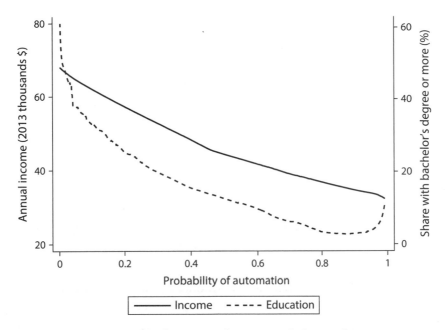

FIGURE 18: Jobs at Risk of Automation by Income and Educational Attainment
Source: C. B. Frey and M. A. Osborne, 2017, "The Future of Employment:
How Susceptible Are Jobs to Computerisation?," Technological Forecasting and
Social Change 114 (January): 254–80.
Note: The figure plots the probability that an occupation is automatable against the
median annual income and educational attainment in that occupation. It shows that
occupations in which people earn higher wages and have higher educational attainment
are less exposed to automation on average.

A number of studies have emerged since we first published our article, reaching somewhat different conclusions. Research by the OECD, for example, estimates that 14 percent of jobs are at risk of being replaced, with another 32 percent being at risk of significant change.[53] The OECD mistakenly argues that we overestimated the scope of automation by focusing on occupations rather than tasks. What they miss is that we inferred the automatability of occupations on the basis of the tasks they entail. According to our estimates, medical doctors are not at risk of automation even if algorithms are becoming more pervasive in tasks like medical diagnostics. And journalists are not exposed to automation just because AI algorithms are now capable of churning out shorter news stories. Both journalists and medical doctors are safe from automation, according to our estimates, even though they entail some tasks that lend themselves to automation. So why, then, do the OECD's estimates diverge so much from ours? One explanation is that they use less-detailed occupational data. Another is that their model performs less well against our training data set.[54]

However, for all their differences, these studies concur that unskilled jobs are most exposed to automation.[55] When President Barack Obama's Council of Economic Advisers used our estimates to sort by wage levels the occupations most at risk of being automated, they found that 83 percent of workers in occupations that paid less than $20 an hour were at high risk of being replaced, while the corresponding figure for workers in occupations that paid more than $40 per hour was only 4 percent.[56] What this shows is that the labor market prospects of the unskilled will likely continue to deteriorate, unless other forces counteract that trend. We saw in chapter 9 that the first wave of automation of routine work pushed many Americans out of decent middle-class jobs and into low-paying service jobs. Many of these low-skilled jobs are now threatened by automation, too. If anything, the next wave can be expected to put more downward pressure on the wages of the middle class, many members of which are already competing for low-income jobs. In the words of Harvard University's Jason Furman, who chaired Obama's Council of Economic Advisers, "We are already seeing some of this play out—for example, when we go shopping and take our groceries to a kiosk instead of a cashier, or when we call a customer service help line and interact with an automated customer service representative."[57]

Thus, a widespread misconception is that automation is coming for the jobs of the skilled. In the best-selling *Rise of the Robots*, Martin Ford declared that "employment for many skilled professionals—including lawyers, journalists, scientists, and pharmacists—is already being significantly eroded by advancing information technology [so that] acquiring more education and skills will not necessarily offer effective protection against job automation in the future."[58] Though many of the jobs Ford highlights surely involve some tasks that can be automated away, they also involve many more tasks that cannot. For example, when

Dana Remus and Frank Levy recently analyzed lawyers' billing records, they found that if AI and several related applications were all adopted immediately—which seems highly unlikely—this would substitute for roughly 13 percent of their time. Most of their time is spent performing tasks like legal writing, investigating facts, negotiating, appearing in court, and advising clients. As Remus and Levy explain, lawyers' work requires more than prediction: "It requires a lawyer to understand a client's situation, goals, and interests; to think creatively about how best to serve those interests pursuant to law; and sometimes, to push back against a client's proposed course of action and counsel compliance. These are things that frequently require human interaction and emotional intelligence and cannot, at least for the time being, be automated."[59] Reassuringly, our algorithm also predicted that lawyers are at low risk of automation.

Amara's Law

Though the scope of automation is significant, its pace is a different matter. Like Simon's forecast, our predictions were based on merely observing what computers can do. And like Simon, we did not try to predict the pace of change, which depends on many unpredictable factors beyond the technology itself.[60] But we surely didn't expect 47 percent of jobs to be automated anytime soon. We highlighted a wide range of factors that are likely to shape the pace of automation. The bottom line is that all the prototype technologies discussed here will not arrive at the same time. And their adoption will not be frictionless, either. Regulation, consumer preferences, and worker opposition, among many other variables, will shape the speed of adoption. Thus, inflated expectations have typically been followed by disillusionment. As Roy Amara famously observed, "We tend to overestimate the effect of a technology in the short run and underestimate the effect in the long run." Indeed, Amara's Law has been a good guide to the trajectories of technological progress in the past.

Seen through the lens of history, the scope of automation this time around is probably not as staggering as has sometimes been suggested.

In 1870, around 46 percent of the American workforce was still employed in agriculture, while today the agriculture sector absorbs about 1 percent of the labor force (table 1).[61] Tractors played a key role in reducing labor requirements on the farms (see chapters 6 and 8). But while one might have inferred that many farm jobs were at risk of replacement when the gasoline-powered tractor arrived, the speed of adoption would have been much harder to predict.

The hurdles to tractor adoption were many. First, increasingly complex machinery required more skilled operators. Early on, farmers typically waited to purchase tractors, wanting to see how long it took for laborers on other farms to acquire the mechanical skill required. As an article in the *New York Times* reported in 1918, "A tractor is a too good machine to put in the hands of a poor operator. . . . Where to get first-class tractor operators is often more of a puzzle to the buyer than how to get the machine."[62] In the same year, the New York State College of Agriculture announced a three-week course for tractor and truck operators, to bridge the skills gap and accelerate adoption. Second, the adoption of tractors—like other GPTs—moved at different speeds across applications: "The earliest models were suitable only for tillage and harvesting small grains, and only in the late 1920s did the technology begin to generalize for use with row crops such as corn, cotton, and vegetables."[63] Some use cases emerged only in the later stages of mechanical development.

Third, even as tractors became more pervasive, the abundance of cheap labor in the countryside meant that the mechanization of farming did not make economic sense for a long time. As the Second Industrial Revolution produced an ever-growing number of well-paying industrial jobs, however, many Americans left the farms for the cities, which increased incentives to mechanize. But even so, tractors were still not economically viable in many settings. They were mainly used on large farms that relied on wage labor. Many low-income farmers were highly risk averse and preferred to continue to rely on horses rather than investing in expensive tractors—even though this meant that they had to set aside acreage to grow feed. If a tractor could not be bought outright, loan payments were still significant enough to deter adoption. In 1921, the *New*

York Times pointed out that there were still seventeen million horses on American farms but only 246,139 tractors and expressed concern that adoption was lagging. A push, the article argued, was required to increase productivity in agriculture.[64] And a push followed a decade later. One reason that adoption finally accelerated in the 1930s, despite a decade of the Great Depression, were New Deal programs like the Commodity Credit Corporation and the Farm Credit Administration, which reduced the price risk, lowered interest rates, and made cash available to farmers.[65]

Amara's Law applied to the computer revolution, too. Despite widespread automation anxiety in the 1950s and 1960s (chapter 7), computers were too bulky and expensive to find widespread use before the 1980s (see chapter 9). While many businesses marveled at the capabilities of computers, few opened their wallets. As risk-adverse farmers were reluctant to adopt expensive tractors, businesses deemed the cost of computers too high to bear. And they were right in thinking so. When computerization finally took off, unforeseen glitches emerged. In 1987, when Robert Solow puzzled that "we can see the computer age everywhere but in the productivity statistics," an article in the *Wall Street Journal* reported: "Companies are automating in smaller doses now, a strategy that allows bugs to be worked out before huge investments are made."[66] As the director of engineering at AT&T explained, "If you make 30 million boxes of Wheaties a year, you can use automation without many problems, but if you're in a competitive market where the product is changing and its life cycle is short, you better be damned careful."[67]

The performance of the technology is not all that matters. Realizing the productivity gains of computers required complementary organizational, process, and strategic changes. In the early days of automation, the training and retraining of employees often took longer than expected, and many companies did not fully appreciate the obstacles involved in getting machines, computers, and sophisticated software to work together effectively. In a number of studies, the economists Erik Brynjolfsson, Timothy Bresnahan, and Lorin Hitt consistently found that investments in computer technology contributed to firm productivity mainly when complementary organizational changes were made.[68] In the 1980s, the computer revolution centered on productivity improvements in individual tasks, such as word processing and manufacturing operations control. Yet preexisting business processes remained intact for the most part. In 1990, Michael Hammer, a management scholar and former professor of computer science, published his famous essay "Re-engineering Work: Don't Automate, Obliterate." In the article, which appeared in the *Harvard Business Review*, Hammer argued that productivity gains would not come from using automation to make existing processes more efficient.[69] Managers trying to do so had gotten it wrong from the outset. Unleashing the full potential of automation, he declared, required analyzing and redesigning workflows and business processes to improve customer service and cut operational costs. By the mid-1990s, the majority of Fortune 500 companies claimed to have re-engineering plans.[70] It was also around then that computers began to have an impact on productivity.

Just like the switch from group drive to unit drive in the age of mass production, computerization and reorganization were gradual processes that required rethinking how the firm worked. Thus, the productivity puzzle of the late 1980s was not a puzzle to everyone. Economic historians realized that they had heard this story before. Studying the evolution of factory electrification, Oxford University's Paul David noted that it took roughly four decades for electricity to appear in the productivity statistics, after the construction of Thomas Edison's first power station in 1882. As discussed in chapter 6, harnessing the mysterious force of electricity required a complete reorganization of the factory, and the switch to unit drive as the organizing principle took plenty of experimentation—so the productivity gains of electrification did not show up until the 1920s.[71] David went on to predict a similar trajectory for computer-led productivity growth. And he was right on target: the similarities between the 1920s and 1990s are tantalizing. Both decades saw productivity blossom and an explosion in the application of GPTs (electricity in the 1920s and computers in the 1990s).[72] The former, economists agree, was the consequence of the latter. About 70 percent of the productivity acceleration in the years 1996–99, relative to that in the period 1991–95, has been attributed to computer technologies.[73] And the productivity rebound was not narrowly focused on a few sectors but was extremely broad based, with wholesale trade, retail, and services showing sizable gains—which pointed to GPTs at work.[74]

AI has only recently expanded the realm of what computers can do. Thus, there are good reasons to believe that the greatest productivity gains from automation are still to come. Multipurpose robots, as noted above, are already being adopted, but though their contributions to productivity growth have been significant, their use is still largely confined to heavy industry.[75] And AI, more broadly, is still in its infancy. A 2017 survey of three thousand executives by the McKinsey Global Institute found that AI adoption outside of the tech sector is still at an early stage. Few firms have deployed it at scale, declaring that they are uncertain of the business case or return on investment. And a review of more than 160 use cases further showed that AI had been deployed commercially in only 12 percent of the cases.[76]

As is well known, productivity growth has slowed since 2005, but that can happen when technologies are at an experimental stage.[77] Technology improves productivity only after long delays, and it primarily incurs costs in the early stages of development. And after a new discovery is made, it often takes years until prototypes become economically viable in production. Thus, the contribution of new technologies to aggregate economic variables has always been delayed: "The case of self-driving cars discussed earlier provides a more prospective example of how productivity might lag technology. Consider what happens to the current pools of vehicle production and vehicle operation workers when autonomous vehicles are introduced. Employment on production side will initially increase to handle R&D, AI development, and new vehicle engineering."[78] The Brookings Institution, for example, calculates that investments in autonomous driving amounted to roughly $80 billion in the period 2014–17, with only a few first case uses of adoption.[79] Over those three years, this is estimated to have lowered labor productivity by 0.1 percent per year.[80] In this light, it is not all that surprising that economists have found current productivity growth to be a bad predictor of future productivity growth.[81]

It is true that the smartphone and the internet have spread much faster than the electric motor or the tractor once did. Yet it makes little sense to compare the spread of consumer goods and services to technologies being used in production. The latter requires the reconfiguration of production processes while the former does not. What's more, firms faced with the decision to automate or not have to weigh not just the engineering bottlenecks to be overcome. Beyond the technology, they must also consider increased overheads, the availability of sufficiently large markets, the cost of scrapping existing machines, the cost of financing new ones, and (as Harry Jerome pointed out) "the possible opposition of [their] workers, and sometimes adverse public opinion and even restrictive legislation."[82] While one might think that in the age of AI, much less capital expenditure is required for automation to happen, significant complementary investments are required to deploy a machine learning system. As Google's chief economist Hal Varian explains:

> The first requirement is to have a data infrastructure that collects and organizes the data of interest—a data pipeline. For example, a retailer would need a system that can collect data at point of sale, and then upload it to a computer that can then organize the data into a database. This data would then be combined with other data, such as inventory data, logistics data, and perhaps information about the customer. Constructing this data pipeline is often the most labor intensive and expensive part of building a data infrastructure, since different businesses often have idiosyncratic legacy systems that are difficult to interconnect.[83]

And while data may be the new oil, the bottlenecks are often not related just to data but also to skills and training:

> In my experience, the problem is not lack of resources, but is lack of skills. A company that has data but no one to analyze it is in a poor

position to take advantage of that data. If there is no existing exper-tise internally, it is hard to make intelligent choices about what skills are needed and how to find and hire people with those skills. Hiring good people has always been a critical issue for competitive advan-tage. But since the widespread availability of data is comparatively recent, this problem is particularly acute. Automobile companies can hire people who know how to build automobiles since that is part of their core competency. They may or may not have sufficient internal expertise to hire good data scientists, which is why we can expect to see heterogeneity in productivity as this new skill percolates through the labor markets.[84]

For these reasons, Amara's Law will likely to apply to AI, too. Myriad necessary ancillary inventions and adjustment are required for automa-tion to happen. Erik Brynjolfsson, who was among those investigating the role of computer technologies in the productivity boom of the late 1990s, thinks that the trajectory of AI adoption is likely to mirror the past in this regard. In a joint paper with Daniel Rock and Chad Syver-son, two economists, he argues that as happened with computers back in the 1990s, the adoption of AI will require not only improvements in the technology itself, but significant complementary investment and plenty of experimentation to exploit its full potential.[85] During this phase, history tells us, the economy goes through an adjustment process with slow productivity growth.

The Industrial Revolution in Britain was exceedingly similar. As Nicho-las Crafts has shown, James Watt's steam engine delivered its main boost to productivity some eight decades after it was invented.[86] When John Smeaton examined Watt's invention, patented in 1769, he declared that "neither the tools nor the workmen existed that could manufacture so complex a machine with sufficient precision."[87] Complementary skills had to be developed to perfect the technology. But ten years later, the combined genius of Matthew Boulton and Watt saw his engine a commercial success. Writing in 1815, Patrick Colquhoun, a Scottish merchant and statistician, declared: "It is impossible to contemplate the progress of manufactures in Great Britain within the last thirty years without wonder and astonishment. Its rapidity . . . exceeds all credibility. The improvement of the steam engines, but above all the facilities afforded to the great branches of the woollen and cotton manufactories by ingenious machinery, invigorated by capital and skill, are beyond all calculation."[88] Yet water power remained a cheaper source of energy for some time, so that the contribution of the steam engine to productivity growth remained absent.

Had Malthus been given the modern statistical apparatus in 1800, he would not have found much suggestive of the coming productivity boom. In the early stages of technological revolutions, current productivity growth does not tell us much about future productivity growth. We have to examine what is going on inside the labs instead. Malthus was dismissive of this view, and consequently he had no way of seeing what was coming. As he declared in his famous 1798 essay, "The moment we leave past experience as the foundation of our conjectures concerning the future, and, still more, if our conjectures absolutely contradict past experience, we are thrown upon a wide field of uncertainty, and any one supposition is then just as good as another. . . . Persons almost entirely unacquainted with the powers of a machine cannot be expected to guess at its effects."[89]

When Malthus wrote his essay, of course, the world barely knew of Schumpeterian growth. We now also know from past experience that what is going on in the labs is a better guide to the future of productivity at times of accelerating innovation. Great inventions may deliver enormous economic benefits, but often with long time lags. At the same time, we must acknowledge that this approach also has shortcomings. The mere existence of new technology does not tell us whether it will find widespread use. Even if Malthus had looked more to the wave of gadgets that made the Industrial Revolution and had realized the pervasiveness of the first machine age, how could he had known that they would be adopted so eagerly? For most of history, as noted, worker-replacing technologies have been fiercely resisted by angry workmen, leading governments to implement policies to restrict their use due to the fear of social upheaval (see chapter 3). As Malthus was writing, the British government had only recently begun to side with the innovators.

Looking forward, worker resistance and adverse public opinion could slow the pace of change, as it has in the past. And some economists have begun to point to the risk of opposition. As Harvard University's Rebecca Henderson warned at a recent National Bureau of Economic Research conference, "There is a real risk of a public backlash against AI that could dramatically reduce its diffusion rate. . . . Productivity seems likely to sky rocket, while with luck tens of thousands of people will no longer perish in car crashes every year. But 'driving' is one of the largest occupations there is. What will happen when millions of people begin to be laid off? . . . I'm worried about the transition problem at the societal level quite as much as I'm worried about it at the organizational level."[90] Those societal consequences are already being felt. The return of Engels's pause, as discussed above, has fueled the populist backlash, and attitudes toward automation itself are seemingly shifting (see chapter 11). The pervasiveness of AI and citizens' reactions to displacement will jointly determine future productivity growth. Any attempt to analyze the role of globalization in shaping the labor market going forward would be misleading if it overlooked the political economy of trade: the future impact of globalization on labor markets, for example, cannot be analyzed in isolation from the Trump administration's trade war with China. The same likely goes for automation. A worry, as automation progresses, is that resistance will grow. As we have seen, historically when machines threaten to take people's jobs and governments fear instability as a consequence, implementation can often be blocked for entirely political reasons.

If Amara's Law ceases to hold, it will likely be due to the return of Luddite sentiment.

Work and Leisure

Will there be enough jobs if automation is allowed to progress uninterrupted? In the public mind, there is a widespread dystopian belief that the rise of brilliant machines will ruin working people's lives by causing

wages to fall and unemployment to rise. By contrast, an equally common utopian belief is that technology will herald a new age of leisure,
where people will prefer to work less and play more. Neither of these
beliefs is new. And over the long run, both have so far been proven
wrong, or at least vastly exaggerated. Though there have clearly been
episodes when workers have suffered hardships as technology has advanced, fears over end-of-work scenarios have always been overblown,
as has the idea that we would all give up work and live a life of fulfillment
and leisure.

In the 1930 "Economic Possibilities for Our Grandchildren," John
Maynard Keynes famously declared that mechanization was progressing at a rate greater than at any other time in history. Our discovery of
ways to replace people with machines, he suggested, was outrunning
the pace at which new uses for labor could be found—which he held
would lead to widespread technological unemployment. Keynes's essay
was a reflection of the productivity boom of the 1920s, which did indeed
come with some adjustment problems that sparked a revival of the machinery question (see chapter 7). But Keynes was still optimistic about
the long run. Technology, he argued, would solve mankind's economic
problems and deprive us of our purpose of subsistence. Instead, our
main concern would become how to occupy our leisure. In a century,
Keynes predicted, people would enjoy a fifteen-hour workweek.[91]

Keynes was right that mechanization was progressing at a more rapid
pace than had been previously seen, yet things still unfolded quite differently. It is true that people in richer countries have shorter workweeks,
take more vacations, and spend more years in retirement as they live
longer. But the time citizens have decided to take as leisure, as they have
grown richer, has not increased by as much as is commonly believed,
and certainly not by as much as Keynes predicted. That is what the economists Valerie Ramey and Neville Francis found when they traced the
trajectories of work and leisure in America over the past century.[92] True,
in 1900, a typical workweek in manufacturing was around fifty-nine
hours. Yet in 1900, manufacturing still accounted for only about a fifth
of total employment, and industrial laborers worked much longer hours

than those in other sectors of the economy.[93] When government and farm workers are taken into account, Americans in 1900 worked around fifty-three hours per week on average. By 2005, this figure had fallen to roughly thirty-eight hours. However, looking merely at changes in hours per worker misses the fact that a larger share of the population today work than they did a century ago, as a growing percentage of women have entered the workforce (see chapter 6). When they accounted for the growing share of citizens at work, Ramey and Francis found a much less pronounced decline in working hours: average weekly hours worked per person fell by 4.7 hours between 1900 and 2005.[94]

All of this decline, in turn, occurred among the young and the elderly. Among those ages 25–54, in contrast, the average workweek actually got longer, even though weekly hours among men declined. The upsurge was driven entirely by working women. Among the young, the reason for the decline in working hours is straightforward: it followed from more children going to school and additional years spent in school, as farmers realized that their children would need an education to prosper in the age of the Second Industrial Revolution. And the fall in weekly hours among the elderly is no mystery, either. Before the Social Security Act of 1935, which provided a nationwide pension system, most people worked until they dropped; private pension plans were available only to a fraction of the population. As pension coverage gradually increased thereafter, citizens who reached retirement age could suddenly enjoy a life of leisure—which, if anything, served to create more jobs. The demands of a new class of leisured but active citizens caused a massive boom in the construction of retirement homes, golf courses, and shopping centers, and retirement cities like Sun City, Arizona, were built to accommodate the massive exodus from the Northeast to the Sun Belt.

Factoring in weekly hours of paid work, hours of schooling, household work, and so on, Ramey and Francis estimate average lifetime leisure over a century. This entails estimating people's average weekly hours of leisure for each year of life from age fourteen to expected death for different cohorts.[95] In doing so, they show that average weekly leisure increased from 39.3 to 43.1 hours per week between 1890

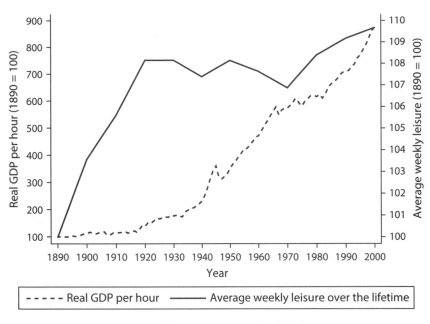

FIGURE 19: The Trajectories of Gross Domestic Product (GDP) per Hour and Average
Weekly Leisure in the U.S, 1890–2000

*Sources: V. A. Ramey and N. Francis, 2009, "A Century of Work and Leisure," American
Economic Journal: Macroeconomics 1 (2): 189–224. See figure 9 for data on GDP per hour.*

and 2000. Most of the increase was due to the welcome fact that people
live longer today. Their findings allow us to shed some light on Keynes's
predictions, whereby he suggested that productivity would increase by
four to eight times over the next century. Despite the unforeseeable
event of World War II, his productivity forecast was quite accurate:
labor productivity is now almost nine times higher than it was in 1900,
yet the time citizens decided to take as leisure had increased by a mere
10 percent by 2000 (figure 19). And after 1930, when Keynes was writing,
labor productivity experienced a fivefold increase while leisure grew
just by 3 percent.[96]

To be sure, Keynes did not overestimate the potential scope of mechanization. He was broadly right in thinking that "we may be able to perform all the operations of agriculture, mining, and manufacture with a
quarter of the human effort to which we have been accustomed."[97] The

economist Robert Heilbroner observed thirty-six years later, in the midst of the 1960s automation debates:

> One can maintain that the labor-displacing effects have run ahead of the job-creating effects within the two historically most important areas of work—the farm and the factory. . . . In mining, we see the same absolute shrinkage as in agriculture, in spite of a huge rise in output (as in agriculture). Eight hundred thousand men entered the earth or worked at the mine surface in 1900; only 600,000 in 1965. . . . Thus it seems beyond dispute that the labor-displacing effect of investment *can* outpace the job-creating effect, and in fact has done so in many key sectors of the economy.[98]

Heilbroner was, of course, well aware that while workers had been replaced from agriculture and mining, they had not become detached from the labor market altogether. On the contrary, the percentage of the population engaged in paid employment had risen as more and more women entered the workforce. As home production became increasingly mechanized, women could have decided to take full advantage of the new entertainment possibilities provided by phonographs, radios, and televisions and enjoyed their newly found leisure time at home. But instead they decided to move into the labor market to take on paid work. More broadly, an average American worker in 2015 who merely wishes to maintain the average income level of 1915 could do so by working just seventeen weeks per year, aided by modern technology.[99] But most citizens do not find this trade-off desirable. Instead, their demand for new goods and services has risen along with productivity. As labor-saving technology has given us the means to do more with less, most of us have preferred to take on other productive tasks instead of opting for more leisure.

The concern going forward, Heilbroner argued, was twofold. This time, it was not just jobs in agriculture and industry that were being affected. Automation, he worried, was rendering employment in the service sector redundant, too. And the demand for services produced by labor, he predicted, would eventually be fully met:

> But here is the critical point: Technology today seems to be invading the services as well as other kinds of work. One secretary, with a machine, can now type and style a writer's work. . . . There is no reason why technology should not penetrate deep into the job skills of the white-collar class. Where, then, will the new labor emigrants go? . . . Suppose that we *can* employ most of the population as psychiatrists, artists, or whatever. I am afraid there is still an upper limit on employment due, very simply, to the prospect of a ceiling on the *total* demand that can be generated for marketable goods and services.[100]

Whether there is a saturation point or not is controversial, but if our "basic needs" have all been met, higher incomes should no longer translate into higher subjective well-being. To that end, the economists Betsey Stevenson and Justin Wolfers examined whether there's a critical income level beyond which the well-being–income relationship diminishes. Analyzing multiple data sets, and using various definitions of basic needs and different measures of well-being, they found no evidence of a satiation point so far. When they compared average levels of subjective well-being and gross domestic product (GDP) per capita across countries, they found that the well-being–income relationship observed among poor countries holds equally in the rich world. This relationship also holds across income levels within countries. In America, for example, there is no evidence of a significant break in the well-being–income relationship, even at annual incomes of half a million dollars.[101] Thus, if there's a satiation point, it is yet to be reached.

In 1966, Herbert Simon wrote a response to Heilbroner's article, arguing that "insofar as they are economic problems at all, the world's problems in this generation and the next are problems of scarcity, not of intolerable abundance."[102] It is tempting to conclude that Simon was right and nothing much has changed since. The dystopian belief that automation must lead to unemployment, and the utopian view that it will bring about a life of leisure, have both seemingly been wrong so far. Looking forward, one observer aptly summarizes the matter as follows:

That shorter hours are inevitable with automation and the kind of technological progress we envisage reflects, in many instances, a fear that unemployment will spread if work is not shared. Just as commonly, however, it reflects a utopianism that the new technology heralds a new day of all play and no work. Whether workers prefer shorter hours to additional income depends upon their judgment as to the relative worth of leisure and income. Progressive gains in productivity and levels of living make this choice easier to make in favor of leisure, but the outcome is hardly predictable. It becomes more and more uncertain as hours input and arduousness of work fall below a point where the physical strain and other detriments of work impinge on health, family life, and full participation in social life, while at the same time the material standards of consumption rise. I do not know what workweek industrial and other workers will choose in the future. It is interesting to note that with substantially full employment in recent years, little reduction in average hours worked has occurred in non-agricultural employment in the United States.... At present, workers seem inclined generally to place a higher value on additional income than on more leisure, but this may not always be the case.[103]

The extract above is from a Bureau of Labor Statistics report that was first published in 1956. I could have used those same words today. After a century of staggering advances in mechanization and soaring productivity, it is quite remarkable how little time Americans take in leisure.

One relatively recent trend is noteworthy, however. Historically, the working poor were the ones who had to put in longer hours to be able to support themselves and their families. As Hans-Joachim Voth has shown, average working hours in Britain increased from fifty per week in 1760 to sixty hours in 1800.[104] This was the time of Engels's pause, when material standards for the working class faltered. It was also around the time when Jane Austen's novels about the elites depicted a society at leisure, whose lives centered on refined conversations and literature. But in recent years, those who would have flocked into the factories in the postwar period have become less likely to work, also relative to the new cognitive elite. The economists Mark Aguiar and Erik Hurst find that the least educated increasingly "enjoy" more leisure hours than symbolic analysts. Data from the American Time Use Surveys also show that college-educated Americans now work about two hours more per day than people who did not go to college.[105] The most convincing explanation for this pattern is simply that it reflects diminishing opportunity in the labor market for the would-be working class. As we saw in chapter 9, as automation has progressed, opportunity for the unskilled has regressed. Faced with falling wages and fading job options, some people may have chosen welfare instead of work, while others have struggled to find a job.

In 1983, when computers finally began to enter the workplace in large numbers, Wassily Leontief remarked: "Think what would happen if all unemployed steel and auto workers were retrained to operate computers. . . . There aren't enough computers to go around. . . . More and more workers will be replaced by machines. I do not see that the new industries can employ everybody who wants a job."[106] One reason why there are still so many jobs is that computers did create new tasks for labor (see chapter 9). But those tasks were primarily created for the highly skilled. This stands in contrast to the period of the Second Industrial Revolution, when technological change spun off new tasks for semiskilled workers, leaving the middle class with more and increasingly well-paying jobs (see chapter 8). The industries of the computer age failed to provide the same opportunities for the middle class as the smokestack industries that preceded them.

It is hard—not to say impossible—to predict exactly what new jobs and tasks AI technologies will bring about in the future. Yet we should put some faith in the observations of Frederic Bastiat. In his brilliant 1850 essay, "That Which Is Seen, and That Which Is Not Seen," he wrote: "In the department of economy, an act, a habit, an institution, a law, gives birth not only to an effect, but to a series of effects. Of these effects, the first only is immediate; it manifests itself simultaneously with its cause—it is seen. The others unfold in succession—they are not seen: it is well for us, if they are foreseen."[107] With regard to machines, substitution is the observable first-order effect. What isn't seen are the new jobs that will be created. Few jobs that exist in America today existed in 1750, at the onset of the Industrial Revolution. And many of today's jobs did not exist in official occupational classifications even in the 1970s, including those of robot engineers, database administrators, and computer support specialists. Almost half of employment growth between 1980 and the Great Recession happened in new types of work.[108]

The unseen will always be unknown, but it seems unlikely that AI technologies will reverse the twentieth-century pattern of rising skill requirements. With some exceptions, the jobs that are least likely to be replaced in the next wave are indeed the jobs of the skilled. And if we look to new industries that did not exist in 2000, most of these industries are related to digital technologies, and the majority of the workers they employ have a college degree (many a degree in science, technology, engineering, or mathematics).[109] Thus, the next wave of automation is likely to have effects similar to those of earlier computer technologies, but it is likely to affect more people. Those who would have taken on jobs in the factories in the postwar era have already seen their job options diminish since the computer revolution. And with retail, construction, transport, and logistics becoming more exposed to automation, too, the options of those people will likely deteriorate even further. Indeed, even if the next three decades mirror the past three, that is not all that comforting since automation recently has pushed up joblessness among groups in the labor market and put downward pressure on the wages of those with no more than a high school degree. In their bestselling book *The Second Machine Age*, Erik Brynjolfsson and Andrew McAfee make a similar observation: "Technological progress is going to leave behind some people, perhaps even a lot of people, as it races ahead. . . . There's never been a better time to be a worker with special skills or the right education, because these people can use technology to create and capture value. However, there's never been a worse time

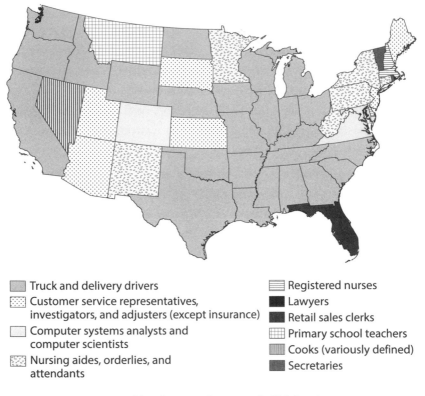

Truck and delivery drivers

Customer service representatives, investigators, and adjusters (except insurance)

Computer systems analysts and computer scientists

Nursing aides, orderlies, and attendants

Registered nurses

Lawyers

Retail sales clerks

Primary school teachers

Cooks (variously defined)

Secretaries

FIGURE 20: Most Common Occupation by U.S. State in 2016
Source: S. Ruggles et al., 2018, IPUMS USA, version 8.0 (dataset), https://usa.ipums.org/usa/.

to be a worker with only 'ordinary' skills and abilities to offer, because computers, robots, and other digital technologies are acquiring these skills and abilities at an extraordinary rate."[110]

Today, the largest single occupation in most American states is that of the truck driver (figure 20). It is true, as the economist Austan Goolsbee has pointed out, that if all 3.5 million truck, bus, and taxi drivers lose their jobs to autonomous vehicles over a fifteen-year period, that would amount to just over nineteen thousand per month: in 2017, 5.1 million Americans were separated from their jobs on a monthly basis, while 5.3 million jobs were generated on average. In this scenario, autonomous vehicles would increase the separation rate by less than four-tenths of a percent.[111] And this would be very unlikely to happen over a fifteen-year period. Technology adoption is never frictionless, and it will take much longer for taxis to become fully autonomous than long-haul trucks. The worrying part is that the alternative options for large groups in the labor market are continuously worsening. Even assuming that replaced truck drivers are reabsorbed with relative ease in the ceaseless churn in the labor market, we have to ask ourselves into which jobs and at what wages? And if their options look unattractive, will they choose to work at all?

A truck driver in the Midwest is not likely to become a software engineer in Silicon Valley. He might take up work as a janitor. Or he might find work in grounds maintenance, keeping parks, houses, and businesses attractive. (Both of these jobs, our estimates suggest, will not be exposed to the next wave of automation.) If he became a janitor he would trade a $41,340 job (2016 annual median income) as a truck driver for a $24,190 job. If he manages to become a ground maintenance worker, he would make $26,830 per year. Or he might get a job as a social care worker, earning $46,890 per year. But that would require him to get a college degree.

Leontief once joked that if horses had been given the right to vote, their disappearance from farms would have been less likely. Though the American middle class has hardly suffered the same fate as that of the farm horse, we would not expect Americans to simply accept falling wages. People might willingly accept automation if it reduces their incomes temporarily. But if their earnings seem unlikely to recover for years or even decades, they are more likely to resist it. Indeed, if individuals are unhappy with the verdict of the market, they can either try to block the technology or demand more redistribution through nonmarket mechanisms and political activism. We saw in chapter 3 that the Luddites and other groups vehemently opposed the spread of machines that threatened their livelihoods. In addition to rioting, they petitioned Parliament, appealing to the government to restrict the introduction of worker-replacing technologies—but their case was hopeless because they lacked political influence. Today, working people do not just have higher expectations for what governments must provide. They also have political rights.

13

THE ROAD TO RICHES

From time to time, people have thought that technological progress was about to come to an end. As the industries that constituted the key drivers of the Industrial Revolution—textiles, rail transport, and steam engineering—started to slow at the end of the nineteenth century, some observers asserted that the capitalist system had broken down.[1] In a similar spirit, during the Great Depression, Marxist critics declared that capitalism was incapable of achieving sustained growth, and non-Marxist writers, like the economist Alvin Hansen, predicted that the U.S. economy was in for a period of secular stagnation in part due to faltering innovation: "When a revolutionary new industry like the railroad or the automobile . . . reaches maturity and ceases to grow, as all industries finally must, the whole economy must experience a profound stagnation. . . . And when giant new industries have spent their force, it may take a long time before something else of equal magnitude emerges."[2]

Robert Gordon has recently put forward an equally bleak outlook on the future of growth in *The Rise and Fall of American Growth*.[3] He argues that contemporary breakthroughs in artificial intelligence (AI), mobile robotics, drones, and other by-products of the computer revolution cannot match the great inventions of the early twentieth century. There is no way of knowing whether future productivity growth will reach rates seen in the golden years, but in light of the technologies on the horizon (see chapter 12), it seems to me that productivity will pick up again, provided that innovation is allowed to progress uninterrupted.

The trouble is that many of those technologies are of the replacing sort, so that they will exert further pressure on the wages of the unskilled (see figure 18).

Before the dawn of automation, more than half of working American adults were employed in blue-collar and clerical jobs, which supported middle-class living for citizens with no more than a high school degree. Over the past thirty years, the number of these jobs has steadily declined, causing many of those who did not go to college to seek employment in low-income service jobs (see chapter 9). AI now threatens also to replace humans in many of those jobs, which used to be a safe haven for the unskilled—worsening their employment prospects even further. In this light, the concern is not that productivity growth will fail to pick up. The more serious challenge, it seems to me, exists not in technology itself but in the area of political economy. As the brilliant David Landes observed, "Even assuming that the ingenuity of scientists and engineers will always generate new ideas to relay the old . . . there is no assurance that those men charged with utilizing these ideas will do so intelligently [and] there is no assurance that noneconomic exogenous factors—above all, man's incompetence in dealing with his fellow-man—will not reduce the whole magnificent structure to dust."[4]

If left unaddressed, the increasing divide between winners and losers from automation could have serious social costs that go far beyond those borne by the individuals whose jobs are directly affected (see chapter 10). Already, the growing economic divide has translated into a greater political divide that is challenging the very fabric of liberal democracy (see chapter 11). In the twentieth century, steadily growing incomes were taken as a given, and people still expect their material standards to improve. But in the age of automation, it has become harder for governments to deliver on that promise, as growth in middle-class wages has fallen behind that in productivity. The populist backlash, in large part, reflects the failure of governments to make the gains from growth more widely shared. Indeed, the wages of noncollege workers have been in decline for more than thirty years—a long-standing process that was unmasked by the Great Recession (see chapter 11). As Francis Fukuyama puts it, "The future of democracy in developed countries will depend on their ability to deal with the problem of a disappearing middle class."[5]

Major social maladjustments lie ahead, and how best to respond to them is not a straightforward issue. If the negative impacts of automation are overemphasized, exaggerated fears over its deleterious effects may be aroused. But if their significance is underestimated, it is likely that precautions to minimize the individual and social costs will be neglected, and as a result people might quite rationally oppose replacing technologies.[6] If history is any guide to workers' reactions to the next wave of automation, it is telling that the Industrial Revolution was a time when many citizens fell between the cracks of transition and consequently, technological change was vehemently resisted (see chapter 5). On several occasions, the British government clashed with angry machine-smashing craftsmen, whom progress was forced upon. Yet resistance was not as vehement everywhere. The Old Poor Law helped ease the transition to the modern world. The economic historians Avner Greif and Murat Iyigun have shown that there was less popular resistance to technological change, and less social disorder, in parts of England where welfare institutions provided more generous support to the poor.[7] There were also some contemporaries, even if they were few, who realized the importance of compensating the losers to progress to avoid social and political upheaval. In his 1797 book on poverty, Sir Frederick Eden rightly pointed out that machines "promote the general wealth," but he added that they "throw many industrious individuals out of work; and thus create distresses that are sometimes exceedingly calamitous." He declared that poor relief must be used so that the machines' "inconvenience to individuals will be softened and mitigated, indeed, as far as it is practical." He contended that failing to adequately do so would lead to stagnation, as people would then oppose machines the way they had in the preindustrial era.[8]

The rise and fall of the poor laws, discussed in chapter 11, reflected a shift in political power from the landed classes to the new urban elite, whose members saw little gain in helping people stay in the countryside. Instead, they needed workers for their factories. But the poor laws' fall was also a consequence of the widespread belief that technology could not improve the human lot. Industrialization was advocated for in the national interest and unleashed to make sure that Britain did not lose ground to its rival nations in trade. And even though Malthusian forces in Britain had long disappeared, Malthusian logic was still thriving. Thomas Malthus's contemporaries and the generation of political economists after him believed that population growth would always undo economic growth in per capita terms. One implication of this belief was that any attempt to redistribute income to make the benefits of industrialization more widely shared was always doomed (see chapter 2). Both Malthus and David Ricardo vehemently opposed poor relief, which they believed would just encourage the poor to have more children rather than helping them.[9] We now know better.

In the twentieth century, governments assumed a wider responsibility for alleviating some of the adjustment costs imposed on the workforce. The labor movement, including its political branch, de facto accepted technology as the engine of growth but insisted on establishing a welfare system to provide credible assurance to all members of society that their personal income would not fall below a certain lower bound, making personal losses more narrowly constrained. And the newly generated wealth from industrialization allowed for more social spending, making it easier for society to compensate the less well off. As noted, one important reason why the socialist revolution that Karl Marx prophesied did not happen is that technology began to work in workers' interests, and consequently laborers quite rightly came to regard it as the engine of their good fortune. The adoption of the steam engine and, later, electrification created new and better-paying jobs for workers, who eventually acquired the skills required to run the machines. But another reason is that governments diffused the threat of revolution from below by expanding the franchise, creating a welfare state, and building an educational system that eased adjustment to the accelerating pace of change. Thus, quite naturally, the coming AI revolution has prompted calls for a capitalist reinvention of similar magnitude.

What Can Be Done?

Historically, the worst times for labor have been those characterized by both worker-replacing technological change and slow productivity growth. If AI technologies turn out to be as brilliant as some of us think, we should be more optimistic about the long run. As Daron Acemoglu and Pascual Restrepo have pointed out, brilliant technologies are much preferable for labor than mediocre ones because as they make us richer, they create more demand for other goods and services produced by humans.[10] Indeed, wages grew faster between 1995 and 2000, when computers prompted a brief productivity boom, than in the preceding and succeeding years. But while high productivity growth is always preferable to slow growth, growth in wages may fall behind that in productivity if technology is of the replacing sort, and some workers might see their incomes vanish in the process—even as new jobs are created elsewhere in the economy. That is what has happened in recent years, and it is also what happened during the classic years of industrialization.[11]

National unemployment in America today stands at 4 percent. Work is seemingly not about to come to an end, despite the rise of the robots. Instead, automation has manifested itself in falling wages for large swaths of the population, leading some to drop out of the workforce. The rising percentage of workers that are now outside the workforce, who are not accounted for in the unemployment rate, is particularly troubling. In *Men without Work*, Nicholas Eberstadt estimates that if the current trend continues, 24 percent of men ages 25–45 will be out of work by 2050. Joblessness is especially prevalent among men without a college degree, who lack the skills to compete in the ever-higher-tech economy.[12] They are the ones who have seen their earning potential diminish due to automation, and because they lack the necessary skills, they have been excluded from the new and emerging well-paying jobs (see chapter 9).

If current trends continue in the coming years, the divide between the winners and losers to automation will become even wider. And there are good reasons to think that it will. Looking at the automatability of existing jobs, we have seen that most occupations that require a college degree remain hard to automate, while many unskilled jobs—like those of cashiers, food preparers, call center agents, and truck drivers—seem set to vanish, though how soon is highly uncertain. But there are also unskilled jobs that remain outside the realms of AI. Many in-person service jobs—like those of fitness trainers, hairstylists, concierges, and massage therapists—that center on complex social interactions remain safe from automation.[13]

There is no way of knowing exactly what jobs the future will bring. At the advent of the Industrial Revolution, nobody could have foretold that many Englishmen would become telegraphers, locomotive engineers, and railroad repairmen. Today, futurologists are just as ill equipped to predict the jobs that AI will create. Official employment statistics are always behind the curve when it comes to capturing new occupations, which are not included in the data until they have reached a critical mass in terms of the number of people in them. But other sources, like LinkedIn data, allow us at least to nowcast some emerging jobs. Among them are the jobs of machine learning engineers, big data architects, data scientists, digital marketing specialists, and Android developers.[14] But we also find jobs like Zumba instructors and Beachbody coaches.[15]

In a world that is becoming increasingly technologically sophisticated, rising returns on skills are unlikely to disappear and likely to intensify. Like computers, AI seems set to spawn more skilled jobs for labor, in the process creating more demand for in-person service jobs that remain hard to automate. As noted above, much of recent job creation has centered on the so-called labor multiplier. Computers have created jobs for software engineers and programmers, which in turn have raised the demand for in-person service jobs in the places where they work and live (see chapter 10). Thus, where skilled jobs are abundant, the unskilled earn better wages, too. In San Jose, California, fitness trainers and aerobics instructors made $57,230 on average in 2017. In Flint, Michigan, they averaged $35,550 annually. Of course, direct comparisons are complicated by a variety of factors. It is true that the cost of living in the Bay Area is higher than it is in Flint. But it is just as true that amenities are more plentiful, health outcomes and public services are better, and crime rates are lower.

Automation then represents a double whammy. Where machines have replaced middle-class workers, the demand for local services has also suffered. The growing divide between the skilled and the unskilled has been amplified by the staggering divergence between skilled and unskilled places. The Bay Area has prospered from the miracles of software engineering, while labor in the Rust Belt has suffered from the implementation of new technologies that were invented elsewhere. And in many places, where middle-class jobs have dried up, vanishing incomes have brought a range of social problems like rising crime, faltering marriages, and deteriorating health (see chapter 10). Many of these problems, as we all know, are negatively correlated with rates of intergenerational mobility. They could have long-lasting effects on communities as they also have adverse consequences for the next generation. In this light, the appeal of populism is not hard to understand. It gives voice to the anger of those who have been excluded from the engines of growth and trapped in places of despair.

The message of this book is that we have been here before. We should recall Maxine Berg's noting of the "unprecedented demands for mobility, both geographical and occupational," that accompanied the Industrial Revolution. We should remember that machines "meant, or at least threatened, unemployment, an unemployment which at best was transitional between and within sectors of the economy." But above all, we should bear in mind that "the conceptual changes in political economy over the period are also very closely connected to class struggle [which was evident] in the very seriousness attached by political economists to the 1826 anti-machinery riots in Lancashire and to the 1830 agricultural riots."[16]

Engels's pause eventually came to an end, as enabling technologies came to the rescue and workers acquired new skills. But by that time, three generations of ordinary Englishmen had seen living standards decline. Governments today can mercifully assume wider responsibility for the social costs brought by technological change. Indeed, the growing percentage of men in their prime who are not working and the steady decline in the earnings capacity of those with no more than a high school degree suggest that we must think carefully about short-run dynamics as AI-enabled automation progresses. As productivity growth makes the pie larger, everyone could in principle be made better off. The challenge lies in the sphere of politics, not in that of technology. Given the enormous potential for AI to make us richer on the one hand, and the specter of disruptions to labor on the other hand, governments must carefully manage the short run, which was a lifetime for many during the classic years of industrialization.

As the former secretary of the treasury Lawrence Summers puts it, "Little is certain. But we will do better going forward than backward [which] means embracing rather than rejecting technological progress. . . . This will be a major debate that I suspect will define a large part of the politics of the industrial world over the next decade."[17] To avoid the technology trap, governments must pursue policies to kickstart productivity growth while helping workers adjust to the onrushing wave of automation. Addressing the social costs of automation will require major reforms in education, providing relocation vouchers to help people move, reducing barriers to switching jobs, getting rid of zoning restrictions that spur social and economic divisions, boosting the incomes of low-income households through tax credits, providing wage insurance for people who lose their jobs to machines, and investing more in early childhood education to mitigate the adverse consequences for the next generation. In the next sections, we shall look at what can be done in more detail.

Education

If people race alongside the machine, they are less likely to rage against it. And historically, education has been the way workers have adjusted to accelerating technological change. The seminal 2008 book by the economists Claudia Goldin and Lawrence Katz, *The Race between Education and Technology*, shows that the solid performance of the U.S. economy and the expansion of education over the first three-quarters of the twentieth century were not coincidental. The former was, at least in part, the consequence of the latter. The fact that the twentieth century was dominated by America and was the human capital century, the authors write, was not a historical accident: "Economic growth in the more modern period requires educated workers, managers, entrepreneurs, and citizens. Modern technologies must be invented, innovated, put in place, and maintained. They must have capable workers at the helm. Rapid technological advance, measured in various ways, has characterized the twentieth century. Because the American people were the most educated in the world, they were in the best position to invent, be entrepreneurial, and produce goods and services using advanced technologies."[18]

We saw in chapter 8 that the race between technology and education did a good job of explaining much of what was going on in the American labor market up until 1980, when technological change increasingly became of the replacing type. Yet replacing technological change made education more rather than less important. As discussed in chapter 9, people have adjusted very differently to automation depending on their educational background. As middle-income jobs for the semiskilled began to dry up, people falling into low-paying service jobs or out of the workforce were overwhelmingly citizens without a college degree. Those with a college education, in contrast, were more likely to move up in the ranks.

Unskilled work is not coming to an end, but as noted, low-skilled jobs are more exposed to future automation, while occupations that require a college degree remain relatively safe. And though it remains to be seen what the jobs of the future will be, and exactly what skills they will require, we do know what some of the barriers to acquiring new skills are. Arguably the greatest policy challenge is that in study after study, children from disadvantaged backgrounds are shown to have consistently lower educational attainment. As is well known, deficits in basic skills like math and reading, which surface in the early years of life, mean that children typically fail to catch up with their peers in later grades. One reason that such deficits arise in the first place is that children from low-income families often lack the intellectual stimulation from in-home reading and daily conversation that are so common in families where one or both parents have completed college. We also know that parents in the top fifth of the income distribution spend seven times more on enriching extracurricular activities and educational materials for their children—like books, computers, and music lessons—than those in the bottom fifth.[19] In this light, it stands to reason that as automation causes the incomes of many parents to vanish, it also diminishes the future prospects of their children. The economist Jeffrey Sachs and colleagues have indeed argued that AI threatens not just to reduce the jobs, wages, and savings of the current generation, but also to impoverish future generations as a consequence.[20]

To level the playing field, governments are advised to invest more heavily in early childhood education. Gaps in knowledge and ability between children from disadvantaged backgrounds and their relatively advantaged peers open early on and tend to persist throughout life. A proactive approach through investments in high-quality early child-hood programs will therefore be more effective and economically preferable than efforts to bridge the gap later on. And preschool education for children from poor families pays for itself. Those are the findings of James Heckman, winner of the Nobel Prize in Economics, and colleagues—whose research shows dramatic long-term effects of early interventions, with a rate of return on investment of 7–10 percent per annum through much improved educational outcomes, better health, productivity, and reduced crime.[21] Other work by Arthur Reynolds and colleagues, published in *Science*, reaches a similar conclusion. In their study, they traced the fates of more than fourteen hundred participants in the Chicago-based Child-Parent Center Education Program over twenty-five years. Their findings show that relative to the control group, program participants did strikingly better in terms of educational attainment, income, substance abuse, and crime, with the strongest enduring effects for males and children of high school dropouts.[22] As things stand, the overall societal costs of the opportunity gap, though hard to estimate, are striking by all accounts. Economists have put the aggregate annual costs of child poverty to the U.S. economy at $500 billion per year, which is equivalent to almost 4 percent of the gross domestic product. These costs stem from low productivity growth, higher crime rates, and cascading health expenditures.[23]

It is true that none of these studies account for the fact that the opportunity gap almost certainly will have some bearing on the future rate of innovation. In a pathbreaking study, the economist Alexander Bell and colleagues set out to analyze why some Americans are more likely than others to become inventors. Drawing upon data on 1.2 million inventors from patent records, the authors found that children from low-income families are much less likely to become inventors even if they exhibit the same ability as high-income children, as measured by test scores.[24] This innovation gap grows in later grades—which, the authors argue, "is because low-income children steadily fall behind their high-income peers over time, perhaps because of differences in their schools and childhood environments."[25]

What's more, the opportunity gap is not just bad for the economy. It is bad for democracy, too. College-educated people ages 20–25 are much more likely to engage in activities such as discussing politics, contacting public officials, voluntary work, and so on. More than twice as many people with no more than a high school education are completely detached from all forms of civic life, compared to those who went to college. And in terms of democratic participation, those who went to college are between two to three times more likely to vote in national elections.[26] Perhaps more worryingly still, as the political scientists Kay Schlozman, Sidney Verba, and Henry Brady demonstrate, political engagement has also become increasingly intergenerational, so that children tend to inherit their parents' degree of political participation. In other words, having well-educated or wealthy parents shapes not only children's job prospects but also their level of engagement in the political sphere.[27] This leads to a well-known dilemma. As Robert Dahl noted, "If you are deprived of an equal voice in the government, the chances are quite high that your interests will not be given the same attention as the interests of those who do have a voice. If you have no voice, who will speak up for you?"[28] Indeed, the political disenfranchisement of the unskilled, whose interests are no longer represented by mainstream politics, has made the discontents caused by automation harder to address (see chapter 11).

Retraining

How can we help those already in the labor market whose jobs are threatened by AI? Training people out of unemployment is a popular idea and a common response to rapid technological change. In the 1960s, when automation anxiety was at a high, retraining became a national priority. In his 1962 State of the Union address to Congress, President John F. Kennedy urged Congress to enact "the Manpower Training and Development Act [MDTA], to stop the waste of able-bodied men and women who want to work, but whose only skill has been replaced by a machine, or moved with a mill, or shut down with a mine."[29] The MDTA, which was signed into law on March 15, 1962, was the first federal manpower program, designed to train and retrain thousands of workers left behind by automation, though it was soon expanded to train people more broadly. In the period 1963–71, almost two million Americans enrolled in the program. How did they fare thereafter? The economist Orley Ashenfelter set out to evaluate the MDTA in 1978 and found that the answer was not straightforward. The program initially focused on the most easily retrained and was later geared toward more disadvantaged workers, many of whom dropped out. Though Ashenfelter found some evidence that workers earned higher incomes after participating in the program, he concluded that it was difficult to see if the benefits warranted the costs.[30]

Since the MDTA, federal policy makers have enacted an array of employment and training programs. Many of these are difficult to evaluate, because forgone earnings during training are hard to account for; data on costs for most training programs are sparse; and for the most part, studies trace outcomes only over a few years, meaning that we cannot know to what extent any effects on earnings fade out over time.[31] In a recent review of the literature, the economists Burt Barnow and Jeffrey Smith concluded, "Taken together, the recent evidence presents a mixed but somewhat disheartening picture."[32] While the policy conclusion is not that we should dismiss the idea of retraining people later in life, it would be unwise to put too much faith in large-scale training efforts without proof of concept. We must pursue a strategy of trial and error and learn from practical experience what works where. And there are surely some interesting ideas out there that go beyond training programs. Lifelong Learning Accounts, for example, which already exist in Maine and Washington, provide tax incentives for people to invest in their own training and target low-income earners. Eligible citizens may contribute up to $2,500 annually and can receive a refundable tax credit equal to 50 percent of the first $500 contributed and 25 percent of the next $2,000 for any given tax year. Workers can then use these funds at different stages in their career for training purposes, in response to replacement or to advance their careers more generally. Before scaling up such efforts, however, they need to be carefully evaluated.

Major reforms to transform education and training might also be required more broadly. As Harvard University's Clayton Christensen has forcefully argued, there is no particular reason why people with different learning requirements should have to conform to rigid academic programs that run for a specified period of time. The factory-based education model, which emerged in the aftermath of the Industrial Revolution, gradually expanded across many dimensions with more hours spent in school, more subjects covered, and more years of schooling. And that was a good thing. But if people have to consistently update their skills later in life, more flexible approaches to education will be needed. The learning process could be broken down so that instead of completing a standardized academic program, students could choose from a menu of skills and competencies they wish to acquire. Massive Open Online Courses (MOOCs), for example, can now be used to provide modularized education for people wanting to update their skills. And people can complete courses at their own pace.

Wage Insurance

It must also be acknowledged that retraining is not going to be the answer for everyone. People who experience dislocation late in life and see their skills rendered redundant might find it easier to take on a low-skilled job, even if it pays less. As noted above, displaced worker studies consistently show that many end up in jobs that pay less well than the jobs they previously held, and this is especially true of older people. Retraining and unemployment insurance do little to help displaced workers whose new jobs mean a significant pay cut. However, wage insurance—which compensates workers if they are forced to move to a job with a lower salary—would help ensure that fewer people are left worse off by automation. And it would make unskilled work pay more, relative to joblessness, which would likely reduce nonworking rates among the unskilled (see chapter 9). In America, wage insurance currently exists as a component of Trade Adjustment Assistance, a federal program designed to reduce the negative effects of imports felt by workers in some sectors. But the program is restricted to workers over fifty who do not earn over $50,000 a year. At the very least, it should be expanded to cover other sources of dislocation, like automation, which can lead to a permanent drop in people's income. In the words of the economist Robert LaLonde, "Whereas private markets offer insurance for storms and fire, no such insurance is available when a middle-aged worker loses a job and suffers a permanent drop in wages. There is a market failure here, and government should correct it."[33]

Tax Credits

In the popular press, universal basic income (UBI) has become a widely discussed way of limiting individual losses resulting from automation and deindustrialization. Of course, there are arguments in favor of UBI that have nothing to do with technological change, but this is not the place to dwell on them. The question here is whether it provides a good way of addressing the discontents brought about by the rise of the robots. In essence, UBI—which is closely tied to Milton Friedman's old idea of a negative income tax—would give people a minimum income regardless of whether they worked or not. They could then earn additional income if they decided to work. The way it was originally conceived, UBI would replace other existing welfare programs. If introduced this way, the downside is that it would increase inequality, unless people were willing to accept a significant increase in tax rates. Because existing welfare programs are designed to help those in need, at the lower end of the income distribution, UBI (which, as the name suggests, would be received by everyone) would effectively redistribute income back to those at the higher end. But more fundamentally, the welfare state emerged the way it did because most citizens felt uncomfortable transferring resources to those not in need.[34] UBI, in other words, would require a significant shift in attitudes and politics, which is highly unlikely given the sharpening economic and political divide in the past few decades. Growing economic segregation has meant that people rarely have firsthand knowledge of the realities other people face, which has led to diminishing cross-class loyalty (see chapter 11).

A change in attitude might come if people are faced with a serious threat of AI-driven mass unemployment. But for the time being, there is little to suggest that widespread joblessness is imminent. As discussed above, AI is a long way from being able to replace workers in all domains, and the new technologies on the horizon will not all arrive at the same time, nor will they be adopted overnight. What's more, as the historical record makes abundantly clear, fears that work will disappear have always turned out to be false alarm. If we think that this time is different, we should at least be able to explain why. Yet when we look at previous episodes of automation anxiety, like those of the 1830s, 1930s, 1960s, and 2010s, it is striking how much technology has advanced, but how little the debate has progressed. When I was researching this book, I struggled to find a single argument for why this time should be different that had not been made in earlier debates about automation.

Another false claim is that UBI is preferable to the welfare state because people do not like work. Back in 1970, for example, Walter Reuther, a vocal proponent of UBI, looked forward to the day when workers could spend less time on the job and devote themselves to music, painting, and scientific research. As we grew richer, he argued, we would work less and spend more time doing more self-fulfilling things. Yet most people find fulfillment and meaning in their work, whereas time-use studies show that the unskilled, who have seen their prospects in the labor market deteriorate, spend much of their time in front of the television, despite many studies showing that there is a negative correlation between television consumption and individual well-being.[35] Contrary to the anthropologist David Graeber's witty essay on "bullshit jobs," in which he claims that most people spend their working lives doing work they perceive to be meaningless, large-scale survey evidence shows the exact opposite.[36] And a wide range of studies across many countries and periods of time has consistently shown that people who work are happier than those who do not.[37] As Ian Goldin puts it, "Individuals gain not only income, but meaning, status, skills, networks and friendships through work. Delinking income and work, while rewarding people for staying at home, is what lies behind social decay."[38]

Thus, rather than subsidizing everyone, whether they work or not and regardless of their income, which is what UBI would do, it makes more sense to specifically target low-income groups in the labor market who have seen their earnings capacity fall. In contrast to UBI, which for the reasons outlined above remains controversial, such policies have broad support. In a recent op-ed in the *Washington Post*, for example, the economist Glenn Hubbard, who served as chairman of the Council of Economic Advisers in the George W. Bush administration, pointed out that "the economic-growth-lifts-all-boats camp needs to confront the question of what happens when growth alone fails to generate inclusion."[39] Hubbard argued for the introduction of a set of vouchers targeting low-income individuals, to be used for their training and their children's education. He also argued for expanding the eligibility for the Earned Income Tax Credit (EITC).

The EITC is a negative income tax that is available only to working low-income individuals and already has a good track record. Scholars have found that those receiving it have seen dramatic increases in take-home income. They have found that its expansion helped put single parents back into the workforce. And the children of people receiving the credit have benefited enormously, in terms of both well-being and educational attainment, as reflected in higher math and reading scores as well as higher college enrollment rates.[40] Thus, unsurprisingly, American states with more generous EITC policies also have higher rates of intergenerational mobility.[41] As the sociologist Lane Kenworthy summarizes the research, "Government cash transfers of just a few thousand dollars could give a significant lifelong boost to the children who need it most."[42]

For these reasons, the EITC should be expanded. First, making it more generous for low-income households with children would help level the playing field in terms of increasing the chance that children from disadvantaged backgrounds can move up in the ranks. Second, there is a good case for extending it to citizens without qualifying children, for whom the subsidy is currently minimal. As noted, the divide between college-educated people and others is likely to continue to grow. Thus, governments must make low-paying jobs pay more to improve incentives to work and reduce inequality, and the EITC or an equivalent provides a credible way of doing so. And if the past is any guidance, some of its cost will be offset by rising labor force participation among the unskilled.

Regulation

A different set of policies is needed to make it easier to move between jobs. Regulatory barriers to job switching are bad for productivity, wages, and equality. Of course, there are good reasons to insist on people like doctors and nurses to be licensed to practice. Yet the U.S. government practice of allowing only licensed practitioners to receive pay in a growing number of professions is worrying. To become a hair shampooer in Tennessee, for example, one must complete seventy days of training and pass two exams. Across America, the share of workers requiring a license to perform their jobs legally expanded from 10 percent in 1970 to almost 30 percent in 2008.[43] Because obtaining a license often requires considerable investments in human capital and licensing fees, Americans who lose their jobs to machines are less likely to switch into licensed occupations, and workers in those jobs are less likely to switch out. Licensing requirements also often vary considerably across states—and even more so across countries—meaning that those in licensed jobs frequently have to make additional investments in obtaining a license when they move. Hence, unsurprisingly, economists have found that places with more people in licensed occupations also have higher rates of joblessness.[44]

In addition, the use of noncompete clauses—in which the employee agrees not to take a similar job at a competing firm for a prespecified period of time after leaving their current firm—has increased in many American states, providing another hurdle for engineers, scientists, and professionals seeking to move to expanding firms. The ability to change jobs easily is frequently mentioned as a prime reason for the success of Silicon Valley. Indeed, the departure of Gordon Moore and Robert Noyce from Fairchild Semiconductor to found Intel in 1968 was a critical moment in the area's history. As is well known, worker mobility is much higher in the computer industry in California in general, and in Silicon Valley in particular, than elsewhere in America.[45] One widely accepted explanation credits the California Civil Code of 1872, which outlawed all covenants in employment contracts—thus ensuring that Moore and Noyce could set up Intel after leaving Fairchild.[46]

The economist Steven Klepper has found that the same sort of dynamism underpinned the success of Detroit's automobile industry in its heyday. In this regard, the rise of Detroit had much in common with the rise of Silicon Valley.[47] Noncompete clauses were long prohibited in Michigan, too. But the Michigan Antitrust Reform Act of 1985, repealed Michigan's—and thus Detroit's—ban on the enforcement of such clauses. My research with Thor Berger shows that the technological dynamism of Michigan declined in response, as fewer workers shifted into new computer-related jobs relative to those in other similar states that did not see any change in legislation.[48] In other words, the repeal exacerbated the decline of Detroit as an innovation hub. Thus, while it is unclear how much the rolling back of excessive occupational licensing practices and the abandonment of noncompete clauses would help, it is certainly worth finding out.

Relocation

The computer revolution has been a double-edged sword for American cities (see chapter 10). Cities with skilled populations were better able to take advantage of skill-intensive computer technologies and have subsequently prospered. In contrast, many of America's problems are concentrated geographically, where middle-class jobs have been replaced by robots. Looking forward, even if new and improved substitutes for face-to-face interactions are developed, they cannot substitute for spontaneous encounters that require physical proximity. Any digital communication must always be planned on at least one end, which means that the type of random interactions that occur in a workplace cannot happen at distance. Rather, the value of proximity will probably increase as AI makes production more skill intensive. Thus, the curse of geography is likely to intensify.

Historically, migration was the mechanism by which cities adjusted to trade and technology shocks. Workers moved to areas where new industries, spawned by the Second Industrial Revolution, created an abundance of well-paying, semiskilled manufacturing jobs. In the Great Migration, millions of Americans left the South for flourishing smokestack cities like Chicago and Buffalo. And agricultural workers left the farms for booming cities like Pittsburgh and Detroit. More and more people moving to higher-productivity areas served to equalize incomes across regions. But migration is no longer the equalizer it once was. While symbolic analysts are still highly mobile, the unskilled have become less likely to migrate since the dawn of the computer revolution (see chapter 10). One reason might be financial. Even if skilled cities provide better employment opportunities, moving is an investment that requires liquidity up front. Thus, as Enrico Moretti has convincingly argued, there is a case for subsidizing relocation.[49] Mobility vouchers could pay for themselves by shifting the unemployed into paid employment elsewhere, while serving to equalize incomes across space. Some will argue that mobility vouchers might serve to accelerate the exodus from communities in decline, leaving parts of America in an even more dire state, but even those who stayed put would likely benefit in terms of having a better chance of finding a job.

Housing and Zoning

Another dilemma is that as skilled cities are becoming more attractive, rising housing prices makes them less affordable. To counteract this, the housing supply must be expanded where new jobs are being created. This will require getting rid of some zoning restrictions, such as minimum lot sizes, height limits, prohibitions on multifamily housing, lengthy permitting processes, and so on, which effectively cap the number of people who can live in thriving places. Because dynamic places like New York and the Bay Area have adopted stringent restrictions on new housing supply, they have effectively limited the number of workers who can participate in the growth created by tech industries. The consequence has been that tech companies find it more difficult to hire due to the rising cost of housing. But more importantly still, an unemployed unskilled worker in Flint who finds a job in Boston cannot afford to live there. As discussed above, the next wave of automation will render many low-skilled jobs redundant, but there are still a variety of in-person service jobs that remain exceedingly hard to automate. Those jobs, it stands to reason, will emerge in skilled cities, where people can afford such services.

The combined effect of zoning restrictions has been slower economic growth, fewer jobs, lower wages, and higher inequality across the nation. Economists have estimated that in the absence of such restrictions on housing supply, the American economy would be 9 percent larger today, which would mean an additional $6,775 in annual income for the average American worker.[50] Abolishing land use restrictions would also have welcome side effects. The breathtaking rise in wealth inequality that has been documented by Thomas Piketty stems almost entirely from housing.[51] Inflated house prices due to land-use restrictions are surely part of the reason, and the abolition of those restrictions must therefore be part of the solution.[52]

Removing barriers to the expansion and development of skilled cities would help social mobility, too. As Raj Chetty, Nathaniel Hendren, and Lawrence Katz have shown, a person who moves from Oakland to San Francisco at the age of nine will, as an adult, gain more than half of the income differential between the two locations.[53] Because zoning restrictions are not distributed randomly but are much more prevalent in high-income cities and neighborhoods, they put people born into less affluent communities at a further disadvantage. Zoning, in other words, has priced lower-income families out of the places with more social capital and better schools.

Another benefit would be more innovation. Children growing up in places with more inventors, who are thus more exposed to innovation in their early years, are much more likely to become inventors themselves. This, we know, also has an impact on the types of inventions that they are likely to produce. Those growing up in Silicon Valley are more likely to drive innovation in computing, while those who spend their early years in places specializing in medical devices, like Minneapolis, for example, are more likely to invent related technologies.[54]

Connectivity

Transportation infrastructure that linked high-paying labor markets to areas where housing is cheap would also allow more people to tap into strong local economies. Subways or high-speed rail that connected declining places (where jobs have dried up and housing is cheap) with expanding ones (where jobs are abundant and housing is expensive) would serve to level incomes across space. And such connections would boost local service economies in decline, as people spend large parts of their incomes locally. In this light, economists have pointed to the potential benefits of current efforts to connect California's low-income cities—like Sacramento, Stockton, Modesto, and Fresno—with the Bay Area through high-speed rail.[55] Many Californians could remain in Fresno, where housing is cheap, and commute to San Francisco for work.

In the future, new transportation technologies could also be used to connect places that are much farther apart. Hyperloop technology, which uses a sealed system of tubes to allow people to travel free of air resistance or friction, offers the potential of reaching distant locations at staggering speeds. Hyperloop Transportation Technologies, for example, recently signed agreements with the Illinois Department of Transportation to examine the feasibility of connecting Cleveland and Chicago through a number of different corridors.[56] The commute currently takes around 5.5 hours by car or 7.1 hours by public transportation in one direction. The Hyperloop, if successful, is expected to bring this commute down to twenty-eight minutes. All of a sudden, it could become feasible to commute long distances to work.

Industrial Renewal

A regrettably less promising approach has been the effort to revitalize cities in decline through place-based policies, which target local industry rather than individuals. It is true that some of these policies have been successful in attracting new jobs, but the costs of doing so have been significant. For example, distressed urban and rural areas that were designated empowerment zones in the 1990s increased local employment through grants, tax credits for business, and other benefits, but each new job is estimated to have cost over $100,000.[57] And while large-scale projects to revive communities have generated sustained growth locally, these projects seems to have attracted resources at the expense of other places. The greatest push of this kind in American history was the Tennessee Valley Authority (TVA) Act of 1933, which became law in the midst of the Great Depression. The TVA set out to rapidly modernize the Tennessee Valley's economy, harnessing promising new technologies like electricity to attract the manufacturing industry. The push included large-scale public infrastructure projects including dams, an extensive road network, and a 650-mile navigation canal. For the valley, the push was surely a good thing: the positive employment contributions were seen as late as the turn of the new millennium, when the area was still growing more rapidly than comparable areas—though the effects have begun to fade. Yet manufacturing jobs created in the valley were offset by losses elsewhere. This is a troubling finding, because federal and local governments are estimated to spend some $95 billion per year on place-based policy programs, which is much more than is spent on unemployment insurance.[58]

Big pushes that focus on investment in physical capital are also likely to yield much lower benefits locally today, as manufacturing has become more automated. A more promising way forward for places that are falling behind is to divert resources to investment in human capital. Economists have shown that the presence of a college or university increases the supply of skilled workers not only by educating them, but also by attracting more college-educated people from elsewhere.[59] For example, the Land-Grant College Act of 1862 (also known as the Morrill Act) established several land-grant universities and is estimated to have increased labor productivity by 57 percent over a period of eighty years.[60] And needless to say, people take their human capital with them if they move. Physical capital stays put.

Final Thoughts

In the mid-nineteenth century, Karl Marx and Friedrich Engels predicted that continued mechanization would mean the continued impoverishment of the working class, just around the time when Britain finally escaped from Engels's pause. They were right about the past: many Englishmen had been left worse off by the Industrial Revolution. However, they were wrong in thinking that continued progress would lead in the same direction. Like so many others, they were fooled by the mysterious force of technology.

There have been long periods when things did not work out well for labor. But even those episodes came to an end. The thesis of this book is not that current economic trends must continue indefinitely. On the contrary, there are good reasons to be optimistic about an AI-induced productivity revival, which, besides making us richer on average, would help offset some of the negative effects replacing technologies have on parts of the labor force. But if history is any guide, that could take many years or even decades. And while it is possible that we are at the cusp of a wave of enabling technologies that could reinstate labor in new jobs more broadly, that is unlikely to provide much relief to people in the middle class unless they have the right skills. Even if we assume that AI will spawn gigantic new industries, as the automobile did a century ago, Henry Ford's invention of the assembly line broke complex operations down into simple tasks that could be performed by a person with a fifth-grade education. For more than thirty years now, technological change has created few new jobs that do not require a college degree. In a world that is becoming increasingly technologically sophisticated, new jobs are unlikely to open up for those who would have flocked into the factories before the dawn of automation.

The economic order that gave rise to a broad middle class has withered, along with the middle-class politics that rested on it. Until the Great Recession the pressures from automation on middle-income households were masked by subsidized credit, which counterbalanced falling wages among non-college-educated workers and left consumption broadly unaffected. The housing boom also meant that abundant construction jobs helped offset some of the job losses in manufacturing—until the burst of the housing bubble.[61] In other words, the recession unmasked the steady decline in the wages of the middle class, which helps explain the relatively recent rise of populism.

Looking forward, the divide between the winners and losers from automation can be expected to grow further. The next wave is not coming just for manufacturing jobs, but also for many unskilled jobs in transportation, retail, logistics, and construction. Thus, while there are good reasons to be optimistic about the long run, such optimism is only possible if we successfully manage the short-term dynamics. People who lose out to automation will quite rationally oppose it, and if they do, the short-term effects cannot be seen in isolation from the long run. In light of the long history of resistance to technology that threatens people's skills and the recent backlash against globalization, automation cannot be seen as an inexorable fact of life. It is true that unlike the Luddites of the nineteenth century, people now have seen how in the twentieth century technology made everyone richer. As mechanization progressed during the first three-quarters of the twentieth century, wages rose at all levels. But if technology fails to lift all boats in the coming years, broad acceptance of technological change cannot be taken for granted. People have higher expectations than at the time of Engels's pause. They have the right to vote. And they are already demanding change.

No single government policy can address the full spectrum of societal challenges brought by automation. Regrettably, proposing apparently easy solutions to a complex set of problems may win elections in the short term, but reality catches up sooner or later. Moderate conservatives and liberals face a tricky balancing act, because exaggerating the effects of automation might prompt fears of mass unemployment and lead to the wrong policy responses, the growth of populist parties, and possibly a backlash against technology itself. At the same time, however, if governments gloss over the social costs of automation, their credibility will diminish. For a long time, governments chose to overlook the costs of globalization and focus on the benefits. Those benefits were indeed significant, but the failure to deal with the individual and societal costs ended up costing mainstream politics its credibility. Governments must avoid making the same mistake with automation. And the stakes could not be higher.

Some readers might still think that we are entering a new era in which machines take all of the jobs, and of course, there is no way of knowing if that is true. But for now, there is little to suggest that this time is different: our current trajectories look exceedingly similar to those in the classic years of industrialization, and we all know what happened after that. Even assuming that this time is different, however, still means that the challenges ahead lie in the area of political economy, not in technology. In a world where technology creates few jobs and enormous wealth, the challenge is a distributional one. The bottom line is that regardless of what the future of technology holds, it is up to us to shape its economic and societal impact.

ACKNOWLEDGMENTS

If this book could be regarded as an invention, it would surely be a recombinant one. It draws upon a vast body of research to which numerous scholars have contributed. I guess my own journey to writing it began in my school years, when my father, Christopher, got back from a business trip with two new books for me. The first was Joel Mokyr's *The Lever of Riches*. The second was Clayton Christensen's *The Innovator's Dilemma*. Their work showed me that long-term prosperity derives from technological innovation. But it also made abundantly clear that progress often comes with economic and societal disruption. My lifelong interest in the subject is thanks to my father.

Over the past four years of writing, I have accumulated many debts. This book could not have been written without generous financial support from Citigroup. I'm especially indebted to Andrew Pitt and Robert Garlick at Citi, whose genuine intellectual curiosity made this project possible. Special thanks also go to Sarah Caro, my editor at Princeton University Press, for her guidance and many thoughtful comments. Chinchih Chen has done a fabulous job of providing diligent research assistance. And my long-standing friend Thor Berger has read many different versions of this manuscript, for which I'm enormously thankful. I'm also grateful to Ian Goldin, Logan Graham, Jane Humphries, Frank Levy, Jonas Ljungberg, Joel Mokyr, Michael Osborne, and Anil Prashar for reading all or part of this manuscript and providing invaluable comments.

Above all, my family has long been gracefully supportive of my many professional preoccupations, including this one. They are the ones who have kept me sane.

APPENDIX

FIGURE 5

Constructed following R. C. Allen, 2009b, "Engels' Pause: Technical Change, Capital Accumulation, and Inequality in the British Industrial Revolution," *Explorations in Economic History* 46 (4): 418–35, appendix I, using the sources below:

- Gross domestic product (GDP) factor cost estimate from C. H. Feinstein, 1998, "Pessimism Perpetuated: Real Wages and the Standard of Living in Britain during and after the Industrial Revolution," *Journal of Economic History* 58 (3): 625–58; B. Mitchell, 1988, *British Historical Statistics* (Cambridge: Cambridge University Press), 837, for 1830–1900.
- Real output per capita from N. F. Crafts, 1987, "British Economic Growth, 1700–1850: Some Difficulties of Interpretation," *Explorations in Economic History* 20 (4): 245–68.
- Average full-employment weekly earnings for the United Kingdom for 1770–1882 from Feinstein 1998, appendix table 1, 652–53; average full-employment weekly earnings for the United Kingdom for 1883–1900 from Feinstein, 1990, "New Estimates of Average Earnings in the United Kingdom," *Economic History Review* 43 (4): 592–633.
- Cost of living index for 1770–1869 from R. C. Allen, 2007, "Pessimism Preserved: Real Wages in the British Industrial

Revolution" (Working Paper 314, Department of Economics, Oxford University), appendix 1.

- Great Britain/United Kingdom cost of living index for 1870–1900 from C. H. Feinstein, 1991, "A New Look at the Cost of Living," in *New Perspectives on the Late Victorian Economy*, edited by J. Foreman-Peck (Cambridge: Cambridge University Press), 151–79.
- I have converted the wage index for 1882 onward from Feinstein 1990, based on 1880–81, the benchmark year in C. H. Feinstein, 1998, "Pessimism Perpetuated: Real Wages and the Standard of Living in Britain during and after the Industrial Revolution," *Journal of Economic History* 58 (3): 625–58. The nominal wage for 1770–1881 is derived from Feinstein 1998 and for 1882–1900, it is derived from Feinstein 1990.
- Following Allen 2009b, I used the growth rate of real output per capita from N. F. Crafts 1987, table 1, to extrapolate backward to 1770.
- All GDP, wage, and population data have been collected from R. Thomas and N. Dimsdale, 2016, "Three Centuries of Data–Version 3.0" (London: Bank of England), https://www.bankofengland.co.uk/statistics/research-datasets.

FIGURE 9

Constructed following R. J. Gordon, 2016, *The Rise and Fall of American Growth: The U.S. Standard of Living since the Civil War* (Princeton, NJ: Princeton University Press), figure 8-7, using the sources below:

- 1929–2016 U.S. real GDP data, 1870–2016 production workers' hourly compensation (nominal dollars), and the 1870–1928 GDP deflator are collected from L. Johnston and S. H. Williamson, 2018, "What Was the U.S. GDP Then?," http://www.measuringworth.org/usgdp//.
- 1870–1929 nominal gross national product (GNP) is collected from N. S. Balke and R. J. Gordon, 1989, "The Estimation of

Prewar Gross National Product: Methodology and New Evidence," *Journal of Political Economy* 97 (1): 38–92, table 10.

- Total civilian man-hours for 1870–1947 are from J. W. Kendrick, 1961, *Productivity Trends in the United States* (Princeton, NJ: Princeton University Press), table A-X.
- Total civilian man-hours for 1948–66 are from J. W. Kendrick, 1973, *Postwar Productivity Trends in the United States, 1948–1969* (Cambridge, MA: National Bureau of Economic Research [NBER] Books), table A-10.
- Total private average weekly hours of production and nonsupervisory employees' data for the years 1967–75 are from Bureau of Labor Statistics, 2015. "Employment, Hours, and Earnings from the Current Employment Statistics Survey" (Washington, DC: U.S. Department of Labor).
- Average weekly hours at work in all industries and in nonagricultural industries for the years 1976–2016 are from Bureau of Labor Statistics, 2015, "Labor Force Statistics from the Current Population Survey" (Washington, DC: U.S. Department of Labor).

FIGURE 14

Constructed following B. Milanovic, 2016b, *Global Inequality: A New Approach for the Age of Globalization* (Cambridge, MA: Harvard University Press), figure 2-1, using the sources below:

- U.S. Gini for 1774 and 1860 from P. H. Lindert and J. G. Williamson, 2012, "American Incomes 1774–1860" (Working Paper 18396, National Bureau of Economic Research, Cambridge, MA), tables 6 and 7; for 1935, 1941, and 1944 from S. Goldsmith, G. Jaszi, H. Kaitz, and M. Liebenberg, 1954, "Size Distribution of Income Since the Mid-Thirties," *Review of Economics and Statistics* 36 (1): 1–32; for 1947–49 from E. Smolensky and R. Plotnick, 1993, "Inequality and Poverty in the United States: 1900 to 1990" (Paper 998-93, University of Wisconsin Institute for Research on

Poverty, Madison); and for 1950–2015 from B. Milanovic 2016a, "All the Ginis (ALG) Dataset," https://datacatalog.worldbank. org/dataset/all-ginis-dataset, Version October 2016."

- United Kingdom/England Gini for 1688, 1759, and 1801–3 from B. Milanovic, P. H. Lindert, and J. G. Williamson, 2010, "Pre-Industrial Inequality," *Economic Journal* 121 (551): 255–72, table 2; for 1867, 1880, and 1913 from P. H. Lindert and J. G. Williamson, 1983, "Reinterpreting Britain's Social Tables, 1688–1913," *Explorations in Economic History* 20 (1): 94–109, table 2; for 1938–59 from P. H. Lindert, 2000a, "Three Centuries of Inequality in Britain and America," in *Handbook of Income Distribution*, ed. A.B. Atkinson and F. Bourguignon, table 1; and for 1961–2014 from Milanovic 2016a.

NOTES

Preface

1. J. Gramlich, 2017, "Most Americans Would Favor Policies to Limit Job and Wage Losses Caused by Automation," Pew Research Center, http://www.pewresearch.org/fact-tank/2017/10/09/most-americans-would-favor-policies-to-limit-job-and-wage-losses-caused-by-automation/.

2. K. Roose, 2018, "His 2020 Campaign Message: The Robots Are Coming," *New York Times*, February 18.

3. C. B. Frey and M. A. Osborne, 2017, "The Future of Employment: How Susceptible Are Jobs to Computerisation?," *Technological Forecasting and Social Change* 114 (January): 254–80.

4. B. DeLong, 1998, "Estimating World GDP: One Million BC–Present" (Working paper, University of California, Berkeley).

5. D. Acemoglu and P. Restrepo, 2018a, "Artificial Intelligence, Automation and Work" (Working Paper 24196, National Bureau of Economic Research, Cambridge, MA).

6. Quoted in G. Allison, 2017, *Destined for War: Can America and China Escape Thucydides's Trap?*, Boston: Houghton Mifflin Harcourt, chapter 2, Kindle.

7. D. S. Landes, 1969, *The Unbound Prometheus: Technological Change and Development in Western Europe from 1750 to the Present* (Cambridge: Cambridge University Press), introduction.

8. Quoted in Roose, 2018, "His 2020 Campaign Message."

9. R. Foorohar, 2018, "Why Workers Need a 'Digital New Deal' to Protect against AI," *Financial Times*, February 18.

Introduction

1. "Lamplighters Quit; City Dark in Spots," 1907, *New York Times*, April 25.

2. B. Reinitz, 1924, "The Descent of Lamp-Lighting: An Ancient and Honorable Profession Fallen into the Hands of Schoolboys," *New York Times*, May 4.

3. B. Reinitz, 1929, "New York Lights Now Robotized," *New York Times*, April 28.

4. W. D. Nordhaus, 1996, "Do Real-Output and Real-Wage Measures Capture Reality? The History of Lighting Suggests Not," in *The Economics of New Goods*, ed. T. F. Bresnahan and

R. J. Gordon (Chicago: University of Chicago Press), 27–70. On the early uses of electric light, see D. E. Nye, 1990, *Electrifying America: Social Meanings of a New Technology, 1880–1940* (Cambridge, MA: MIT Press), chapter 1.

5. Lamplighters and Electricity," 1906, *Washington Post*, July 1.

6. J. A. Schumpeter, [1942] 1976, *Capitalism, Socialism and Democracy*, 3d ed. (New York: Harper Torchbooks), 76.

7. Quoted in R. J. Gordon, 2014, "The Demise of U.S. Economic Growth: Restatement, Rebuttal, and Reflections" (Working Paper 19895, National Bureau of Economic Research, Cambridge, MA), 23.

8. D. Comin and M. Mestieri, 2018, "If Technology Has Arrived Everywhere, Why Has Income Diverged?," *American Economic Journal: Macroeconomics* 10 (3): 137–78.

9. Quoted in Nye, 1990, *Electrifying America*, 150.

10. Robert Gordon has called the period 1870–1970 the "special century" in American history (2016, *The Rise and Fall of American Growth: The U.S. Standard of Living Since the Civil War* [Princeton, NJ: Princeton University Press]).

11. S. Landsberg, 2007, "A Brief History of Economic Time," *Wall Street Journal*, June 9.

12. E. Hobsbawm, 1968, *Industry and Empire: From 1750 to the Present Day* (New York: New Press), chap. 3, Kindle.

13. Hobsbawm has called the period 1789–1848 the "dual revolution" (1962, *The Age of Revolution: Europe 1789–1848* [London: Weidenfeld and Nicolson], preface, Kindle. The term refers to the political changes of the French Revolution fused with the technological changes of the Industrial Revolution.

14 T. Hobbes, 1651, *Leviathan*, chapter 13, https://ebooks.adelaide.edu.au/h/hobbes/thomas/h68l/chapter13.html.

15. A. Deaton, 2013, *The Great Escape: Health, Wealth, and the Origins of Inequality* (Princeton, NJ: Princeton University Press).

16. W. Blake, 1810, "Jerusalem," https://www.poetryfoundation.org/poems/54684/jerusalem-and-did-those-feet-in-ancient-time.

17. For more on living standards during the Industrial Revolution, see chapter 5.

18. For more on the causes of the living standards crisis, see chapter 5.

19. J. Brown, 1832, *A Memoir of Robert Blincoe: An Orphan Boy; Sent From the Workhouse of St. Pancras, London at Seven Years of Age, to Endure the Horrors of a Cotton-Mill* (London: J. Doherty).

20. D. S. Landes, 1969, *The Unbound Prometheus: Technological Change and Development in Western Europe from 1750 to the Present* (Cambridge: Cambridge University Press), 7.

21. For examples of preindustrial resistance, see chapter 1. For a discussion of why British governments began to side with the innovators, see chapter 3.

22. Quoted in E. Brynjolfsson, 2012, *Race Against the Machine* (MIT lecture), slide 2, http://ilp.mit.edu/images/conferences/2012/IT/Brynjolfsson.pdf.

23. Bruce Stokes, 2017, "Public Divided on Prospects for Next Generation," Pew Research Center Spring 2017 Global Attitudes Survey, June 5, http://www.pewglobal.org/2017/06/05/2-public-divided-on-prospects-for-the-next-generation/.

24. R. Chetty et al., 2017, "The Fading American Dream: Trends in Absolute Income Mobility Since 1940," *Science* 356 (6336): 398–406.

25. For more on disappearing middle-income jobs, see chapter 9.

26. For more on communities where jobs have disappeared, see chapter 10.

27. C. B. Frey, T. Berger, and C. Chen, 2018, "Political Machinery: Did Robots Swing the 2016 U.S. Presidential Election?," *Oxford Review of Economic Policy* 34 (3): 418–42. For how automation has increased support for nationalist and radical-right parties in Europe, see M. Anelli, I. Colantone, and P. Stanig, 2018, "We Were the Robots: Automation in Manufacturing and Voting Behavior in Western Europe" (working paper, Bocconi University, Milan, Italy).

28. E. Hoffer, 1965, "Automation Is Here to Liberate Us," *New York Times*, October 24.

29. "Danzig Bars New Machinery Except on Official Permit," 1933, *New York Times*, March 14.

30. Quoted in "Nazis to Curb Machines as Substitutes for Men," 1933, *New York Times*, August 6.

31. P. R. Krugman, 1995, *Peddling Prosperity: Economic Sense and Nonsense in the Age of Diminished Expectations* (New York: Norton), 56.

32. On the distinction between productivity-enhancing and worker-replacing technological changes, see H. Jerome, 1934, "Mechanization in Industry" (Working Paper 27, National Bureau of Economic Research, Cambridge, MA), 27–31.

33. Ibid., 65.

34. D. Acemoglu and P. Restrepo, 2018a, "Artificial Intelligence, Automation and Work" (Working Paper 24196, National Bureau of Economic Research, Cambridge, MA).

35. Ibid.

36. J. Bessen, 2015, *Learning by Doing: The Real Connection between Innovation, Wages, and Wealth* (New Haven, CT: Yale University Press), chapter 6.

37. Schumpeter, [1942] 1976, *Capitalism, Socialism and Democracy*, 85.

38. Quoted in D. Akst, 2013, "What Can We Learn from Past Anxiety over Automation?," *Wilson Quarterly*, Summer, https://wilsonquarterly.com/quarterly/summer-2014-where-have-all-the-jobs-gone/theres-much-learn-from-past-anxiety-over-automation/.

39. Quoted in J. Mokyr, 2001, "The Rise and Fall of the Factory System: Technology, firms, and households since the Industrial Revolution," in Carnegie-Rochester Conference Series on Public Policy 55 (1): 20.

40. Ibsen. H., 1919, *Pillars of Society* (Boston: Walter H. Baker & Co.), https://archive.org/details/pillarsofsocietyooibse/page/36.

41. The adoption of the printing press in the empire had to wait until 1727. Even in the late nineteenth century, Ottoman book production was primarily undertaken by scribes. The consequences of the long absence of the printing press become obvious when we examine regional disparities in literacy rates. In 1800, about 2–3 percent of the population of the Ottoman Empire was literate, compared to 60 percent of adult males and 40 percent of adult females in Britain (D. Acemoglu and J. A. Robinson, 2012, *Why Nations Fail: The Origins of Power, Prosperity and Poverty* [New York: Crown Business], 207–8).

42. Ibid., 80.

43. For efforts by the ruling classes to block replacing technologies, see chapters 1 and 3.

44. J. Mokyr, 2002, *The Gifts of Athena: Historical Origins of the Knowledge Economy* (Princeton, NJ: Princeton University Press), 232.

45. J. Mokyr, 1992b, "Technological Inertia in Economic History," *Journal of Economic History* 52 (2): 331–32.

46. D. S. Landes, 1969, *The Unbound Prometheus: Technological Change and Development in Western Europe from 1750 to the Present* (Cambridge: Cambridge University Press), 8.

47. Quoted in C. Curtis, 1983, "Machines vs. Workers," *New York Times*, February 8.

48. P. H. Lindert and J. G. Williamson, 2016, *Unequal Gains: American Growth and Inequality Since 1700* (Princeton, NJ: Princeton University Press), 194.

Part 1

Epigraph: The first epigraph, from a royal edict to resolve conflicts in the town of Thorn (or Toruń) in 1523, is quoted in S. Ogilvie, 2019, *The European Guilds: An Economic Analysis* (Princeton, NJ: Princeton University Press), 390.

1. J. Diamond, 1993, "Ten Thousand Years of Solitude," *Discover*, March 1, 48–57.

2. D. Cardwell, 2001, *Wheels, Clocks, and Rockets: A History of Technology* (New York: Norton), 186.

Chapter 1

1. B. Russell, 1946, *History of Western Philosophy and Its Connection with Political and Social Circumstances: From the Earliest Times to the Present Day* (New York: Simon & Schuster), 25.

2. P. Bairoch, 1991, *Cities and Economic Development: From the Dawn of History to the Present* (Chicago: University of Chicago Press) 17–18.

3. D. R. Headrick, 2009, *Technology: A World History* (New York: Oxford University Press), 32–33.

4. D. Cardwell, 2001, *Wheels, Clocks, and Rockets: A History of Technology* (New York: Norton), 16–17.

5. P. Mantoux, 1961, *The Industrial Revolution in the Eighteenth Century: An Outline of the Beginnings of the Modern Factory System in England*, trans. M. Vernon (London: Routledge), 189.

6. Quoted in F. Klemm, 1964, *A History of Western Technology* (Cambridge, MA: MIT Press), 51.

7. Early accounts suggested that classical civilizations didn't achieve much meaningful technological progress. See, for example, M. I. Finley, 1965, "Technical Innovation and Economic Progress in the Ancient World," *Economic History Review* 18 (1): 29–45, and 1973, *The Ancient Economy* (Berkeley: University of California Press); H. Hodges, 1970, *Technology in the Ancient World* (New York: Barnes & Noble); D. Lee, 1973, "Science, Philosophy, and Technology in the Greco-Roman World: I," *Greece and Rome* 20 (1): 65–78. But more recently, scholars have argued that these accounts understate the civiliations' technological achievements. See K. D. White, 1984, *Greek and Roman Technology* (Ithaca, NY: Cornell University Press); J. Mokyr, 1992a, *The Lever of Riches: Technological Creativity and Economic Progress* (New York: Oxford University

Press); Cardwell, 2001, *Wheels, Clocks, and Rockets*; K. Harper, 2017, *The Fate of Rome: Climate, Disease, and the End of an Empire* (Princeton, NJ: Princeton University Press).

8. Finley, 1973, *The Ancient Economy*.

9. Mokyr, 1992a, *The Lever of Riches*, 20.

10. Harper, 2017, *The Fate of Rome*, 1.

11. Many of these technologies, however, were borrowed from earlier civilizations like the Babylonians or Egyptians.

12. Mokyr, 1992a, *The Lever of Riches*, 20.

13. The Samos aqueduct, the first of its kind, was built around 600 B.C. by the Greek engineer Eupalinus of Megara.

14. Mokyr, 1992a. *The Lever of Riches*, 20.

15. R. J. Forbes, 1958, *Man: The Maker* (New York: Abelard-Schuman), 73.

16. H. Heaton, 1936, *Economic History of Europe* (New York: Harper and Brothers), 58.

17. K. D. White, 1984, *Greek and Roman Technology*.

18. Mokyr, 1992a, *The Lever of Riches*, 27.

19. A. C. Leighton, 1972, *Transport and Communication in Early Medieval Europe AD 500–1100* (Newton Abbot: David and Charles Publishers).

20. On the importance of Archimedes to Galileo's work, see Cardwell, 2001, *Wheels, Clocks, and Rockets*, 83.

21. J. G. Landels, 2000, *Engineering in the Ancient World* (Berkeley: University of California Press), 201.

22. Price, D. de S., 1975, *Science Since Babylon* (New Haven, CT: Yale University Press), 48.

23. B. Gille, 1986, *History of Techniques*, vol. 2: *Techniques and Sciences* (New York: Gordon and Breach Science Publishers). See also Mokyr, 1992a, *The Lever of Riches*, 194.

24. J. D. Bernal, 1971, *Science in History*, vol. 1: *The Emergence of Science* (Cambridge, MA: MIT Press), 222.

25. Quoted in D. Acemoglu and J. A. Robinson, 2012, *Why Nations Fail: The Origins of Power, Prosperity and Poverty* (New York: Crown Business), 165.

26. For other examples of Roman rulers blocking replacing technologies, see ibid., 164–66.

27. A. P. Usher, 1954, *A History of Mechanical Innovations* (Cambridge, MA: Harvard University Press), 101.

28. P. Temin, 2006, "The Economy of the Early Roman Empire," *Journal of Economic Perspectives* 20 (1): 133–51, and 2012, *The Roman Market Economy* (Princeton, NJ: Princeton University Press).

29. Mokyr, 1992a, *The Lever of Riches*, 29.

30. Ibid., 31.

31. On Roman roads, see Cardwell, 2001, *Wheels, Clocks, and Rockets*, 33.

32. Mokyr, 1992a, *The Lever of Riches*, 31.

33. Cardwell, 2001. *Wheels, Clocks, and Rockets*, 48.

34. On the three-field system, see Mokyr, 1992a, *The Lever of Riches*, 31.

35. L. White, 1962, *Medieval Technology and Social Change* (New York: Oxford University Press), 43.

36. Although some Roman plows had wheels, the complete heavy plow didn't make its appearance until the sixth century.

37. L. White, 1962, *Medieval Technology and Social Change*.

38. Ibid.

39. The collar's being placed on the horse's neck instead of its shoulders meant that heavy strain almost choked it. On Richard Lefebvre des Noëttes and advances in horse technology, see Mokyr, 1992a, *The Lever of Riches*, 36–38.

40. For an analysis of the economics of horse technology relative to that of the ox, see J. Langdon, 1982, "The Economics of Horses and Oxen in Medieval England," *Agricultural History Review* 30 (1): 31–40.

41. See Mokyr, 1992a, *The Lever of Riches*, 36–38.

42. On the Domesday book, see M. T. Hodgen, 1939, "Domesday Water Mills," *Antiquity* 13 (51): 261–79.

43. Cardwell, 2001, *Wheels, Clocks, and Rockets*, 49.

44. L. White, 1962, *Medieval Technology and Social Change*, 89.

45. On Burchard and Pope Celestine III, see E. J. Kealey, 1987, *Harvesting the Air: Windmill Pioneers in Twelfth-Century England* (Berkeley: University of California Press), 180.

46. Usher, 1954, *A History of Mechanical Innovations*, 209.

47. L. Boerner and B. Severgnini, 2015, "Time for Growth" (Economic History Working Paper 222/2015, London School of Economics and Political Science).

48. L. Boerner and B. Severgnini, 2016, "The Impact of Public Mechanical Clocks on Economic Growth," *Vox*, October 10, https://voxeu.org/article/time-growth.

49. J. Le Goff, 1982, *Time, Work, and Culture in the Middle Ages* (Chicago: University of Chicago Press).

50. L. Mumford, 1934, *Technics and Civilization* (New York: Harcourt, Brace and World), 14.

51. On markets and clocks, see Boerner and Severgnini, 2015, "Time for Growth."

52. There was significant productivity growth in watch making from the late seventeenth century onward, but the industry was tiny. See M. Kelly and C. Ó Gráda, 2016, "Adam Smith, Watch Prices, and the Industrial Revolution," *Quarterly Journal of Economics* 131 (4): 1727–52.

53. On the price of books, see J. Van Zanden, 2004, "Common Workmen, Philosophers and the Birth of the European Knowledge Economy" (paper for Global Economic History Network Conference, Leiden, September 16–18).

54. Cardwell, 2001, *Wheels, Clocks, and Rockets*, 55.

55. On the numbers of books published, see ibid., 49.

56. G. Clark, 2001. "The Secret History of the Industrial Revolution" (Working paper, University of California, Davis), 60.

57. J. E. Dittmar, 2011, "Information Technology and Economic Change: The Impact of the Printing Press," *Quarterly Journal of Economics* 126 (3): 1133–72.

58. Quoted in F. J. Swetz, 1987, *Capitalism and Arithmetic: The New Math of the 15th Century* (La Salle, IL: Open Court), 20.

59. Dittmar, 2011, "Information Technology and Economic Change," 1140.

60. Quoted in W. Endrei and W. v. Stromer, 1974, "Textiltechnische und hydraulische Erfindungen und ihre Innovatoren in Mitteleuropa im 14. / 15. Jahrhundert," *Technikgeschichte* 41:90.

See also S. Ogilvie, 2019, *The European Guilds: An Economic Analysis* (Princeton, NJ: Princeton University Press), 390.

61. S. Füssel, 2005, *Gutenberg and the Impact of Printing* (Aldershot, UK: Ashgate).

62. U. Neddermeyer, 1997, "Why Were There No Riots of the Scribes?," *Gazette du Livre Médiéval* 31 (1): 1–8.

63. Quoted in ibid., 7.

64. Ibid., 8.

65. Mokyr, 1992a, *The Lever of Riches*, 57.

66. Quoted in B. Gille, 1969, "The Fifteenth and Sixteenth Centuries in the Western World," in *A History of Technology and Invention: Progress through the Ages*, ed. M. Daumas and trans. E. B. Hennessy (New York: Crown), 2:135–36.

67. On Bauer, Zonca, and Drebbel, see Mokyr, 1992a, *The Lever of Riches*, chapter 4.

68. Ibid., 58.

69. On the steam engine, see R. C. Allen, 2009a, *The British Industrial Revolution in Global Perspective* (Cambridge: Cambridge University Press), chapter 7.

70. F. Reuleaux, 1876, *Kinematics of Machinery: Outlines of a Theory of Machines*, trans. A.B.W. Kennedy (London: Macmillan), 9.

71. On Galileo's theory of mechanics, see D. Cardwell, 1972, *Turning Points in Western Technology: A Study of Technology, Science and History* (New York: Science History Publications).

72. On the machine maker, see Cardwell, 2001, *Wheels, Clocks, and Rockets*, 44.

73. For underground transport, horse-driven treadmills were introduced.

74. On advances in mining, new husbandry, and the seed drill, see Mokyr, 1992a, *The Lever of Riches*, chapter 4.

75. On the worker-replacing effects of the gig mill, see A. Randall, 1991, *Before the Luddites: Custom, Community and Machinery in the English Woollen Industry, 1776–1809* (Cambridge: Cambridge University Press), 120.

76. Quoted in Acemoglu and Robinson, 2012, *Why Nations Fail*, 176.

77. For more on examples of resistance to replacing technologies, see L. A. White, 2016, *Modern Capitalist Culture* (New York: Routledge), 77.

78. For more on the Leiden riots, see R. Patterson, 1957, "Spinning and Weaving," in *A History of Technology*, vol. 3, *From the Renaissance to the Industrial Revolution, c. 1500–c. 1750*, ed. C. Singer, E. J. Holmyard, A. R. Hall, and T. I. Williams (New York: Oxford University Press), 167.

79. Acemoglu and Robinson, 2012, *Why Nations Fail*, 197.

80. I. A. Gadd and P. Wallis, 2002, *Guilds, Society, and Economy in London 1450–1800* (London: Centre for Metropolitan History).

81. S. Ogilvie, 2019, *The European Guilds*, 5.

82. K. Desmet, A. Greif, and S. Parente, 2018, "Spatial Competition, Innovation and Institutions: The Industrial Revolution and the Great Divergence" (Working Paper. 24727, National Bureau of Economic Research, Cambridge, MA); J. Mokyr, 1998, "The Political Economy of Technological Change," in *Technological Revolutions in Europe: Historical Perspectives*, ed. K. Bruland and M. Berg (Cheltenham, UK: Edward Elgar), 39–64.

83. S. R. Epstein, 1998, "Craft Guilds, Apprenticeship and Technological Change in Pre-industrial Europe," *Journal of Economic History* 58 (3): 684–713.

84. Ibid., 696.

85. Ogilvie, 2019, *The European Guilds*, 415

86. Ibid., 410.

87. C. Dent, 2006, "Patent Policy in Early Modern England: Jobs, Trade and Regulation," *Legal History* 10 (1): 79–80.

88. In particular, the Thirty Years' War put pressure on governments to constantly modernize their armies.

89. Q. Wright, 1942, *A Study of War* (Chicago: University of Chicago Press), 1:215.

90. C. Tilly, 1975, *The Formation of National States in Western Europe* (Princeton, NJ: Princeton University Press), 42.

91. N. Ferguson, 2012, *Civilization: The West and the Rest* (New York: Penguin), 37.

92. N. Rosenberg and L. E. Birdzell, 1986, *How the West Grew Rich: The Economic Transformation of the Western World* (London: Basic), 138.

93. On the age of instruments, see Mokyr, 1992, *The Lever of Riches*, chapter 4.

94. Cardwell, 2001, *Wheels, Clocks, and Rockets*, 107.

Chapter 2

1. G. Clark, 2008. *A Farewell to Alms: A Brief Economic History of the World* (Princeton, NJ: Princeton University Press), 39.

2. Quoted in ibid.

3. D. Cannadine, 1977, "The Landowner as Millionaire: The Finances of the Dukes of Devonshire, c. 1800–c. 1926," *Agricultural History Review* 25 (2): 77–97.

4. P. H. Lindert, 2000b, "When Did Inequality Rise in Britain and America?," *Journal of Income Distribution* 9 (1): 11–25.

5. H. A. Taine, 1958, *Notes on England, 1860–70*, trans. E. Hyams (London: Strahan), 181. See also Cannadine, 1977, "The Landowner as Millionaire."

6. See P. H. Lindert, 1986, "Unequal English Wealth since 1670," *Journal of Political Economy* 94 (6): 1127–62.

7. T. Piketty, 2014, *Capital in the Twenty-First Century* (Cambridge, MA: Harvard University Press), figure 3.1.

8. See, for example, C. Boix, and F. Rosenbluth, 2014, "Bones of Contention: The Political Economy of Height Inequality," *American Political Science Review* 108 (1): 1–22.

9. J. Diamond, 1987, "The Worst Mistake in the History of the Human Race," *Discover*, May 1, 64–66.

10. See J. J. Rousseau, [1755] 1999, *Discourse on the Origin of Inequality* (New York: Oxford University Press).

11. See, for example, P. Eveleth and J. M. Tanner, 1976, *Worldwide Variation in Human Growth*, Cambridge Studies in Biological & Evolutionary Anthropology (Cambridge: Cambridge University Press).

12. G. J. Armelagos and M. N. Cohen, *Paleopathology at the Origins of Agriculture* (Orlando, FL: Academic Press).

13. C. S. Larsen, 1995, "Biological Changes in Human Populations with Agriculture," *Annual Review of Anthropology* 24 (1): 185–213.

14. A. Mummert, E. Esche, J. Robinson, and G. J. Armelagos, 2011. "Stature and Robusticity During the Agricultural Transition: Evidence from the Bioarchaeological Record," *Economics and Human Biology* 9 (3): 284–301.

15. Larsen, 1995, "Biological Changes in Human Populations with Agriculture."

16. K. Marx and F. Engels, [1848] 1967, *The Communist Manifesto*, trans. Samuel Moore (London: Penguin), 55.

17. On population pressure, see E. Boserup, 1965. *The Condition of Agricultural Growth: The Economics of Agrarian Change under Population Pressure* (London: Allen and Unwin).

18. J. Diamond, 1987, "The Worst Mistake in the History of the Human Race."

19. M. L. Bacci, 2017, *A Concise History of World Population* (London: John Wiley and Sons).

20. During the time period 1–1500, higher land productivity appears to have had significant effects on population density but insignificant effects on the standard of living. See Q. Ashraf and O. Galor, 2011, "Dynamics and Stagnation in the Malthusian Epoch," *American Economic Review* 101 (5): 2003–41.

21. For an overview, see J. Mokyr and H. J. Voth, 2010, "Understanding Growth in Europe, 1700–1870: Theory and Evidence," in *The Cambridge Economic History of Modern Europe*, ed. S. Broadberry and K. O'Rourke (Cambridge: Cambridge University Press), 1:7–42.

22. O. Galor and D. N. Weil, 2000, "Population, Technology, and Growth: From Malthusian Stagnation to the Demographic Transition and Beyond," *American Economic Review* 90 (4): 806–28; G. Clark, 2008, *A Farewell to Alms*.

23. For example, Ronald Lee and Michael Anderson cast doubt on the idea that the world was still Malthusian after 1500, showing that very little of the long-term variation in either fertility or mortality can be explained by wage patterns (2002, "Malthus in State Space: Macroeconomic-Demographic Relations in English History, 1540 to 1870," *Journal of Population Economics* 15 [2]: 195–220). Esteban Nicolini also found that after 1650, the fertility effect became much weaker (2007, "Was Malthus Right? A VAR Analysis of Economic and Demographic Interactions in Pre-Industrial England," *European Review of Economic History* 11 [1]: 99–121).

24. Using per capita gross domestic product (GDP) estimates for England, Alessandro Nuvolari and Mattia Ricci find that the period 1250–1580 was a Malthusian phase with no positive growth. During the period 1580–1780, however, Malthusian constraints appear to have relaxed, leading to a positive growth rate. (Nuvolari and Ricci, 2013, "Economic Growth in England, 1250–1850: Some New Estimates Using a Demand Side Approach," *Rivista di Storia Economica* 29 [1]: 31–54.)

25. R. C. Allen, 2009, "How Prosperous Were the Romans? Evidence from Diocletian's Price Edict (AD 301)," in *Quantifying the Roman Economic: Methods and Problems*, ed. Alan Bowman and Andrew Wilson (Oxford: Oxford University Press), 327–45.

26. J. Bolt and J. L. Van Zanden, 2014, "The Maddison Project: Collaborative Research on Historical National Accounts," *Economic History Review* 67 (3): 627–51.

27. A notable exception to stagnant growth outside the North Sea area are the per capita GDP estimates for northern Italy: per capita GDP there is suggested to have nearly doubled in

the period 1–1300. However, there are good reasons to believe that these estimates are over-stated, as has been suggested by several scholars (Bolt and Van Zanden, 2014, "The Maddison Project"; W. Scheidel, and S. J. Friesen, 2009, "The Size of the Economy and the Distribution of Income in the Roman Empire," *Journal of Roman Studies* 99 (March): 61–91). In the period 1300–1800, per capita GDP in northern Italy is estimated to have declined.

28. A. Maddison, 2005, *Growth and Interaction in the World Economy: The Roots of Modernity* (Washington: AEI Press), 21.

29. See J. De Vries, 2008, *The Industrious Revolution: Consumer Behavior and the Household Economy, 1650 to the Present* (Cambridge: Cambridge University Press).

30. See S. D. Chapman, 1967, *The Early Factory Masters: The Transition to the Factory System in the Midlands Textile Industry* (Exeter, UK: David and Charles).

31. F. F. Mendels, 1972, "Proto-industrialization: The First Phase of the Industrialization Process," *Journal of Economic History* 32 (1): 241–61.

32. P. H. Lindert and J. G. Williamson, 1982, "Revising England's Social Tables 1688–1812," *Explorations in Economic History* 19 (4): 385–408.

33. A. Maddison, 2002, *The World Economy: A Millennial Perspective* (Paris: Organisation for Economic Co-operation and Development).

34. Based on data from the 1086 Domesday Book and the numbers published by Gregory King in 1688, Graeme Snooks has estimated that the British economy grew at an annual rate of 0.29 percent in per capita terms (1994, "New Perspectives on the Industrial Revolution," in *Was the Industrial Revolution Necessary?*, ed. G. D. Snooks [London: Routledge], 1–26).

35. D. Defoe, [1724] 1971, *A Tour through the Whole Island of Great Britain* (London: Penguin), 432.

36. A. Smith, [1776] 1976, *An Inquiry into the Nature and Causes of the Wealth of Nations* (Chicago: University of Chicago Press), 365–66.

37. As noted above, a larger share of this wealth was appropriated by a fraction of the population. But although everyone didn't gain equally from growth, most workers lived well above subsistence levels. Based on King's 1688 social table of Britain, Allen estimates that the poorest group—consisting of cottagers, paupers, and vagrants—earned just about enough to buy a bare-bones subsistence basket. This group was probably no better off than hunter-gatherers several millennia before, but they accounted for less than a fifth of Britain's population. Other groups were substantially better off: Manufacturing workers, agricultural laborers, building craftsmen, miners, soldiers, sailors, and domestic servants (who made up 35 percent of the population) earned almost three times a subsistence income. And the largest category (consisting of shopkeepers, manufacturers, and farmers) earned as much as five times a subsistence income, while the wealthiest (including the landed classes and the bourgeoisie) could afford about twenty subsistence baskets (R. C. Allen, 2009a, *The British Industrial Revolution in Global Perspective* [Cambridge: Cambridge University Press], table 2.5).

38. Defoe, [1724] 1971, *A Tour through the Whole Island of Great Britain*, 338.

39. On downward social mobility, see G. Clark and G. Hamilton, 2006, "Survival of the Richest: The Malthusian Mechanism in Pre-Industrial England," *Journal of Economic History* 66 (3): 707–36.

40. Smith, [1776] 1976, *An Inquiry into the Nature and Causes of the Wealth of Nations*, 432.

41. M. Doepke and F. Zilibotti, 2008, "Occupational Choice and the Spirit of Capitalism," *Quarterly Journal of Economics* 123 (2): 747–93.

42. D. N. McCloskey, 2010, *The Bourgeois Virtues: Ethics for an Age of Commerce* (Chicago: University of Chicago Press).

43. Marx and Engels, [1848] 1967, *The Communist Manifesto*, 35.

44. F. Crouzet, 1985, *The First Industrialists: The Problems of Origins* (Cambridge: Cambridge University Press).

45. Defoe, [1724] 1971, *A Tour through the Whole Island of Great Britain*.

46. Crouzet, 1985, *The First Industrialists*, 4.

Chapter 3

1. On Schumpeterian versus Smithian growth in the preindustrial world, see J. Mokyr, 1992a, *The Lever of Riches: Technological Creativity and Economic Progress* (New York: Oxford University Press).

2. J. A. Schumpeter, 1939, *Business Cycles* (New York: McGraw-Hill), 1:161–74.

3. T. Malthus, [1798] 2013, *An Essay on the Principle of Population*, Digireads.com, 279, Kindle.

4. H. J. Habakkuk, 1962, *American and British Technology in the Nineteenth Century: The Search for Labour Saving Inventions* (Cambridge: Cambridge University Press), 22.

5. S. Lilley, 1966, *Men, Machines and History: The Story of Tools and Machines in Relation to Social Progress* (Paris: International Publishers).

6. In Czech, *robota* means forced labor of the kind that serfs had to perform on their master's lands and is derived from *rab*, meaning slave.

7. A. Young, 1772, *Political Essays Concerning the Present State of the British Empire* (London: printed for W. Strahan and T. Cadell).

8. On cheap labor and mechanization, see R. Hornbeck and S. Naidu, 2014, "When the Levee Breaks: Black Migration and Economic Development in the American South," *American Economic Review* 104 (3): 963–90.

9. R. C. Allen, 2009a, *The British Industrial Revolution in Global Perspective* (Cambridge: Cambridge University Press).

10. Robert Allen follows in the footsteps of the economist Sir John Habakkuk, who argued that the scarcity of labor in antebellum America, together with its abundance of land, led to high wages that in turn produced efforts to substitute machines for workers (1962, *American and British Technology in the Nineteenth Century*).

11. Edward Anthony Wrigley has also argued that productivity soared during the Industrial Revolution because of the abundance of coal at the disposal of the British worker. He suggests that the switch from an organic economy to an energy-rich one was at the heart of the Industrial Revolution (2010, *Energy and the English Industrial Revolution* [Cambridge: Cambridge University Press]).

12. For revised wages of spinners, see J. Humphries and B. Schneider, forthcoming, "Spinning the Industrial Revolution," *Economic History Review*. For evidence suggesting that real wages in England in the period 1650–1800 were lower than previously thought, see

J. Z. Stephenson, 2018, "'Real' Wages? Contractors, Workers, and Pay in London Building Trades, 1650–1800," *Economic History Review* 71 (1): 106–32.

13. Mokyr, 1992a, *The Lever of Riches*, 151.

14. J. Diamond, 1998, *Guns, Germs and Steel: A Short History of Everybody for the Last 13,000 Years* (New York: Random House), chapter 13.

15. For a detailed summary of supply-side hurdles to innovation, see Mokyr, 1992a, *The Lever of Riches*, chapter 7.

16. J. Mokyr, 2011, *The Enlightened Economy: Britain and the Industrial Revolution, 1700–1850* (London: Penguin), Kindle.

17. M. Weber, 1927, *General Economic History* (New Brunswick, NJ: Transaction Books).

18. B. Russell, 1946, *History of Western Philosophy and Its Connection with Political and Social Circumstances: From the Earliest Times to the Present Day* (New York: Simon & Schuster), 110.

19. Mokyr, 1992a, *The Lever of Riches*, 196.

20. L. White, 1967, "The Historical Roots of Our Ecologic Crisis," *Science* 155 (3767): 1205.

21. Mokyr, 1992a, *The Lever of Riches*, 203.

22. Mokyr, 2011, *The Enlightened Economy*, introduction.

23. See, for example, D. C. North and B. R. Weingast, 1989, "Constitutions and Commitment: The Evolution of Institutions Governing Public Choice in Seventeenth-Century England," *Journal of Economic History* 49 (4): 803–32; D. C. North, 1991, "Institutions," *Journal of Economic Perspectives* 5 (1): 97–112.

24. D. Acemoglu, S. Johnson, and J. Robinson, 2005, "The Rise of Europe: Atlantic Trade, Institutional Change, and Economic Growth," *American Economic Review* 95 (3): 546–79.

25. On commercial partnerships and the prevention of royal monopolies see R. Davis, 1973, *English Overseas Trade 1500–1700* (London: Macmillan), 41; R. Cameron, 1993, *A Concise Economic History of the World from Paleolithic Times to the Present*, 2nd ed. (New York: Oxford University Press), 127; Acemoglu, Johnson, and Robinson, 2005, "The Rise of Europe," 568.

26. See, for example, W. C. Scoville, 1960, *The Persecution of Huguenots and French Economic Development, 1680–1720* (Berkeley: University of California Press).

27. Of course, parliaments didn't exist only in Britain and the Dutch Republic. Beginning in twelfth-century Spain, they gradually spread across Western Europe. Medieval parliaments were independent bodies that represented various social groups, including members of three estates (the nobility, the clergy, and, in some instances, the peasantry), and provided checks on the crown by granting taxes and taking an active role in the legislative process. However, after their initial rise and relative success during the late Middle Ages, monarchs often refused to convene their parliament and found various ways to limit their powers.

28. J. L. Van Zanden, E. Buringh, and M. Bosker, 2012, "The Rise and Decline of European Parliaments, 1188–1789," *Economic History Review* 65 (3): 835–61.

29. Acemoglu, Johnson, and Robinson, 2005, "The Rise of Europe," 546–79.

30. On the Bill of Rights, see G. W. Cox, 2012, "Was the Glorious Revolution a Constitutional Watershed?," *Journal of Economic History* 72 (3): 567–600.

31. On the Whig coalition, see D. Stasavage, 2003, *Public Debt and the Birth of the Democratic State: France and Great Britain 1688–1789* (Cambridge: Cambridge University Press).

32. Mokyr, 1992a, *The Lever of Riches*, 243.

33. On diversified wealth, see D. Acemoglu and J. A. Robinson, 2006, "Economic Backwardness in Political Perspective," *American Political Science Review* 100 (1): 115–31.

34. For a detailed account of how political elites might block technological progress out of fear of political replacement, see ibid.

35. Quoted in Mokyr, 2011, *The Enlightened Economy*, chap. 3.

36. Quoted in ibid.

37. Quoted in P. Mantoux, 1961, *The Industrial Revolution in the Eighteenth Century: An Outline of the Beginnings of the Modern Factory System in England*, trans. M. Vernon (London: Routledge), 135.

38. Quoted in ibid., 134.

39. Ibid., 30–31.

40. Quoted in D. Acemoglu and J. A. Robinson, 2012, *Why Nations Fail: The Origins of Power, Prosperity and Poverty* (New York: Crown Business), 219.

41. Ibid., 221.

42. "Machinery Causes a Riot," 1895, *New York Times*, November 25.

43. Acemoglu and Robinson, *Why Nations Fail*, 197.

44. A. Randall, 1991, *Before the Luddites: Custom, Community and Machinery in the English Woollen Industry, 1776–1809* (Cambridge: Cambridge University Press).

45. According to Francis Aiden Hibbert, new industries did not fall under the Apprenticeship Act of the Statute of Artificers, meaning that their existence weakened the guilds (1891, *The Influence and Development of English Guilds* [New York: Sentry], 129).

46. K. Desmet, A. Greif, and S. Parente, 2018, "Spatial Competition, Innovation and Institutions: The Industrial Revolution and the Great Divergence" (Working Paper 24727, National Bureau of Economic Research, Cambridge, MA).

47. C. MacLeod, 1998. *Inventing the Industrial Revolution: The English Patent System, 1660–1800* (Cambridge: Cambridge University Press), 160.

48. H. B. Morse, 1909, *The Guilds of China* (London: Longmans, Green and Co.), 1.

49. Quoted in Desmet, Greif, and Parente, "Spatial Competition, Innovation and Institutions," 37–38.

50. Quoted in ibid., 38.

51. Ibid., 39.

52. Mokyr, 1992a, *The Lever of Riches*, 257.

53. Quoted in Mantoux, 1961, *The Industrial Revolution in the Eighteenth Century*, 403.

54. J. Horn, 2008, *The Path Not Taken: French Industrialization in the Age of Revolution, 1750–1839* (Cambridge, MA: MIT Press) chapter 4, Kindle.

55. See E. P. Thompson, 1963, *The Making Of The English Working Class* (London: Gollancz, Vintage Books).

56. On the French machinery riots, see J. Horn, 2008, *The Path Not Taken*, chap. 4.

57. Ibid., 8.

58. F. Machlup, 1962, *The Production and Distribution of Knowledge in the United States* (Princeton, NJ: Princeton University Press), 166.

59. Desmet, Greif, and Parente, 2018, "Spatial Competition, Innovation and Institutions," 15–16.

Part 2

1. Marx's chapter "The Division of Labour and Manufacture" in *Das Kapital* well captures the extreme divisions of labor that existed, and it is followed by a chapter titled "Machinery and the Factory System" ([1867] 1999, *Das Kapital*, trans. S. Moore and E. Aveling [New York: Gateway], chapter 15, Kindle).

2. W. W. Rostow, 1960, *The Stages of Growth: A Non-Communist Manifesto* (Cambridge: Cambridge University Press).

3. D. Phyllis and W. A. Cole, 1962, *British Economic Growth, 1688–1959: Trends and Structure* (Cambridge: Cambridge University Press); N. F. Crafts, 1985, *British Economic Growth during the Industrial Revolution* (New York: Oxford University Press); N. F. Crafts and C. K. Harley, 1992, "Output Growth and the British Industrial Revolution: A Restatement of the Crafts-Harley View," *Economic History Review* 45 (4): 703–30.

4. B. Mitchell, 1975, *European Historical Statistics, 1750–1970* (London: Macmillan), 438.

5. T. S. Ashton, 1948, *An Economic History of England: The Eighteenth Century* (London: Routledge), 58.

6. M. W. Flinn, 1966, *The Origins of the Industrial Revolution* (London: Longmans), 15.

Chapter 4

1. D. Cardwell, 1972, *Turning Points in Western Technology: A Study of Technology, Science and History* (New York: Science History Publications).

2. A. Ure, 1835, *The Philosophy of Manufactures* (London: Charles Knight), 14.

3. Quoted in P. Mantoux, 1961, *The Industrial Revolution in the Eighteenth Century: An Outline of the Beginnings of the Modern Factory System in England*, trans. M. Vernon (London: Routledge [first published in 1928]), 39.

4. On the domestic system, see ibid., 54–61.

5. For a detailed account suggesting that the rise of the factory was a technological event, see J. Mokyr, 2001, "The Rise and Fall of the Factory System: Technology, Firms, and Households since the Industrial Revolution," *Carnegie-Rochester Conference Series on Public Policy*, 55(1): 1–45.

6. M. W. Flinn, 1962, *Men of Iron: The Crowleys in the Early Iron Industry* (Edinburgh: Edinburgh University Press), 252.

7. On cotton yarn manufacturing, see R. C. Allen, 2009a, *The British Industrial Revolution in Global Perspective* (Cambridge: Cambridge University Press), chapter 8, Kindle.

8. Mantoux, 1961, *The Industrial Revolution in the Eighteenth Century*, 234.

9. It was in the same year that Adam Smith published *The Wealth of Nations* that the industrial undertakings that would eventually make Britain a truly wealthy nation took off.

10. Quoted in Mantoux, 1961, *The Industrial Revolution in the Eighteenth Century*, 213.

11. Ibid., 14.

12. On the labor savings of Arkwright's inventions, see Allen, 2009a, *The British Industrial Revolution*, chapter 8.

13. R. C. Allen, 2009d, "The Industrial Revolution in Miniature: The Spinning Jenny in Britain, France, and India," *Journal of Economic History* 69 (4): 901–27.

14. J. Humphries, 2013, "The Lure of Aggregates and the Pitfalls of the Patriarchal Perspective: A Critique of the High Wage Economy Interpretation of the British Industrial Revolution," *Economic History Review* 66 (3): 709.

15. Ure, 1835, *The Philosophy of Manufactures*, 23.

16. On pauper apprentices, see J. Humphries, 2010, *Childhood and Child Labour in the British Industrial Revolution* (Cambridge: Cambridge University Press), 246.

17. Humphries, 2013, "The Lure of Aggregates and the Pitfalls of the Patriarchal Perspective," 710.

18. Mantoux, 1961, *The Industrial Revolution in the Eighteenth Century*, 241–44.

19. J. Bessen, 2015, *Learning by Doing: The Real Connection between Innovation, Wages, and Wealth* (New Haven, CT: Yale University Press), 75. Though Bessen's calculations are for American factories, the power loom likely had similar labor-saving effects in Britain.

20. K. Marx, [1867] 1999, *Das Kapital*, trans. S. Moore and E. Aveling (New York: Gateway), chapter 15, section 1, Kindle.

21. Like Savery, Watt conceived of numerous applications for his engine. The patent that he took out in 1784 makes clear that it was not an invention intended for a specific purpose. In the words of Marx, he thought of it as "an agent universally applicable in mechanical industry" (ibid). Some of the many applications he listed in his patent filing would have to wait, but eventually they were realized in practice. The steam hammer, for example, was introduced about half a century later. And other later applications even went beyond his envisioned uses. While in doubt about the use of steam in shipping, the Boulton & Watt company later displayed steam engines for ocean steamers at the Crystal Palace Exhibition of 1851, some three decades after Watt's death.

22. G. N. Von Tunzelmann, 1978, *Steam Power and British Industrialization to 1860* (Oxford: Oxford University Press).

23. J. Kanefsky and J. Robey, 1980, "Steam Engines in 18th-Century Britain: A Quantitative Assessment," *Technology and Culture* 21 (2): 161–86.

24. N. F. Crafts, 2004, "Steam as a General Purpose Technology: A Growth Accounting Perspective," *Economic Journal* 114 (495): 338–51.

25. J. Hoppit, 2008, "Political Power and British Economic Life, 1650–1870," in *The Cambridge Economic History of Modern Britain*, vol. 1, *Industrialisation, 1700–1870*, ed. R. Floud, J. Humphries, and P. Johnson (Cambridge: Cambridge University Press), 370–71.

26. J. Mokyr, 2011, *The Enlightened Economy: Britain and the Industrial Revolution, 1700–1850* (London: Penguin), chapter 10, Kindle.

27. T. Leunig, 2006, "Time Is Money: A Re-Assessment of the Passenger Social Savings from Victorian British Railways," *Journal of Economic History* 66 (3): 635–73.

28. For a history of the Darby family and the Coalbrookdale Iron Company, see Allen, 2009a, *The British Industrial Revolution*, chapter 9.

29. Ibid.

30. Quoted in J. Langton and R. J. Morris, 2002, *Atlas of Industrializing Britain, 1780–1914* (London: Routledge), 88.

31. G. R. Hawke, 1970, *Railways and Economic Growth in England and Wales, 1840–1870* (Oxford: Clarendon Press of Oxford University Press).

32. Leunig, 2006, "Time Is Money."

33. On the social savings of the turnpikes, see C. Bogart, 2005, "Turnpike Trusts and the Transportation Revolution in 18th Century England," *Explorations in Economic History* 42 (4): 479–508.

34. Of course, none of these estimates capture the benefits of steam-powered transportation in full, as steam also transformed water travel. Even as early as 1821, there were 188 steamers operating in Britain. In their absence, goods would have had to have been hauled around, although for shorter routes canals were often more suitable than coasters. In addition, not long after the first railroad began operation, steam began to revolutionize ocean travel. Isambard Kingdom Brunel's *Great Western*—which in 1838 became the first ocean steamship to cross the Atlantic— was a landmark achievement on a par with the Rocket. Yet because longer journeys required ships to carry vast amounts of coal as cargo, it took almost half a century for steam to displace sail. It was not until the end of the nineteenth century that the coal requirements of steam engines had fallen enough for steamships to cover the distance between China and Britain.

35. E. Baines, 1835, *History of the Cotton Manufacture in Great Britain* (London: H. Fisher, R. Fisher, and P. Jackson), 5.

Chapter 5

1. B. Disraeli, 1844, *Coningsby* (a Public Domain Book), 187, Kindle.

2. F. Engels, [1844] 1943, *The Condition of the Working-Class in England in 1844*. Reprint, London: Allen & Unwin, 100; 25–26.

3. See D. Defoe, [1724] 1971, *A Tour through the Whole Island of Great Britain* (London: Penguin), 432.

4. D. S. Landes, 1969, *The Unbound Prometheus: Technological Change and Development in Western Europe from 1750 to the Present* (Cambridge: Cambridge University Press), 128.

5. "The Present Condition of British Workmen," 1834, accessed December 15, 2018, https://deriv.nls.uk/dcn9/7489/74895330.9.htm.

6. On the urban wage premium, see J. G. Williamson, 1987, "Did English Factor Markets Fail during the Industrial Revolution?," *Oxford Economic Papers* 39 (4): 641–78.

7. On the trends in output, see N. F. Crafts and C. K. Harley, 1992, "Output Growth and the British Industrial Revolution: A Restatement of the Crafts-Harley View," *Economic History Review* 45 (4): 703–30.

8. C. H. Feinstein, 1998, "Pessimism Perpetuated: Real Wages and the Standard of Living in Britain during and after the Industrial Revolution," *Journal of Economic History* 58 (3): 625–58; R. C. Allen, 2009b, "Engels' Pause: Technical Change, Capital Accumulation, and Inequality in the British Industrial Revolution," *Explorations in Economic History* 46 (4): 418–35. The real wage series of Gregory Clark suggests that it was not till the 1820s that real wages advanced beyond their level in the middle of the eighteenth century (2005, "The Condition of the Working Class in England, 1209–2004," *Journal of Political Economy*, 113 [6] 1307–40). After 1820, according to Clark, they rose more rapidly than estimated by Allen or Feinstein. However, this is hard to square with what we know from data on consumption and heights, as well as contemporary accounts.

9. On working hours, see H. Voth, 2000, *Time and Work in England 1750–1830* (Oxford: Clarendon Press of Oxford University Press).

10. On the rate of profit, see Allen, 2009b, "Engels' Pause."

11. On the top 5 percent income share, see P. H. Lindert, 2000b, "When Did Inequality Rise in Britain and America?," *Journal of Income Distribution* 9 (1): 11–25.

12. G. Clark, M. Huberman, and P. H. Lindert, 1995, "A British Food Puzzle, 1770–1850," *Economic History Review* 48 (2): 215–37. As noted, however, recent studies show that there is no puzzle, as real wages stagnated and fell among the lower ranks.

13. S. Horrell, 1996, "Home Demand and British Industrialisation," *Journal of Economic History* 56 (September): 561–604.

14. R. H. Steckel, 2008, "Biological Measures of the Standard of Living," *Journal of Economic Perspectives* 22 (1): 129–52. The idea that observed height data can be used to approximate the elusive standard of living was first proposed by Robert Fogel (1983, "Scientific History and Traditional History," in *Which Road to the Past?*, ed. R. W. Fogel and G. R. Elton (New Haven, CT: Yale University Press), 5–70.

15. R. C. Floud, K. Wachter, and A. Gregory, 1990, *Height, Health, and History: Nutritional Status in the United Kingdom, 1750–1980* (Cambridge: Cambridge University Press), chapter 4; J. Komlos, 1998, "Shrinking in a Growing Economy? The Mystery of Physical Stature during the Industrial Revolution," *Journal of Economic History* 58 (3): 779–802.

16. On the environmental versus the poverty view, see J. Mokyr, 2011, *The Enlightened Economy: Britain and the Industrial Revolution, 1700–1850* (London: Penguin), chapter 10, Kindle.

17. See, for example, J. G. Williamson, 2002, *Coping with City Growth during the British Industrial Revolution* (Cambridge: Cambridge University Press).

18. S. Szreter and G. Mooney, 1998, "Urbanization, Mortality, and the Standard of Living Debate: New Estimates of the Expectation of Life at Birth in Nineteenth-Century British Cities," *Economic History Review* 51 (1): 84–112.

19. J. Komlos and B. A'Hearn, 2017, "Hidden Negative Aspects of Industrialization at the Onset of Modern Economic Growth in the US," *Structural Change and Economic Dynamics* 41 (June): 43.

20. F. M. Eden, 1797, *The State of the Poor; or, An History of the Labouring Classes in England* (London: B. and J. White), 3:848.

21. D. Ricardo, [1817] 1911, *The Principles of Political Economy and Taxation* (Repr., London: Dent).

22. Jean-Baptiste Say, for example, argued that cost reductions caused by labor-saving technology would result in price reductions and thus growing demand, implying that it was only a matter of time until displaced workers were reemployed elsewhere. Although Ricardo later revisited his model, he still did not believe—along with Malthus and Marx—that industrialization could improve real wages over the long run.

23. E. C. Gaskell, 1884, *Mary Barton* (London: Chapman and Hall), 104.

24. See K. Marx, [1867] 1999, *Das Kapital*, trans. S. Moore and E. Aveling (New York: Gateway), chapter 15, section 4, Kindle; C. Dickens, [1854] 2017, *Hard Times* (Amazon Classics), chapter 5, Kindle.

25. As Kay-Shuttleworth noted, "While the engine works, the people must work. Men, women and children are thus yokefellows with iron and steam. . . . The persevering labour of

the operative must rival the mathematical precision, the incessant motion, and the exhaustless power of the machine" (1832, *The Moral and Physical Condition of the Working Classes Employed in the Cotton Manufacture in Manchester* Manchester: Harrisons & Crosfield).

26. On the factory and its discontents, see P. Gaskell, 1833, *The Manufacturing Population of England, Its Moral, Social, and Physical Conditions* (London: Baldwin and Cradock), 16.

27. Landes, 1969, *The Unbound Prometheus*, 2.

28. P. Gaskell, 1833, *The Manufacturing Population of England*, 12 and 341.

29. Marx, [1867] 1999, *Das Kapital* , chapter 15, section 5.

30. C. Babbage, 1832, *On the Economy of Machinery and Manufactures* (London: Charles Knight), 266–67.

31. A. Ure, 1835, *The Philosophy of Manufactures* (London: Charles Knight), 220.

32. E. Baines, 1835, *History of the Cotton Manufacture in Great Britain* (London: H. Fisher, R. Fisher, and P. Jackson), 452.

33. Ibid., 460.

34. Ibid., 435.

35. J. Humphries and B. Schneider, forthcoming, "Spinning the Industrial Revolution," *Economic History Review*.

36. J. Humphries, 2010, *Childhood and Child Labour in the British Industrial Revolution* (Cambridge: Cambridge University Press), 342.

37. R. C. Allen, forthcoming, "The Hand-Loom Weaver and the Power Loom: A Schumpeterian Perspective," *European Review of Economic History*.

38. Humphries, 2010, *Childhood and Child Labour*.

39. Allen, forthcoming, "The Hand-Loom Weaver and the Power Loom."

40. D. Bythell, 1969, *The Handloom Weavers: A Study in the English Cotton Industry during the Industrial Revolution* (Cambridge: Cambridge University Press), 139.

41. C. Nardinelli, 1986, "Technology and Unemployment: The Case of the Handloom Weavers," *Southern Economic Journal* 53 (1): 87–94.

42. On technological versus cyclical unemployment, see ibid.

43. J. Fielden, 2013, *Curse of the Factory System* (London: Routledge).

44. On urban migration, see J. Humphries and T. Leunig, 2009, "Was Dick Whittington Taller Than Those He Left Behind? Anthropometric Measures, Migration and the Quality of Life in Early Nineteenth-Century London," *Explorations in Economic History* 46 (1): 120–31; J. Long, 2005, "Rural-Urban Migration and Socioeconomic Mobility in Victorian Britain," *Journal of Economic History* 65 (1): 1–35; M. Anderson, 1990, "The Social Implications of Demographic Change," in *The Cambridge Social History of Britain, 1750–1950*, vol. 2: *People and Their Environment*, ed. F.M.L. Thompson (Cambridge: Cambridge University Press), 1–70; H. R. Southall, 1991, "The Tramping Artisan Revisits: Labour Mobility and Economic Distress in Early Victorian England," *Economic History Review* 44 (2): 272–96. For an overview of urban migration during the Industrial Revolution, see P. Wallis, 2014, "Labour Markets and Training," in *The Cambridge Economic History of Modern Britain*, vol. 1, *Industrialisation, 1700–1870*, ed. R. Floud, J. Humphries, and P. Johnson (Cambridge: Cambridge University Press), 178–210.

45. A. Ure, 1835, *The Philosophy of Manufactures* (London: Charles Knight), 20.

46. Quoted in P. Gaskell, 1833, *The Manufacturing Population of England*, 174.

47. The early jenny was "awkward" for adults but managed "with dexterity"' by children ages 9–12 (M. Berg, 2005, *The Age of Manufactures, 1700–1820: Industry, Innovation and Work in Britain* [London: Routledge], 146).

48. C. Tuttle, 1999, *Hard at Work in Factories and Mines: The Economics of Child Labor during the British Industrial Revolution* (Boulder, CO: Westview Press), 110.

49. Ure, 1835, *The Philosophy of Manufactures*, 144.

50. On the upsurge in child labor, see Tuttle, 1999, *Hard at Work in Factories and Mines*, 96 and 142. See also Wallis, 2014, "Labour Markets and Training," 193.

51. P. Mantoux, 1961, *The Industrial Revolution in the Eighteenth Century: An Outline of the Beginnings of the Modern Factory System in England*, trans. M. Vernon (London: Routledge), 410.

52. Quoted in S. Smiles, 1865, *Lives of Boulton and Watt* (Philadelphia: J. B. Lippincott), 227. See also Mokyr, 2011, *The Enlightened Economy*, chapter 15.

53. Baines, 1835, *History of the Cotton Manufacture in Great Britain*, 452.

54. L. Shaw-Taylor and A. Jones, 2010, "The Male Occupational Structure of Northamptonshire 1777–1881: A Case of Partial De-Industrialization?" (working paper, Cambridge University).

55. M. Berg, 1976, "The Machinery Question," PhD diss., University of Oxford, 2.

56. Mantoux, 1961, *The Industrial Revolution in the Eighteenth Century*, 408.

57. Old Bailey Proceedings, 6th July 1768, Old Bailey Proceedings Online, version 8.0, 01 January 2019, www.oldbaileyonline.org.

58. On Limehouse, see Mantoux, 1961, *The Industrial Revolution in the Eighteenth Century*, 401–8.

59. Ibid.

60. T. C. Hansard, 1834, *General Index to the First and Second Series of Hansard's Parliamentary Debates: Forming a Digest of the Recorded Proceedings of Parliament, from 1803 to 1820* (New York: Kraus Reprint Co.).

61. R. Jackson, 1806, *The Speech of R. Jackson Addressed to the Committee of the House of Commons Appointed to Consider of the State of the Woollen Manufacture of England, on Behalf of the Cloth-Workers and Sheermen of Yorkshire, Lancashire, Wiltshire, Somersetshire and Gloucestershire* (London: C. Stower), 11.

62. Quoted in Mantoux, 1961, *The Industrial Revolution in the Eighteenth Century*, 408.

63. J. Horn, 2008, *The Path Not Taken: French Industrialization in the Age of Revolution, 1750–1830* (Cambridge, MA: MIT Press), chapter 4, Kindle.

64. *Annual Registrar or a View of the History, Politics, and Literature for the Year 1811*, 1811 (London: printed for Baldwin, Cradock, and Joy), 292.

65. On Liverpool and Kenyon, see Berg, 1976, "The Machinery Question," 76.

66. Horn, 2008, *The Path Not Taken*, chapter 4.

67. On the machines destroyed, see B. Caprettini and H. Voth, 2017, "Rage against the Machines: Labour-Saving Technology and Unrest in England, 1830–32" (working paper, University of Zurich).

68. E. Hobsbawm and G. Rudé, 2014, *Captain Swing* (New York: Verso), 265–79.

69. Caprettini and Voth, 2017, "Rage against the Machines."

70. D. Acemoglu and P Restrepo, 2018a, "Artificial Intelligence, Automation and Work" (Working Paper 24196, National Bureau of Economic Research, Cambridge, MA).

71. Allen, 2009b, "Engels' Pause."

72. E. S. Phelps, 2015, *Mass Flourishing: How Grassroots Innovation Created Jobs, Challenge, and Change* (Princeton, NJ: Princeton University Press), 47.

73. Quoted in ibid., 46.

74. O. Galor, 2011, "Inequality, Human Capital Formation, and the Process of Development," in *Handbook of the Economics of Education*, ed. Hanushek, E.A., Machin, S.J. and Woessmann, L. Amsterdam: Elsevier), 4:441–93.

75. For an overview of trends in human capital, see Wallis, 2014, "Labour Markets and Training," 203.

76. M. Sanderson, 1995, *Education, Economic Change and Society in England 1780–1870* (Cambridge; Cambridge University Press); D. F. Mitch, 1992, *The Rise of Popular Literacy in Victorian England: The Influence of Private Choice and Public Policy* (Philadelphia: University of Pennsylvania Press).

77. N. F. Crafts, 1985, *British Economic Growth during the Industrial Revolution* (Oxford: Oxford University Press), 73.

78. Landes, 1969, *The Unbound Prometheus*, 340. David Mitch also shows that the jobs of the early Industrial Revolution did not require much education or even literacy (1992, *The Rise of Popular Literacy in Victorian England*). However, in the late nineteenth century, literacy became increasingly desirable, according to job advertisements (D. F. Mitch, 1993, "The Role of Human Capital in the First Industrial Revolution," in *The British Industrial Revolution: An Economic Perspective*, ed. J. Mokyr [Boulder, CO: Westview Press, 241–80.]).

79. Tuttle, 1999, *Hard at Work in Factories and Mines*, 96 and 142; Wallis, 2014, "Labour Markets and Training," 193.

80. C. Goldin and K. Sokoloff, 1982, "Women, Children, and Industrialization in the Early Republic: Evidence from the Manufacturing Censuses," *Journal of Economic History* 42 (4): 741–74.

81. L. F. Katz and R. A. Margo, 2013, "Technical Change and the Relative Demand for Skilled Labor: The United States in Historical Perspective" (Working Paper 18752, National Bureau of Economic Research, Cambridge, MA), 3.

82. P. Gaskell, 1833, *The Manufacturing Population of England*, 182.

83. See G. Clark, 2005, "The Condition of the Working Class in England."

84. However, the skill premium in itself is not necessarily suggestive of the demand for skills, as it also depends on supply factors: a skill premium will emerge only if the demand for human capital outpaces its supply. And the supply of skills increased throughout the century.

85. G. Clark, 2005. "The Condition of the Working Class in England."

86. J. Bessen, 2015, *Learning by Doing: The Real Connection between Innovation, Wages, and Wealth* (New Haven, CT: Yale University Press), chapter 6.

87. Mokyr, 2011, *The Enlightened Economy*, chapter 15.

88. See D. H. Aldcroft and M. J. Oliver, 2000, *Trade Unions and the Economy: 1870–2000*, (Aldershot, UK: Ashgate Publishing).

Part 3

Epigraph: The second epigraph is from P. Zachary, 1996, "Does Technology Create Jobs, Destroy Jobs, or Some of Both?," *Wall Street Journal*, June 17.

1. J. Horn, 2008, *The Path Not Taken: French Industrialization in the Age of Revolution, 1750–1830* (Cambridge, MA: MIT Press).

2. On guild restrictions in Prussia, see T. Lenoir, 1998, "Revolution from Above: The Role of the State in Creating the German Research System, 1810–1910," *American Economic Review* 88 (2): 22–27.

3. On education and industrialization in Prussia, see S. O. Becker, E. Hornung, and L. Woessmann, 2011, "Education and Catch-Up in the Industrial Revolution," *American Economic Journal: Macroeconomics* 3 (3): 92–126.

4. On catch-up growth, see A. Gerschenkron, 1962, *Economic Backwardness in Historical Perspective: A Book of Essays* (Cambridge, MA: Belknap Press of Harvard University Press).

5. P. H. Lindert, 2004, *Growing Public*), vol. 1, *The Story: Social Spending and Economic Growth Since the Eighteenth Century* (Cambridge: Cambridge University Press), table 1.2.

6. M. Alexopoulos and J. Cohen, 2016, "The Medium Is the Measure: Technical Change and Employment, 1909–1949," *Review of Economics and Statistics* 98 (4): 792–810.

7. D. Acemoglu and P. Restrepo, 2018b, "The Race between Man and Machine: Implications of Technology for Growth, Factor Shares, and Employment," *American Economic Review* 108 (6): 1489

Chapter 6

1. Quoted in G. Tucker, 1837, *The Life of Thomas Jefferson, Third President of the United States: With Parts of His Correspondence Never Before Published, and Notices of His Opinions on Questions of Civil Government, National Policy, and Constitutional Law* (Philadelphia: Carey, Lea and Blanchard), 2:226.

2. A. de Tocqueville, 1840, *Democracy in America*, trans. H. Reeve (New York: Alfred A. Knopf), 2:191.

3. E. W. Byrn, 1900, *The Progress of Invention in the Nineteenth Century* (New York: Munn and Company), 1.

4. R. J. Gordon, 2016, *The Rise and Fall of American Growth: The U.S. Standard of Living Since the Civil War* (Princeton, NJ: Princeton University Press), 150.

5. D. Hounshell, 1985, *From the American System to Mass Production, 1800–1932: The Development of Manufacturing Technology in the United States* (Baltimore, MD: Johns Hopkins University Press), 307.

6. Ibid.

7. Quoted in B. Bryson, 2010, *At Home: A Short History of Private Life* (Toronto: Doubleday Canada), 29.

8. N. Rosenberg, 1963, "Technological Change in the Machine Tool Industry, 1840–1910," *Journal of Economic History* 23 (4): 414–43.

9. Quoted in D. Hounshell, 1985, *From the American System to Mass Production*, 19.

10. Ibid., 17–19.

11. Quoted in ibid., 233.

12. On electricity and working conditions, see D. E. Nye, 1990, *Electrifying America: Social Meanings of a New Technology, 1880–1940* (Cambridge, MA: MIT Press), 232.

13. Quoted in T. C. Martin, 1905, "Electrical Machinery, Apparatus, and Supplies," in *Census of Manufactures, 1905* (Washington, DC: United States Bureau of the Census), 170.

14. P.A. David and G. Wright, 1999, *Early Twentieth Century Productivity Growth Dynamics: An Inquiry into the Economic History of Our Ignorance* (Oxford: Oxford University Press).

15. E. Clark, 1925, "Giant Power Transforming America's Life," *New York Times*, February 22.

16. Ibid.

17. V. Smil, 2005, *Creating the Twentieth Century: Technical Innovations of 1867–1914 and Their Lasting Impact* (New York: Oxford University Press), 53.

18. Nye, 1990, *Electrifying America*, 232.

19. P. A. David, 1990, "The Dynamo and the Computer: An Historical Perspective on the Modern Productivity Paradox," *American Economic Review* 80 (2): 355–61.

20. W. D. Devine Jr., 1983, "From Shafts to Wires: Historical Perspective on Electrification," *Journal of Economic History* 43 (2): 347–72.

21. H. Jerome, 1934, "Mechanization in Industry" (Working Paper 27, National Bureau of Economic Research, Cambridge, MA), 48.

22. D. E. Nye, 2013, *America's Assembly Line* (Cambridge, MA: MIT Press), 23.

23. F. C. Mills, 1934, introduction to "Mechanization in Industry," by H. Jerome (Cambridge, MA: National Bureau of Economic Research), xxi.

24. Jerome, 1934, "Mechanization in Industry," 104–5.

25. Quoted in J. Greenwood, A. Seshadri, and M. Yorukoglu, 2005, "Engines of Liberation," *Review of Economic Studies* 72 (1): 109.

26. Strasser, S. (1982). *Never Done: A History of American Housework.* (New York: Pantheon), 57.

27. Gordon, 2016, *The Rise and Fall of American Growth*), 123.

28. Quoted in "Farm Woman Works Eleven Hours a Day," 1920, *New York Times*, July 6.

29. Quoted in Nye, 1990, *Electrifying America*, 270.

30. J. Greenwood, A. Seshadri, and M. Yorukoglu, 2005, "Engines of Liberation," *Review of Economic Studies* 72 (1): 109–33.

31. Calculations are based on the Muncie, Indiana, median family income level (see Gordon, 2016, *The Rise and Fall of American Growth*, 121.)

32. "The Electric Home: Marvel of Science," 1921, *New York Times*, April 10.

33. S. Lebergott, 1993, *Pursuing Happiness: American Consumers in the Twentieth Century* (Princeton, NJ: Princeton University Press).

34. V. A. Ramey, 2009, "Time Spent in Home Production in the Twentieth-Century United States: New Estimates from Old Data," *Journal of Economic History* 69 (1): 1–47.

35. R. S. Cowan, 1983, *More Work for Mother: The Ironies of Household Technology from the Open Hearth to the Microwave* (New York: Basic).

36. "French's Conical Washing Machine and Young Women at Service," 1860, *New York Times*, August 29.

37. "New Rules for Servants: Pittsburgh Housekeepers Insist on a Full Day's Work," 1921, *New York Times*, January 16.

38. J. Mokyr, 2000, "Why 'More Work for Mother?' Knowledge and Household Behavior, 1870–1945," *Journal of Economic History* 60 (1): 1–41.

39. Nye, 1990, *Electrifying America*, 18.

40. Gordon, 2016, *The Rise and Fall of American Growth*, 227.

41. Greenwood, Seshadri, and Yorukoglu, 2005, "Engines of Liberation."

42. V. E. Giuliano, 1982, "The Mechanization of Office Work," *Scientific American* 247 (3): 148–65.

43. On the term "pink collar," see A. J. Cherlin, 2013, *Labor's Love Lost: The Rise and Fall of the Working-Class Family in America* (New York: Russell Sage Foundation), 119.

44. A. J. Field, 2007, "The Origins of US Total Factor Productivity Growth in the Golden Age," *Cliometrica* 1 (1): 89. See also A. J. Field, 2011, *A Great Leap Forward: 1930s Depression and U.S. Economic Growth* (New Haven, CT: Yale University Press).

45. G. P. Mom and D. A. Kirsch, 2001, "Technologies in Tension: Horses, Electric Trucks, and the Motorization of American Cities, 1900–1925," *Technology and Culture* 42 (3): 489–518.

46. Gordon, 2016, *The Rise and Fall of American Growth*, 227.

47. In the period 1850–80, 80 percent of the residents of Philadelphia were still walking to work.

48. Gordon, 2016, *The Rise and Fall of American Growth*, 56–57.

49. When the first issue of the periodical came out, the automobile industry was not even sufficiently important to be listed in the census under a separate heading.

50. G. Norcliffe, 2001, *The Ride to Modernity: The Bicycle in Canada, 1869–1900* (Toronto: University of Toronto Press).

51. M. Twain, 1835, "Taming the Bicycle," The University of Adelaide Library, last updated March 27, 2016, https://ebooks.adelaide.edu.au/t/twain/mark/what_is_man/chapter15.html.

52. R. H. Merriam, 1905, "Bicycles and Tricycles," in *Census of Manufactures, 1905* (Washington, DC: United States Bureau of the Census), 289.

53. Daimler, for example, fitted his small air-cooled motor to a bicycle.

54. Quoted in Hounshell, 1985, *From the American System to Mass Production*, 214.

55. Martin, 1905, "Electrical Machinery, Apparatus, and Supplies," 20.

56. Quoted in Hounshell, 1985, *From the American System to Mass Production*, 214.

57. K. Kaitz, 1998, "American Roads, Roadside America," *Geographical Review* 88 (3): 372.

58. On automobiles and infrastructure in the U.S., see Gordon, 2016, *The Rise and Fall of American Growth*, 156–59.

59. Quoted in ibid., 167.

60. In the words of Ralph Epstein, "It is sometimes said that the automobile has caused good roads; sometimes, that the construction of good roads has caused the great development of the automobile industry. Both statements are true; here, as so often in economic matters, cause and effect have constantly interacted" (1928, *The Automobile Industry* [Chicago: Shaw], 17).

61. J. J. Flink, 1988, *The Automobile Age* (Cambridge, MA: MIT Press), 33.

62. On the economics and adoption of the Model T, see Gordon, 2016, *The Rise and Fall of American Growth*, 165.

63. Epstein, 1928, *The Automobile Industry*, 16.

64. Wayne Rasmussen writes: "In general the task for which steam engines proved to be most useful was threshing grain. The engines were too heavy and cumbersome for most other farm work. The peak in the manufacture of steam engines for agriculture came in 1913, when 10,000 of them were made" (1982, "The Mechanization of Agriculture," *Scientific American* 247 [3]: 82).

65. On tractor adoption, see R. E. Manuelli and A. Seshadri, 2014, "Frictionless Technology Diffusion: The Case of Tractors," *American Economic Review* 104 (4): 1368–91.

66. W. J. White, 2001, "An Unsung Hero: The Farm Tractor's Contribution to Twentieth-Century United States Economic Growth" (PhD diss., Ohio State University).

67. One of the most important changes for farmers was the operation of highway common carriers, which transported the major portion of milk to the cities and covering distances of up to seventy miles (see International Chamber of Commerce, 1925, "Report of the American Committee on Highway Transport, June, 1925" [Washington, D.C.: American Section, International Chamber of Commerce], 5).

68. On the widening radius of farm operations, see H. R. Tolley and L. M. Church, 1921, "Corn-Belt Farmers' Experience with Motor Trucks," United States Department of Agriculture, Bulletin No. 931, February 25.

69. Field, 2011, *A Great Leap Forward*, table 2.5 and table 2.6.

70. The idea that military research and development during World War II had substantial positive effects that drove American productivity during the subsequent decades remains controversial but is at odds with evidence provided by indicators of technological progress. The number of new technology books did not exceed its 1941 level until the late 1950s (M. Alexopoulos and J. Cohen, 2011, "Volumes of Evidence: Examining Technical Change in the Last Century through a New Lens," *Canadian Journal of Economics/Revue Canadienne d'économique* 44 [2]: 413–50). The military buildup, which gained traction only after Pearl Harbor, was attacked in December 1941, was thus seemingly accompanied by a slowdown in innovation as productive resources were allocated to the American war machine.

71. For an overview of advances in trucking and other transportation technologies in the early twentieth century, see W. Owen, 1962, "Transportation and Technology," *American Economic Review* 52 (2): 405–13.

72. Quoted in R. F. Weingroff, 2005, *Designating the Urban Interstates*, Federal Highway Administration Highway History, https://www.fhwa.dot.gov/infrastructure/fairbank.cfm.

73. M. I. Nadiri and T. P. Mamuneas, 1994, "Infrastructure and Public R&D Investments, and the Growth of Factor Productivity in U.S. Manufacturing Industries" (Working Paper 4845, National Bureau of Economic Research, Cambridge, MA).

74. D. M. Bernhofen, Z. El-Sahli, and R. Kneller, 2016, "Estimating the Effects of the Container Revolution on World Trade," *Journal of International Economics* 98: 36–50.

75. G. Horne, 1968, "Container Revolution Hailed by Many, Feared by Others," *New York Times*, September 22.

76. Ibid.

77. "The Humble Hero: Containers Have Been More Important for Globalisation Than Freer Trade," 2013, *Economist*, May 18, https://www.economist.com/finance-and-economics /2013/05/18/the-humble-hero.

78. R. H. Richter, 1958, "Dockers Demand Container Curbs," *New York Times*, November 27.

79. Ibid.

80. On the federal court's dismissal, see D. F. White, 1976, "High Court Review Sought in Case Involving Jobs for Longshoremen," *New York Times*, October 17.

81. Jerome, 1934, "Changes in Mechanization," 152.

82. J. Lee, 2014, "Measuring Agglomeration: Products, People, and Ideas in U.S. Manufacturing, 1880–1990" (working paper, Harvard University).

83. Ibid.

84. Alexopoulos and Cohen, 2016, "The Medium Is the Measure."

85. D. L. Lewis, 1986, "The Automobile in America: The Industry," *Wilson Quarterly* 10 (5): 50.

Chapter 7

1. W. Green, 1930, "Labor Versus Machines: An Employment Puzzle," *New York Times*, June 1.

2. F. Engels, [1844] 1943, *The Condition of the Working-Class in England in 1844*. Reprint, London: Allen & Unwin, 100.

3. A keyword search for "technological unemployment" in the *New York Times* archives yields 13 hits in the 1920s and 356 in the 1930s, as the term became increasingly popular.

4. G. R. Woirol, 2006, "New Data, New Issues: The Origins of the Technological Unemployment Debates," *History of Political Economy* 38 (3): 480.

5. J. J. Davis, 1927, "The Problem of the Worker Displaced by Machinery," *Monthly Labor Review* 25 (3): 32.

6. Ibid.

7. Quoted in Woirol, 2006, "New Data, New Issues," 481.

8. I. Lubin, 1929, *The Absorption of the Unemployed by American Industry* (Washington, DC: Brookings Institution), 6.

9. R. J. Myers, 1929, "Occupational Readjustment of Displaced Skilled Workmen," *Journal of Political Economy* 37 (4): 473–89.

10. Another study by Ewan Clague and W. J. Couper of shutdowns of two rubber factories in New Haven and Hartford, Connecticut (closed in 1929 and 1930, respectively), further shows that the bulk of workers fared worse economically in their new jobs (1931, "The Readjustment of Workers Displaced by Plant Shutdowns," *Quarterly Journal of Economics* 45 [2]: 309–46).

11. On mechanization in music, see H. Jerome, 1934, "Mechanization in Industry" (Working Paper 27, National Bureau of Economic Research, Cambridge, MA), chapter 4.

12. See Woirol, 2006, "New Data, New Issues."

13. L. Wolman, 1933, "Machinery and Unemployment," *Nation*, February 22, 202–4.

14. Quoted in "Technological Unemployment," 1930, *New York Times*, August 12.

15. "Durable Goods Industries," 1934, *New York Times*, July 16.

16. M. Alexopoulos and J. Cohen, 2016, "The Medium Is the Measure: Technical Change and Employment, 1909–1949," *Review of Economics and Statistics* 98 (4): 793.

17. F. D. Roosevelt, 1940, "Annual Message to the Congress," January 3, by G. Peters and J. T. Woolley, The American Presidency Project, https://www.presidency.ucsb.edu/documents /annual-message-the-congress.

18. R. M. Solow, 1965, "Technology and Unemployment," *Public Interest* 1 (Fall): 17.

19. A keyword search for the new expression in the *New York Times* archives yields no hits for the 1940s. But in the 1950s, "automation" appeared in 1,252 news stories.

20. See U.S. Congress, 1955, "Automation and Technological Change," Hearings Before the Subcommittee on Economic Stabilization of the Congressional Joint Committee on the Economic Report (84th Cong., 1st sess.), pursuant to sec. 5(a) of Public Law 304, 79th Cong. (Washington, DC: Government Printing Office).

21. Quoted in E. Weinberg, 1956, "An Inquiry into the Effects of Automation," *Monthly Labor Review* 79 (1): 7.

22. Quoted in ibid.

23. D. Morse, 1957, "Promise and Peril of Automation," *New York Times*, June 9.

24. Ibid.

25. "Elevator Operator Killed," 1940, *New York Times*, February 10.

26. "Elevator Units Fight Automatic Lift Ban," 1952, *New York Times*, October 7.

27. "New Devices Gain on Elevator Men: Operators May Be Riding to Oblivion," 1956, *New York Times*, May 27.

28. G. Talese, 1963, "Elevator Men Dwindle in City," *New York Times*, November 30.

29. A. H. Raskin, 1961, "Fears about Automation Overshadowing Its Boons," *New York Times*, April 7.

30. On fears about government jobs, see C. P. Trussell, 1960, "Government Automation Posing Threat to the Patronage System," *New York Times*, September 14.

31. J. F. Kennedy, 1960, "Papers of John F. Kennedy. Pre-Presidential Papers. Presidential Campaign Files, 1960. Speeches and the Press. Speeches, Statements, and Sections, 1958–1960. Labor: Meeting the Problems of Automation," https://www.jfklibrary.org/asset-viewer/archives /JFKCAMP1960/1030/JFKCAMP1960-1030-036.

32. President's Advisory Committee on Labor-Management Policy, 1962, *The Benefits and Problems Incident to Automation and Other Technological Advances* (Washington, DC: Government Printing Office), 2.

33. J. F. Kennedy, 1962, "News Conference 24," https://www.jfklibrary.org/archives/other -resources/john-f-kennedy-press-conferences/news-conference-24.

34. L. B. Johnson, 1964," Remarks Upon Signing Bill Creating the National Commission on Technology, Automation, and Economic Progress," August 19, http://archive.li/F9iX8.

35. H. R. Bowen, 1966, *Report of the National Commission on Technology, Automation, and Economic Progress* (Washington, DC: Government Printing Office), xii.

36. Ibid., 9.

37. G. R. Woirol, 1980, "Economics as an Empirical Science: A Case Study" (working paper, University of California, Berkeley), 188.

38. G. R. Woirol, 2012, "Plans to End the Great Depression from the American Public," *Labor History* 53 (4): 571–77.

39. W. A. Faunce, E. Hardin, and E. H. Jacobson, 1962, "Automation and the Employee," *Annals of the American Academy of Political and Social Science* 340 (1): 62.

40. F. C. Mann, L. K. Williams, 1960, "Observations on the Dynamics of a Change to Electronic Data-Processing Equipment," *Administrative Science Quarterly* 5 (2): 255.

41. W. A. Faunce, 1958a, "Automation and the Automobile Worker," *Social Problems* 6 (1): 68–78, and 1958b, "Automation in the Automobile Industry: Some Consequences for In-Plant Social Structure," *American Sociological Review* 23 (4): 401–7.

42. C. R. Walker, 1957, *Toward the Automatic Factory: A Case Study of Men and Machines* (New Haven, CT: Yale University Press), 192.

43. Faunce, Hardin, and Jacobson, 1962, "Automation and the Employee," 60.

Chapter 8

1. "Burning Farming Machinery," 1879, *New York Times*, August 12.

2. D. Nelson, 1995, *Farm and Factory: Workers in the Midwest, 1880–1990* (Bloomington: Indiana University Press), 18–19.

3. P. Taft and P. Ross, 1969, "American Labor Violence: Its Causes, Character, and Outcome," in *Violence in America: Historical and Comparative Perspectives*, ed. H. D. Graham and T. R. Gurr (London: Corgi), 1:221–301.

4. B. E. Kaufman, 1982, "The Determinants of Strikes in the United States, 1900–1977," *ILR Review* 35 (4): 473–90.

5. P. Wallis, 2014, "Labour Markets and Training," in *The Cambridge Economic History of Modern Britain*, vol. 1: *Industrialisation, 1700–1870*, ed. R. Floud, J. Humphries, and P. Johnson (Cambridge: Cambridge University Press), 186.

6. Quoted in D. Stetson, 1970, "Walter Reuther: Union Pioneer with Broad Influence Far beyond the Field of Labor," *New York Times*, May 11.

7. H. J. Rothberg, 1960, "Adjustment to Automation in Two Firms," in *Impact of Automation: A Collection of 20 Articles about Technological Change, from the* Monthly Labor Review (Washington, DC: Bureau of Labor Statistics), 86.

8. G. B. Baldwin and G. P. Schultz, 1960, "The Effects of Automation on Industrial Relations," in *Impact of Automation: A Collection of 20 Articles about Technological Change, from the* Monthly Labor Review (Washington, DC: Bureau of Labor Statistics), 47–49; J. W. Childs and R. H. Bergman, 1960, "Wage-Rate Determination in an Automated Rubber Plant," in ibid, 56–58; H. J. Rothberg, 1960, "Adjustment to Automation in Two Firms," in ibid, 88–93.

9. U.S. Congress, 1984, "Computerized Manufacturing Automation: Employment, Education, and the Workplace," No. 235 (Washington, DC: Office of Technology Assessment).

10. On improving working conditions, see R. J. Gordon, 2016, *The Rise and Fall of American Growth: The U.S. Standard of Living Since the Civil War* (Princeton, NJ: Princeton University Press), chapter 8.

11. R. Hornbeck, 2012, "The Enduring Impact of the American Dust Bowl: Short- and Long-Run Adjustments to Environmental Catastrophe," *American Economic Review* 102 (4): 1477–507.

12. Ibid.

13. Gordon, 2016, *The Rise and Fall of American Growth*, 270.

14. "Shocking Death in Machinery," 1895, *New York Times*, May 23.

15. "The Calamity," 1911, *New York Times*, March 26.

16. D. E. Nye, 1990, *Electrifying America: Social Meanings of a New Technology, 1880–1940* (Cambridge, MA: MIT Press), 210.

17. U.S. Bureau of the Census, 1960, D785, "Work-injury Frequency Rates in Manufacturing, 1926–1956," and D.786–790, "Work-injury Frequency Rates in Mining, 1924–1956," *Historical Statistics of the United States, Colonial Times to 1957* (Washington, DC: Government Printing Office), https://www.census.gov/library/publications/1960/compendia/hist_stats_colonial-1957.html.

18. Quoted in A. H. Raskin, 1955, "Pattern for Tomorrow's Industry?," *New York Times*, December 18.

19. On automation and health, see O. R. Walmer, 1956, "Workers' Health in an Era of Automation," *Monthly Labor Review* 79 (7): 819–23.

20. Quoted in ibid, 821

21. U.S. Department of Agriculture, 1963, *1962 Agricultural Statistics* (Washington, DC: Government Printing Office).

22. On motor vehicles and saved hours, see A. L. Olmstead and P. W. Rhode, 2001, "Reshaping the Landscape: The Impact and Diffusion of the Tractor in American Agriculture, 1910–1960," *Journal of Economic History* 61 (3): 663–98. See also M. R. Cooper, G. T. Barton, and A. P. Brodell, 1947, "Progress of Farm Mechanization," USDA Miscellaneous Publication 630 (October).

23. Nye, 1990, *Electrifying America*, 15.

24. Jerome, 1934, "Mechanization in Industry," 131.

25. Ibid., 134.

26. In the period 1940–80, 24.5 million new white-collar jobs were added to the American economy, causing the share of white-collar employment to grow by 10.8 percentage points—with clerical work accounting for almost the entire increase. In addition, 19.9 million professional and management jobs were created, accounting for 27.8 percent of the total employment by 1980.

27. Jerome, 1934, "Mechanization in Industry," 173.

28. Gordon, 2016, *The Rise and Fall of American Growth*, table 8-1.

29. Ibid., 257.

30. Quoted in D. L. Lewis, 1986, "The Automobile in America: The Industry," *Wilson Quarterly* 10 (5): 53.

31. On corporate welfare programs, see Nye, 1990, *Electrifying America*, 215.

32. Gordon, 2016, *The Rise and Fall of American Growth*, 279.

33. L. Hartz, 1955, *The Liberal Tradition in America: An Interpretation of American Political Thought Since the Revolution* (Boston: Houghton Mifflin Harcourt).

34. J. Cowie, 2016, *The Great Exception: The New Deal and the Limits of American Politics* (Princeton, NJ: Princeton University Press).

35. Pioneering work by H. G. Lewis suggests that the union premium fluctuated between 38 percent around the time of the New Deal and essentially zero in the years just after World War II. Although the union premium reemerged in the 1950s, it accounted for only up to

15 percent of workers' compensation at that time (see H. G. Lewis, 1963, *Unionism and Relative Wages in the U.S.: An Empirical Inquiry* [Chicago: Chicago University Press]). Other studies confirm that the wage advantage of union membership has varied greatly, not just over time but also across occupations and industries (see C. J. Parsley, 1980, "Labor Union Effects on Wage Gains: A Survey of Recent Literature," *Journal of Economic Literature* 18[1]: 1–31; G. E. Johnson, 1975, "Economic Analysis of Trade Unionism," *The American Economic Review* 65 [2]: 23–28).

36. W. K. Stevens, 1968, "Automation Keeps Struck Phone System," *New York Times*, April 20.

37. As James Bessen notes, "Indeed, textile workers saw their wages rise during the latter part of the nineteenth century even though textile unions were small and ineffective. Bessemer steelworkers earned much higher wages than craft ironworkers, and they worked an eight-hour day despite consistent defeats for the unions over the first decades of Bessemer production" (2015, *Learning by Doing: The Real Connection between Innovation, Wages, and Wealth* [New Haven, CT: Yale University Press], 86).

38. Gordon, 2016, *The Rise and Fall of American Growth*, 282.

39. M. Alexopoulos and J. Cohen, 2016, "The Medium Is the Measure: Technical Change and Employment, 1909–1949," *Review of Economics and Statistics* 98(4): 793.

40. On electrical industries, see T. C. Martin, 1905, "Electrical Machinery, Apparatus, and Supplies," in *Census of Manufactures, 1905* (Washington, DC: United States Bureau of the Census), 157–225.

41. On peak industry employment, see J. Bessen, 2018, "Automation and Jobs: When Technology Boosts Employment" (Law and Economics Paper 17-09, Boston University School of Law).

42. At a major manufacturer of radio and television sets, the adoption of new machines in the production of TV receivers led to higher wages. For the new jobs, pay was "set at 5 to 15 percent above the straight-time rates for unskilled assemblers because of some differences in working conditions and increased responsibility." And at one manufacturer of electrical equipment, the adoption of labor-saving technology similarly created new jobs with higher pay. See Rothberg, 1960, "Adjustment to Automation in Two Firms," 80.

43. R. H. Day, 1967, "The Economics of Technological Change and the Demise of the Sharecropper," *American Economic Review* 57 (3): 427–49.

44. Quoted in W. D. Rasmussen, 1982, "The Mechanization of Agriculture," *Scientific American* 247 (3): 87.

45. On rising wages in cities and rural outmigration, see W. Peterson and Y. Kislev, 1986, "The Cotton Harvester in Retrospect: Labor Displacement or Replacement?," *Journal of Economic History* 46 (1): 199–216.

46. R. Hornbeck and S. Naidu, 2014, "When the Levee Breaks: Black Migration and Economic Development in the American South," *American Economic Review* 104 (3): 963–90.

47. Rasmussen, 1982, "The Mechanization of Agriculture," 83.

48. Ibid, 84.

49. On the Mississippi flood, see Hornbeck and Naidu, 2014, "When the Levee Breaks."

50. On the Great Migration, see W. J. Collins and M. H. Wanamaker, 2015, "The Great Migration in Black and White: New Evidence on the Selection and Sorting of Southern Migrants," *Journal of Economic History* 75 (4): 947–92.

51. "Motors on the Farms Replace Hired Labor," 1919, *New York Times*, October 26.

52. N. Kaldor, 1957, "A Model of Economic Growth," *Economic Journal* 67 (268): 591–624.

53. P. H. Lindert and J. G. Williamson, 2016, *Unequal Gains: American Growth and Inequality Since 1700* (Princeton, NJ: Princeton University Press), 194.

54. R. M. Solow, 1956, "A Contribution to the Theory of Economic Growth," *Quarterly Journal of Economics* 70 (1): 65–94; S. Kuznets, 1955, "Economic Growth and Income Inequality," *American Economic Review* 45 (1): 1–28; Kaldor, 1957, "A Model of Economic Growth."

55. Kuznets, 1955, "Economic Growth and Income Inequality."

56. Lindert and Williamson, 2016, *Unequal Gains*.

57. A. de Tocqueville, 1840, *Democracy in America*, trans. H. Reeve (New York: Alfred A. Knopf), 2:646.

58. Quoted in Lindert and Williamson, 2016, *Unequal Gains*, 117.

59. M. Twain and C. D. Warner, [1873] 2001, *The Gilded Age: A Tale of Today* (New York: Penguin).

60. H. J. Raymond, 1859, "Your Money or Your Line," *New York Times*, February 9.

61. M. Klein, 2007, *The Genesis of Industrial America, 1870–1920* (Cambridge: Cambridge University Press), 133–34.

62. Lindert and Williamson, 2016, *Unequal Gains*, tables 5-8 and 5-9.

63. L. F. Katz and R. A. Margo, 2013, "Technical Change and the Relative Demand for Skilled Labor: The United States in Historical Perspective" (Working Paper 18752, National Bureau of Economic Research, Cambridge, MA).

64. Lindert and Williamson, 2016, *Unequal Gains*, table 7-2.

65. I. Fisher, 1919, "Economists in Public Service: Annual Address of the President," *American Economic Review* 9 (1): 10 and 16.

66. T. Piketty, 2014, *Capital in the Twenty-First Century* (Cambridge, MA: Harvard University Press).

67. W. Scheidel, 2018, *The Great Leveler: Violence and the History of Inequality from the Stone Age to the Twenty-First Century* (Princeton, NJ: Princeton University Press).

68. On financial occupations, see Lindert and Williamson, 2016, *Unequal Gains*, figure 8-3.

69. Piketty, 2014, *Capital in the Twenty-First Century*, 506–7.

70. C. Goldin and R. A. Margo, 1992, "The Great Compression: The Wage Structure in the United States at Mid-Century," *Quarterly Journal of Economics* 107 (1): 1–34.

71. H. S. Farber, D. Herbst, I. Kuziemko, and S. Naidu, 2018, "Unions and Inequality over the Twentieth Century: New Evidence from Survey Data" (Working Paper 24587, National Bureau of Economic Research, Cambridge, MA).

72. J. M. Abowd, P. Lengermann, and K. L. McKinney, 2003, "The Measurement of Human Capital in the US Economy" (LEHD Program technical paper TP-2002-09, Census Bureau, Washington).

73. J. Tinbergen, 1975, *Income Distribution: Analysis and Policies* (Amsterdam: North Holland).

74. C. Goldin and L. Katz, 2008, *The Race between Technology and Education* (Cambridge, MA: Harvard University Press).

75. C. Goldin and Margo, 1992, "The Great Compression."

76. Goldin and Katz, 2008. *The Race between Technology and Education*, 303.

77. Ibid., 208–17.

78. Quoted in ibid., 177.

79. Rothberg, 1960, "Adjustment to Automation in Two Firms," 89.

80. E. Weinberg, 1960, "A Review of Automation Technology," *Monthly Labor Review* 83 (4): 376–80.

81. T. Piketty and E. Saez, 2003, "Income Inequality in the United States, 1913–1998," *Quarterly Journal of Economics* 118 (1): 2 and 24.

82. B. Milanovic, 2016b, *Global Inequality: A New Approach for the Age of Globalization* (Cambridge, MA: Harvard University Press).

83. Katz and Margo, 2013, "Technical Change and the Relative Demand for Skilled Labor."

84. Gordon, 2016, *The Rise and Fall of American Growth*, 47.

85. Katz and Margo, 2013, "Technical Change and the Relative Demand for Skilled Labor," 4.

86. S. Thernstrom, 1964, *Poverty and Progress: Social Mobility in a Nineteenth Century City* (Cambridge, MA: Harvard University Press).

87. Gordon, 2016, *The Rise and Fall of American Growth*, 126.

88. Ibid., 379.

89. A. J. Cherlin, 2013, *Labor's Love Lost: The Rise and Fall of the Working-Class Family in America* (New York: Russell Sage Foundation), 115.

90. Speech by John F. Kennedy in Cheyenne, Wyoming, September 23, 1960, https://www.jfklibrary.org/archives/other-resources/john-f-kennedy-speeches/cheyenne-wy-19600923.

Part 4

1. D. Acemoglu and D. H. Autor, 2011, "Skills, Tasks and Technologies: Implications for Employment and Earnings," in *Handbook of Labor Economics*, ed. David Card and Orley Ashenfelter (Amsterdam: Elsevier), 4:1043–171.

Chapter 9

1. P. F. Drucker, 1965, "Automation Is Not the Villain," *New York Times*, January 10.

2. D. A. Grier, 2005, *When Humans Were Computers* (Princeton, NJ: Princeton University Press).

3. On mortgage underwriters, see F. Levy and R. J. Murnane, 2004, *The New Division of Labor: How Computers Are Creating the Next Job Market* (Princeton, NJ: Princeton University Press), 17–19.

4. H. Braverman, 1998, *Labor and Monopoly Capital: The Degradation of Work in the Twentieth Century*, 25th anniversary ed. (New York: New York University Press), 49.

5. N. Wiener, 1988, *The Human Use of Human Beings: Cybernetics and Society* (New York: Perseus Books Group).

6. D. H. Autor and D. Dorn, 2013, "The Growth of Low-Skill Service Jobs and the Polarization of the US Labor Market," *American Economic Review* 103 (5): 1553–97; M. Goos, A.

Manning, and A. Salomons, 2014, "Explaining Job Polarization: Routine-Biased Technological Change and Offshoring," *American Economic Review* 104 (8): 2509–26, and 2009, "Job Polarization in Europe," *American Economic Review* 99 (2): 58–63; M. A. Goos and A. Manning, 2007, "Lousy and Lovely Jobs: The Rising Polarization of Work in Britain," *Review of Economics and Statistics* 89 (1): 118–33.

7. Levy and Murnane, 2004, *The New Division of Labor*, 3.

8. W. D. Nordhaus, 2007, "Two Centuries of Productivity Growth in Computing," *Journal of Economic History* 67 (1): 128–59.

9. J. S. Tompkins, 1958, "Cost of Automation Discourages Stores," *New York Times*, January 26.

10. The first microprocessor, invented in 1971, only paved the way for the IBM PC in 1981. Nordhaus's calculations show that the greatest fall in the cost of computing occurred after the PC arrived.

11. O. Friedrich, 1983, "The Computer Moves In (Machine of the Year)," *Time*, January 3, 15.

12. K. Flamm, 1988, "The Changing Pattern of Industrial Robot Use," in *The Impact of Technological Change on Employment and Economic Growth*, ed. R. M. Cyert and D. C. Mowery (Cambridge, MA: Ballinger Publishing Company), tables 7-1 and 7-6.

13. E. B. Jakubauskas, 1960, "Adjustment to an Automatic Airline Reservation System," in *Impact of Automation: A Collection of 20 Articles about Technological Change, from the* Monthly Labor Review (Washington: Bureau of Labor Statistics), 94.

14. Ibid.

15. Quoted in Levy and Murnane, 2004, *The New Division of Labor*, 4.

16. Quoted in ibid.

17. D. H. Autor, 2015, "Polanyi's Paradox and the Shape of Employment Growth," in *Reevaluating Labor Market Dynamics* (Kansas City: Federal Reserve Bank of Kansas City), 129–177.

18. M. Polanyi, 1966, *The Tacit Dimension* (New York: Doubleday), 4.

19. According to the O-ring production function of Michael Kremer, an improvement in one task in the production of something makes the other tasks more valuable (1993, "The O-Ring Theory of Economic Development," *Quarterly Journal of Economics* 108 [3]: 551–75).

20. Levy and Murnane, 2004, *The New Division of Labor*, 13–14.

21. R. Reich, 1991, *The Work of Nations: Preparing Ourselves for Twenty-First Century Capitalism* (New York: Knopf).

22. E. L. Glaeser, 2013, review of *The New Geography of Jobs*, by Enrico Moretti, *Journal of Economic Literature* 51 (3): 827.

23. H. Moravec, 1988, *Mind Children: The Future of Robot and Human Intelligence* (Cambridge, MA: Harvard University Press), 15.

24. The share of labor hours in service occupations grew by 30 percent between 1980 and 2005. In the three decades before the computer revolution of the 1980s, in contrast, that share had been flat or declining (D. H. Autor and Dorn, 2013, "The Growth of Low-Skill Service Jobs and the Polarization of the US Labor Market").

25. Levy and Murnane, 2004, *The New Division of Labor*, 3. See also D. H. Autor, F. Levy, and R. J. Murnane, 2003, "The Skill Content of Recent Technological Change: An Empirical Exploration," *Quarterly Journal of Economics* 118 (4): 1279–333.

26. A. J. Cherlin, 2014, *Labor's Love Lost: The Rise and Fall of the Working-Class Family in America* (New York: Russell Sage Foundation), 128.

27. Ibid.

28. Douglas Massey defines social class by education, which he sees as the most important resource in our increasingly knowledge-based economy (2007, *Categorically Unequal: The American Stratification System* [New York: Russell Sage Foundation]). Andrew Cherlin also relies on education as the best indicator of social class for the post-1980 period (2014, *Labor's Love Lost*). And Robert Putnam argues along similar lines (2016, *Our Kids: The American Dream in Crisis* [New York: Simon & Schuster]).

29. G. M. Cortes, N. Jaimovich, C. J. Nekarda, and H. E. Siu, 2014, "The Micro and Macro of Disappearing Routine Jobs: A Flows Approach" (Working Paper 20307, National Bureau of Economic Research, Cambridge, MA).

30. D. D. Buss, 1985, "On the Factory Floor, Technology Brings Challenge for Some, Drudgery for Others," *Wall Street Journal*, September 16.

31. G. M. Cortes, N. Jaimovich, and H. E. Siu, 2017, "Disappearing Routine Jobs: Who, How, and Why?," *Journal of Monetary Economics*, 91:69–87.

32. K. G. Abraham and M. S. Kearney, 2018, "Explaining the Decline in the US Employment-to-Population Ratio: A Review of the Evidence" (Working Paper 24333, National Bureau of Economic Research, Cambridge, MA).

33. G. M. Cortes, N. Jaimovich, and H. E. Siu, 2018, "The 'End of Men' and Rise of Women in the High-Skilled Labor Market" (Working Paper 24274, National Bureau of Economic Research., Cambridge, MA).

34. B. A. Weinberg, 2000, "Computer Use and the Demand for Female Workers," *ILR Review* 53 (2): 290–308.

35. D. Acemoglu and P. Restrepo, 2018c, "Robots and Jobs: Evidence from US Labor Markets" (Working paper, Massachusetts Institute of Technology, Cambridge, MA). Economists have found similar effects of robots on labor markets in Britain (A. Prashar, 2018, "Evaluating the Impact of Automation on Labour Markets in England and Wales" [working paper, Oxford University]). In the German context, each additional robot led to two manufacturing jobs being lost, but these were offset by job creation elsewhere (W. Dauth, S. Findeisen, J. Südekum, and N. Woessner, 2017, "German Robots: The Impact of Industrial Robots on Workers" [Discussion Paper DP12306, Center for Economic and Policy Research, London]). That is not really surprising. Technological change inevitably interacts with different labor market institutions in different countries, and the relative strength of German trade unions is likely to go some way toward explaining these differences, as the authors argue. The general pattern across the industrial world, it seems, is that robots have not significantly reduced total employment, only low-skilled workers' employment share. Automation, in other words, has caused employment opportunities for non-college-educated workers to dry up (G. Graetz and G. Michaels, forthcoming, "Robots at Work," *Review of Economics and Statistics*).

36. D. H. Autor and A. Salomons, forthcoming, "Is Automation Labor-Displacing? Productivity Growth, Employment, and the Labor Share," *Brookings Papers on Economic Activity*.

37. J. Bivens, E. Gould, E. Mishel, and H. Shierholz, 2014, "Raising America's Pay" (Briefing Paper 378, Economic Policy Institute, New York), figure A.

38. See M. W. Elsby, B. Hobijn, and A. Şahin, 2013, "The Decline of the US Labor Share," *Brookings Papers on Economic Activity* 2013 (2): 1–63.

39. L. Karabarbounis and B. Neiman, 2013, "The Global Decline of the Labor Share," *Quarterly Journal of Economics* 129 (1): 61–103

40. M. C. Dao, M. M. Das, Z. Koczan, and W. Lian, 2017, "Why Is Labor Receiving a Smaller Share of Global Income? Theory and Empirical Evidence" (Working Paper No. 17/169, International Monetary Fund, Washington, DC), 11.

41. B. Milanovic, 2016b, *Global Inequality: A New Approach for the Age of Globalization* (Cambridge, MA: Harvard University Press), 54.

42. L. F. Katz and R. A. Margo, 2013, "Technical Change and the Relative Demand for Skilled Labor: The United States in Historical Perspective (Working Paper 18752, National Bureau of Economic Research, Cambridge, MA).

43. Autor and Salomons, forthcoming, "Is Automation Labor-Displacing?"

44. E. Weinberg, 1960, "Experiences with the Introduction of Office Automation," *Monthly Labor Review* 83 (4): 376–80.

45. Ibid.

46. J. Bessen, 2015, *Learning by Doing: The Real Connection between Innovation, Wages, and Wealth* (New Haven, CT: Yale University Press), 111.

47. Ibid.

Chapter 10

1. P. Gaskell, 1833, *The Manufacturing Population of England, its Moral, Social, and Physical Conditions* (London: Baldwin and Cradock), 6.

2. Ibid., 9.

3. W. J. Wilson, 1996, "When Work Disappears," *Political Science Quarterly* 111 (4): 567.

4. R. D. Putnam, 2016, *Our Kids: The American Dream in Crisis* (New York: Simon & Schuster), 7.

5. Ibid.

6. Ibid., 20.

7. C. Murray, 2013, *Coming Apart: The State of White America, 1960–2010* (New York: Random House Digital, Inc.), 47.

8. Ibid., 193.

9. W. J. Wilson, 2012, *The Truly Disadvantaged: The Inner City, the Underclass, and Public Policy* (Chicago: University of Chicago Press).

10. R. Chetty, N. Hendren, P. Kline, and E. Saez, 2014, "Where Is the Land of Opportunity? The Geography of Intergenerational Mobility in the United States," *Quarterly Journal of Economics* 129 (4): 1553–623; R. Chetty and N. Hendren, 2018, "The Impacts of Neighborhoods on Intergenerational Mobility II: County-Level Estimates," *Quarterly Journal of Economics* 133 (3): 1163–228.

11. See, for example, G. Becker, 1968, "Crime and Punishment: An Economic Approach," *Journal of Political Economy* 76 (2): 169–217; I. Ehrlich, 1996, "Crime, Punishment, and the Market for Offenses," *Journal of Economic Perspectives* 10 (1): 43–67, and 1973, "Participation in Illegitimate Activities: A Theoretical and Empirical Investigation," *Journal of Political Economy* 81 (3): 521–65.

12. C. Vickers and N. L. Ziebarth, 2016, "Economic Development and the Demographics of Criminals in Victorian England," *Journal of Law and Economics* 59 (1): 191–223.

13. E. D. Gould, B. A. Weinberg, and D. B. Mustard, 2002, "Crime Rates and Local Labor Market Opportunities in the United States: 1979–1997," *Review of Economics and Statistics* 84 (1): 45–61.

14. A. J. Cherlin, 2013, *Labor's Love Lost: The Rise and Fall of the Working-Class Family in America* (New York: Russell Sage Foundation), figure 1.2.

15. D. H. Autor, D. Dorn, and G. Hanson, forthcoming, "When Work Disappears: Manufacturing Decline and the Falling Marriage-Market Value of Men" *American Economic Review: Insights*.

16. L. S. Jacobson, R. J. LaLonde, and D. G. Sullivan, 1993, "Earnings Losses of Displaced Workers," *American Economic Review* 83 (4): 685–709.

17. D. Sullivan and T. Von Wachter, 2009, "Job Displacement and Mortality: An Analysis Using Administrative Data," *Quarterly Journal of Economics* 124 (3): 1265–1306.

18. A. Case and A. Deaton, 2015, "Rising Morbidity and Mortality in Midlife among White Non-Hispanic Americans in the 21st Century," *Proceedings of the National Academy of Sciences* 112 (49): 15078–83.

19. On technology and trade as possible causes of the mortality upsurge, see A. Case and A. Deaton, 2017, "Mortality and Morbidity in the 21st Century," *Brookings Papers on Economic Activity* 1: 397. Yet the mortality puzzle documented by Case and Deaton remains an American phenomenon. As they note, trade and technology have had adverse impacts on labor markets elsewhere, too, but in Europe, for example, mortality rates are still falling across the board. If automation and globalization are behind the recent rise in mortality, institutions across the Atlantic must have done a better job at moderating its negative effects.

20. On unemployment and well-being, see, for example, R. D. Tella, R. J. MacCulloch, and A. J. Oswald, 2003, "The Macroeconomics of Happiness," *Review of Economics and Statistics* 85 (4): 809–27.

21. A. E. Clark, E. Diener, Y. Georgellis, and R. E. Lucas, 2008, "Lags and Leads in Life Satisfaction: A Test of the Baseline Hypothesis," *Economic Journal* 118 (529): 222–43.

22. A. E. Clark and A. J. Oswald, 1994, "Unhappiness and Unemployment," *Economic Journal* 104 (424): 655.

23. D. S. Massey, J. Rothwell, and T. Domina, 2009, "The Changing Bases of Segregation in the United States," *Annals of the American Academy of Political and Social Science* 626 (1): 74–90.

24. See, for example F. Cairncross, 2001, *The Death of Distance: 2.0: How the Communications Revolution Will Change Our Lives* (New York: Texere Publishing).

25. A. Toffler, 1980, *The Third Wave* (New York: Bantam Books).

26. T. L. Friedman, 2006, *The World is Flat: The Globalized World in the Twenty-first Century* (London: Penguin).

27. E. L. Glaeser, 1998, "Are Cities Dying?," *Journal of Economic Perspectives* 12 (2): 139–60.

28. For an overview of the sources of agglomeration, see E. L. Glaeser and J. D. Gottlieb, 2009, "The Wealth of Cities: Agglomeration Economies and Spatial Equilibrium in the United States," *Journal of Economic Literature* 47 (4): 983–1028.

29. E. L. Glaeser, 2013, review of *The New Geography of Jobs*, by Enrico Moretti, *Journal of Economic Literature* 51 (3): 832.

30. E. Moretti, 2012, *The New Geography of Jobs* (Boston: Houghton Mifflin Harcourt), 1–2.

31. Ibid., 3–4.

32. T. Berger and C. B. Frey, 2016, "Did the Computer Revolution Shift the Fortunes of U.S. Cities? Technology Shocks and the Geography of New Jobs," *Regional Science and Urban Economics* 57:38–45.

33. T. Berger and C. B. Frey, 2017a, "Industrial Renewal in the 21st Century: Evidence from US Cities," *Regional Studies* 51 (3): 404–13.

34. E. L. Glaeser, 1998, "Are Cities Dying?," 149–50.

35. R. J. Barro and X. Sala-i-Martin, 1992, "Convergence," *Journal of Political Economy* 100 (2): 223–51.

36. P. Ganong and D. Shoag, 2017, "Why Has Regional Income Convergence in the U.S. Declined?," *Journal of Urban Economics* 102 (November): 76–90.

37. G. Duranton and D. Puga, 2001, "Nursery Cities: Urban Diversity, Process Innovation, and the Life Cycle of Products," *American Economic Review* 91 (5): 1454–77.

38. B. Austin, E. L. Glaeser, and L. Summers, forthcoming, "Saving the Heartland: Place-Based Policies in 21st Century America," *Brookings Papers on Economic Activity*.

39. Ibid.

40. E. Moretti, 2010, "Local Multipliers," *American Economic Review* 100 (2): 373–77.

41. E. L. Glaeser, 2013, review of *The New Geography of Jobs*, 831.

Chapter 11

1. B. Moore Jr., 1993, *Social Origins of Dictatorship and Democracy: Lord and Peasant in the Making of the Modern World* (Boston: Beacon Press), 418.

2. F. Fukuyama, 2014, *Political Order and Political Decay: From the Industrial Revolution to the Globalization of Democracy* (New York: Farrar, Straus and Giroux).

3. In this regard America is a special case, as it never had a feudal system.

4. Fukuyama, 2014, *Political Order and Political Decay*, 407–8.

5. Ibid., 405.

6. W. H. Maehl, 1967, *The Reform Bill of 1832: Why Not Revolution?* (New York: Holt, Rinehart and Winston), 1.

7. T. Aidt and R. Franck, 2015, "Democratization under the Threat of Revolution: Evidence from the Great Reform Act of 1832," *Econometrica* 83 (2): 505–47.

8. D. Acemoglu and J. A. Robinson, 2006, "Economic Backwardness in Political Perspective," *American Political Science Review* 100 (1): 115–31.

9. G. Himmelfarb, 1968, *Victorian Minds* (New York: Knopf).

10. This link, Lindert shows, was less pronounced after 1930, for the simple reason that most advanced economies now differ less in terms of their degree of democracy (P. H. Lindert, 2004, *Growing Public*, vol. 1, *The Story: Social Spending and Economic Growth Since the Eighteenth Century* [Cambridge: Cambridge University Press]).

11. A. de Tocqueville, 1840, *Democracy in America*, trans. H. Reeve (New York: Alfred A. Knopf), 2:237

12. J. S. Hacker and P. Pierson, 2010, *Winner-Take-All Politics: How Washington Made the Rich Richer—and Turned Its Back on the Middle Class* (New York: Simon & Schuster), 77–78.

13. Ibid.

14. Quoted in Lindert, 2004, *Growing Public*, 64.

15. On clientelism, see Fukuyama, 2014, *Political Order and Political Decay*, chapter 9.

16. R. Oestreicher, 1988, "Urban Working-Class Political Behavior and Theories of American Electoral Politics, 1870–1940," *Journal of American History* 74 (4): 1257–86.

17. Lindert, 2004, *Growing Public*, 187.

18. R. D. Putnam, 2016, *Our Kids: The American Dream in Crisis* (New York: Simon & Schuster), 7.

19. R. J. Gordon, 2016, *The Rise and Fall of American Growth: The U.S. Standard of Living Since the Civil War* (Princeton, NJ: Princeton University Press), 503.

20. R. A. Dahl, 1961, *Who Governs? Democracy and Power in an American City* (New Haven, CT: Yale University Press), 1.

21. N. McCarty, K. T. Poole, and H. Rosenthal, 2016, *Polarized America: The Dance of Ideology and Unequal Riches* (Cambridge, MA: MIT Press), 2.

22. L. M. Bartels, 2016, *Unequal Democracy: The Political Economy of the New Gilded Age* (Princeton, NJ: Princeton University Press), 1.

23. Organisation for Economic Co-operation and Development, "Social Expenditure—Aggregated Data," accessed December 22, 2018, https://stats.oecd.org/Index.aspx?DataSetCode =SOCX_AGG.

24. McCarty, Poole, and Rosenthal, 2016, *Polarized America*, 4.

25. Ibid.

26. Bartels, 2016, *Unequal Democracy*, 2.

27. Ibid., 209.

28. M. Geewax, 2005, "Minimum Wage Odyssey: A Yearlong View from Capitol Hill and a Small Ohio Town," *Trenton Times*, November 27.

29. Bartels, 2016, *Unequal Democracy*, chapter 7.

30. G. Lordan and D. Neumark, 2018, "People versus Machines: The Impact of Minimum Wages on Automatable Jobs," *Labour Economics* 52 (June): 40–53.

31. A. J. Cherlin, 2013, *Labor's Love Lost: The Rise and Fall of the Working-Class Family in America* (New York: Russell Sage Foundation), 93 and 143.

32. R. D. Putnam, 2004, in *Democracies in Flux: The Evolution of Social Capital in Contemporary Society*, ed. R. D. Putnam (New York: Oxford University Press).

33. H. S. Farber, D. Herbst, I. Kuziemko, and S. Naidu, 2018, "Unions and Inequality over the Twentieth Century: New Evidence from Survey Data (Working Paper 24587, National Bureau of Economic Research, Cambridge, MA).

34. T. Piketty, 2018, "Brahmin Left vs. Merchant Right: Rising Inequality and the Changing Structure of Political Conflict," (working paper, Paris School of Economics).

35. On political polarization across geographic regions, see D. S. Massey, J. Rothwell, and T. Domina, 2009, "The Changing Bases of Segregation in the United States," *Annals of the American Academy of Political and Social Science* 626 (1): 74–90.

36. A. Goldstein, 2018, *Janesville: An American Story* (New York: Simon & Schuster), 26–27.

37. Ibid.

38. D. C. Mutz, 2018, "Status Threat, Not Economic Hardship, Explains the 2016 Presidential Vote," *Proceedings of the National Academy of Sciences* 115 (19): 4338.

39. M. Lamont, 2009, *The Dignity of Working Men: Morality and the Boundaries of Race, Class, and Immigration* (Cambridge, MA: Harvard University Press).

40. Cherlin, 2013, *Labor's Love Lost*, 53.

41. A. E. Clark and A. J. Oswald, 1996, "Satisfaction and Comparison Income," *Journal of Public Economics* 61 (3): 359–81; A. Ferrer-i-Carbonell, 2005, "Income and Well-Being: An Empirical Analysis of the Comparison Income Effect," *Journal of Public Economics* 89 (5–6): 997–1019; E. F. Luttmer, 2005, "Neighbors as Negatives: Relative Earnings and Well-Being," *Quarterly Journal of Economics* 120 (3): 963–1002.

42. Cherlin, 2013, *Labor's Love Lost*, 170.

43. Ibid., 169 and 172.

44. The evidence for the past four decades shows that immigrants are not responsible for the stagnating or declining wages among the unskilled. This is true both nationally and locally. On the contrary, there is evidence to suggest that immigration may have helped prevent an even further decline in the wages of non-college-educated workers. See G. Peri, 2018, "Did Immigration Contribute to Wage Stagnation of Unskilled Workers?," *Research in Economics* 72 (2): 356–65. Studies have shown that immigration does not crowd out native workers but simply adds to employment, while boosting productivity. The effects on the wages of unskilled natives is close to zero. See G. Peri, 2012, "The Effect of Immigration on Productivity: Evidence from US States," *Review of Economics and Statistics* 94 (1): 348–58.

45. R. Chetty, N. Hendren, P. Kline, and E. Saez, 2014, "Where Is the Land of Opportunity? The Geography of Intergenerational Mobility in the United States," *Quarterly Journal of Economics* 129 (4): 1553–623.

46. Moderate Republican and Democratic legislators alike have been kicked out of Congress: in the period 2002–10, the combined fraction of moderates from both parties declined from 57 percent to 37 percent. See D. H. Autor, D. Dorn, G. Hanson, and K. Majlesi, 2016a, "Importing Political Polarization? The Electoral Consequences of Rising Trade Exposure" (Working Paper 22637, National Bureau of Economic Research, Cambridge, MA).

47. D. H. Autor, D. Dorn, G. Hanson, and K. Majlesi, 2016b, "A Note on the Effect of Rising Trade Exposure on the 2016 Presidential Election," appendix to "Importing Political Polarization? The Electoral Consequences of Rising Trade Exposure" (Working Paper 22637, National Bureau of Economic Research, Cambridge, MA).

48. D. Rodrik, 2016, "Premature Deindustrialization," *Journal of Economic Growth* 21 (1): 1–33; World Bank Group, 2016, *World Development Report 2016*: Digital Dividends (Washington, DC: World Bank Publications).

49. On the effects of technological change being counterbalanced by subsidized credit, see R. G. Rajan, 2011, *Fault Lines: How Hidden Fractures Still Threaten the World Economy* (Princeton, NJ: Princeton University Press).

50. K. K. Charles, E. Hurst, and M. J. Notowidigdo, 2016, "The Masking of the Decline in Manufacturing Employment by the Housing Bubble," *Journal of Economic Perspectives* 30 (2): 179–200.

51. Goldstein, 2018, *Janesville*, 290.

52. T. Gibbons-Neff, 2017, "Feeling Forgotten by Obama, People in This Ohio Town Look to Trump with Cautious Hope," *Washington Post*, January 22.

53. Quoted in "Want to Understand Why Trump Has Rural America Feeling Hopeful? Listen to This Ohio Town," 2017, *Washington Post*, May 11.

54. Ibid.

55. C. B. Frey, T. Berger, and C. Chen, 2018, "Political Machinery: Did Robots Swing the 2016 U.S. Presidential Election?," *Oxford Review of Economic Policy* 34 (3): 418–42.

56. T. Aidt, G. Leon, and M. Satchell, 2017, "The Social Dynamics of Riots: Evidence from the Captain Swing Riots, 1830–31" (Working paper, Cambridge University), 4.

57. Ibid.

58. D. Rodrik, 2017a, "Populism and the Economics of Globalization" (Working Paper 23559, National Bureau of Economic Research, Cambridge, MA), 21.

59. D. Rodrik, 2017b, *Straight Talk on Trade: Ideas for a Sane World Economy* (Princeton, NJ: Princeton University Press), 116.

60. Ibid., 122.

61. Ibid.

62. Ibid., 260.

63. Quoted in A. Oppenheimer, 2018, "Las Vegas Hotel Workers vs. Robots Is a Sign of Looming Labor Challenges," *Miami Herald*, June 1.

64. J. Gramlich, 2017, "Most Americans Would Favor Policies to Limit Job and Wage Losses Caused by Automation," Pew Research Center, http://www.pewresearch.org/fact-tank/2017/10/09/most-americans-would-favor-policies-to-limit-job-and-wage-losses-caused-by-automation/.

65. Acemoglu and Robinson, 2006, "Economic Backwardness in Political Perspective."

66. Ibid., 117.

67. M. Berg, 1976, "The Machinery Question," PhD diss., University of Oxford, 76.

68. Quoted in W. Broad, 1984, "U.S. Factories Reach into the Future," *New York Times*, March 13.

69. Quoted in G. Allison, 2017, *Destined for War: Can America and China Escape Thucydides's Trap?* (Boston: Houghton Mifflin Harcourt), chapter 1, Kindle.

70. P. Druckerman, 2014, "The French Do Buy Books. Real Books," *New York Times*, July 9.

71. G. Rayner, 2017, "Jeremy Corbyn Plans to 'Tax Robots' Because Automation Is a 'Threat' to Workers," *Daily Telegraph*, September 26.

72. Y. Sung-won, 2017, "Korea Takes First Step to Introduce 'Robot Tax,'" *Korea Times*, August 7.

73. B. Merchant, 2018, "The Presidential Candidate Bent on Beating the Robot Apocalypse Will Give Two Americans a $1,000-per-month Basic Income," *Motherboard*, April 19.

74. Quoted in S. Cronwell, 2018, "Rust-Belt Democrats Praise Trump's Threatened Metals Tariffs," *Reuters*, March 2.

75. D. Grossman, 2017, "Highly-Automated Austrian Steel Mill Only Needs 14 People," *Popular Mechanics*, June 22, https://www.popularmechanics.com/technology/infrastructure/a27043/steel-mill-austria-automated/.

76. M. Spence and S. Hlatshwayo, 2012, "The Evolving Structure of the American Economy and the Employment Challenge," *Comparative Economic Studies* 54 (4): 703–38.

77. Quoted in C. Cain Miller, 2017, "A Darker Theme in Obama's Farewell: Automation Can Divide Us," *New York Times*, January 12.

78. R. Rector and R. Sheffield, 2011, "Air Conditioning, Cable TV, and an Xbox: What Is Poverty in the United States Today?" (Washington, DC: Heritage Foundation), 2.

79. J. Mokyr, 2011, *The Enlightened Economy: Britain and the Industrial Revolution, 1700–1850* (London: Penguin), chapter 1, Kindle.

80. J. A. Schumpeter, [1942] 1976, *Capitalism, Socialism and Democracy*, 3d ed. (New York: Harper Torchbooks), 76.

Part 5

1. G. B. Baldwin, and G. P. Schultz, 1960, "The Effects of Automation on Industrial Relations," in *Impact of Automation: A Collection of 20 Articles about Technological Change, from the Monthly Labor Review* (Washington, DC: Bureau of Labor Statistics), 51.

Chapter 12

1. E. Brynjolfsson and A. McAfee, 2017, *Machine, Platform, Crowd: Harnessing Our Digital Future* (New York: Norton), 71–73.

2. C. E. Shannon, 1950, "Programming a Computer for Playing Chess," *Philosophical Magazine* 41 (314): 256–75.

3. C. Koch, 2016, "How the Computer Beat the Go Master," *Scientific American* 27 (4): 20.

4. F. Levy and R. J. Murnane, 2004, *The New Division of Labor: How Computers Are Creating the Next Job Market* (Princeton, NJ: Princeton University Press).

5. E. Brynjolfsson and A. McAfee, 2014, *The Second Machine Age: Work, Progress, and Prosperity in a Time of Brilliant Technologies* (New York: W. W. Norton), chapter 3, Kindle.

6. Koch, 2016, "How the Computer Beat the Go Master," 20.

7. M. Fortunato et al. 2017, "Noisy Networks for Exploration," preprint, submitted, https://arxiv.org/abs/1706.10295.

8. Cisco, 2018, "Cisco Visual Networking Index: Forecast and Trends, 2017–2022," (San Jose, CA: Cisco), https://www.cisco.com/c/en/us/solutions/collateral/service-provider/visual-networking-index-vni/complete-white-paper-c11-481360.html.

9. P. Lyman and H. R. Varian, 2003, "How Much Information?," berkeley. edu/research/projects/how-much-info-2003.

10. A. Tanner, 2007. "Google Seeks World of Instant Translations," *Reuters*, March 27.

11. Y. Wu et al., 2016, "Google's Neural Machine Translation System: Bridging the Gap between Human and Machine Translation," preprint, submitted October 8, https://arxiv.org/pdf/1609.08144.pdf.

12. I. M. Cockburn, R. Henderson, and S. Stern, 2018, "The Impact of Artificial Intelligence on Innovation (Working Paper 24449, National Bureau of Economic Research, Cambridge, MA).

13. E. Brynjolfsson, D. Rock, and C. Syverson, forthcoming, "Artificial Intelligence and the Modern Productivity Paradox: A Clash of Expectations and Statistics," in *The Economics of Artificial Intelligence: An Agenda*, ed. Ajay K. Agrawal, Joshua Gans, and Avi Goldfarb (Chicago: University of Chicago Press), figure 1.

14. "Germany Starts Facial Recognition Tests at Rail Station," 2017, *New York Post*, December 17.

15. N. Coudray et al., 2018, "Classification and Mutation Prediction from Non–Small Cell Lung Cancer Histopathology Images Using Deep Learning," *Nature Medicine* 24 (10): 1559–1567.

16. A. Esteva et al., 2017, "Dermatologist-Level Classification of Skin Cancer with Deep Neural Networks," *Nature* 542 (7639): 115.

17. W. Xiong et al., 2017, "The Microsoft 2017 Conversational Speech Recognition System," Microsoft AI and Research Technical Report MSR-TR-2017-39, August, https://www.microsoft.com/en-us/research/wp-content/uploads/2017/08/ms_swbd17-2.pdf.

18. M. Burns, 2018, "Clinc Is Building a Voice AI System to Replace Humans in Drive-Through Restaurants," *TechCrunch*, https://techcrunch.com/video/clinc-is-building-a-voice-ai-system-to-replace-humans-in-drive-through-restaurants/.

19. D. Gershgorn, 2018, "Google Is Building 'Virtual Agents' to Handle Call Centers' Grunt Work," *Quartz*, July 24, https://qz.com/1335348/google-is-building-virtual-agents-to-handle-call-centers-grunt-work/.

20. Brynjolfsson, Rock, and Syverson, forthcoming, "Artificial Intelligence and the Modern Productivity Paradox."

21. See C. B. Frey and M. A. Osborne, 2017, "The Future of Employment: How Susceptible Are Jobs to Computerisation?," *Technological Forecasting and Social Change* 114 (C): 254–80.

22. B. Mathibela, M. A. Osborne, I. Posner, and P. Newman, 2012, "Can Priors Be Trusted? Learning to Anticipate Roadworks," in IEEE Conference on Intelligent Transportation Systems, 927–32.

23. B. Mathibela, P. Newman, and I. Posner, 2015, "Reading the Road: Road Marking Classification and Interpretation," *IEEE Transactions on Intelligent Transportation Systems* 16 (4): 2080.

24. See C. B. Frey and Osborne, 2017, "The Future of Employment."

25. Rio Tinto, 2017, "Rio Tinto to Expand Autonomous Fleet as Part of $5 Billion Productivity Drive," December 18, http://www.riotinto.com/media/media-releases-237_23802.aspx.

26. A. Agrawal, J. Gans, and A. Goldfarb, 2016, "The Simple Economics of Machine Intelligence," *Harvard Business Review*, November 17, https://hbr.org/2016/11/the-simple-economics-of-machine-intelligence.

27. "A More Realistic Route to Autonomous Driving," 2018, *Economist*, August 2, https://www.economist.com/business/2018/08/02/a-more-realistic-route-to-autonomous-driving.

28. "Tractor Crushes Boy to Death," 1931, *New York Times*, October 12.

29. J. R. Treat et al., 1979, *Tri-Level Study of the Causes of Traffic Accidents: Final Report*, vol. 2: *Special Analyses* (Bloomington, IN: Institute for Research in Public Safety). See also V. Wadhwa, 2017, *The Driver in the Driverless Car: How Our Technology Choices Will Create the Future* (San Francisco: Berrett-Koehler).

30. World Health Organization, 2015, "Road Traffic Deaths," http://www.who.int/gho/road_safety/mortality/en.

31. J. McCurry, 2018, "Driverless Taxi Debuts in Tokyo in 'World First' Trial ahead of Olympics," *Guardian*, August 28.

32. Quoted in F. Levy, 2018, "Computers and Populism: Artificial Intelligence, Jobs, and Politics in the Near Term," *Oxford Review of Economic Policy* 34 (3): 405.

33. Quoted in T. B. Lee, 2016, "This Expert Thinks Robots Aren't Going to Destroy Many Jobs. And That's a Problem," Vox, https://www.vox.com/a/new-economy-future/robert-gordon-interview.

34. Other approaches to the automation of such tasks center on 3-D printing. Roboticists at Nanyang Technological University in Singapore imagine that a robotic swarm of 3-D printers could be used in construction. While this might seem like a distant prospect, engineers have actually managed to create a single-piece concrete structure, using two mobile robots operating concurrently. See X. Zhang et al., 2018, "Large-Scale 3D Printing by a Team of Mobile Robots," *Automation in Construction* 95 (November): 98–106.

35. C. B. Frey and Osborne, 2017, "The Future of Employment," 261.

36. M. Mandel and B. Swanson, 2017, "The Coming Productivity Boom—Transforming the Physical Economy with Information" (Washington, DC: Technology CEO Council), 14.

37. H. Shaban, 2018, "Amazon Is Issued Patent for Delivery Drones That Can React to Screaming Voices, Flailing Arms," *Washington Post*, March 22.

38. D. Paquette, 2018, "He's One of the Only Humans at Work—and He Loves It," *Washington Post*, September 10.

39. Ibid.

40. M. Ryan, C. Metz, and M. Taylor, 2018, "How Robot Hands Are Evolving to Do What Ours Can," *New York Times*, July 30.

41. Ibid.

42. Ibid.

43. Quoted in M. Klein, 2007, *The Genesis of Industrial America, 1870–1920* (Cambridge: Cambridge University Press), 78.

44. Quoted in D. J. Millet, 1972, "Town Development in Southwest Louisiana, 1865–1900," *Louisiana History* 13 (2): 144.

45. "Music over the Wires," 1890, *New York Times*, October 9.

46. E. Clague, 1960, "Adjustments to the Introduction of Office Automation," *Bureau of Labor Statistics Bulletin*, no. 1276, 2.

47. H. Simon, [1960] 1985, "The Corporation: Will It Be Managed by Machines?," in *Management and the Corporation*, ed. M. L. Anshen and G. L. Bach (New York: McGraw-Hill), 17–55.

48. T. Malthus, [1798] 2013, *An Essay on the Principle of Population*, Digireads.com, Kindle, 179.

49. C. B. Frey and Osborne, 2017, "The Future of Employment," 262.

50. The O*NET database contains hundreds of standardized and occupation-specific descriptors for occupations covering the U.S. economy. For a list of occupations involving "originality," see O*NET OnLine, 2018, "Find Occupations: Abilities—Originality," https://www.onetonline.org/find/descriptor/result/1.A.1.b.2.

51. C. B. Frey and Osborne, 2017, "The Future of Employment," 262.

52. These descriptions stem from large-scale surveys of the American workforce, in which workers are asked how often they engage in various tasks. Their responses form part of O*NET OnLine database.

53. L. Nedelkoska and G. Quintini, 2018, "Automation, Skills Use and Training" (OECD Social, Employment and Migration Working Paper 202, Organisation of Economic Co-operation and Development, Paris).

54. One study by researchers at the University of Mannheim suggests that only 9 percent of jobs are exposed to automation. See M. Arntz, T. Gregory, and U. Zierahn, 2016, "The Risk of Automation for Jobs in OECD Countries" (OECD Social, Employment and Migration Working Paper 189, Organisation of Economic Co-operation and Development, Paris). And more recently, a study by the OECD estimates that 14 percent of jobs are at risk of being replaced. See L. Nedelkoska and G. Quintini, 2018, "Automation, Skills Use and Training" (OECD Social, Employment and Migration Working Paper 202, Organisation of Economic Co-operation and Development, Paris). The intuition behind these studies and ours is that we can infer the automatability of jobs by analyzing the tasks they entail. However, instead of relying primarily on tasks, the Mannheim study also incorporated demographic variables such as sex, education, age, and income. Because women and college-educated people, for example, tend to work in occupations that are less exposed to automation, their approach means that a female taxi driver with a PhD is less likely to be displaced by autonomous vehicles than a man who has been driving a taxi for many decades. In practice, however, this seems unlikely to be true. Aware of this problem, the authors of the OECD study followed our approach and relied on tasks rather than worker characteristics. But like the Mannheim study, the OECD study used individual-level data from the Programme for the International Assessment of Adult Competencies (PIAAC) survey instead of occupational averages. This approach allows the authors to distinguish between workers within occupations who might perform slightly different tasks. The drawback is that they have to rely on broader occupational categories, lumping many different occupations together, which means that valuable information is lost, as the OECD study rightly points out (Nedelkoska and Quintini, 2018, "Automation, Skills Use and Training"). What's more, that study regrettably does not provide any detail on any within-occupation variation, which suggests that other things are likely to be more relevant in explaining the differences between their results and ours. Indeed, it is hard to believe that the tasks performed by different truck drivers (or workers in other occupations) vary that greatly. In the end, the only reasonable way to check whether their model or ours is preferable is how well they perform on that training set (the OECD study also used our training data set). A frequently used metric to assess this is the area under the curve (AUC), and by this measure the nonlinear model in our study is much more accurate than their linear model. For a detailed discussion of how and why these estimates differ, see also C. B. Frey and M. Osborne, 2018, "Automation and the Future of Work—Understanding the Numbers," Oxford Martin School, https://www.oxfordmartin.ox.ac.uk/opinion/view/404.

55. See, for example, Arntz, Gregory, and Zierahn, 2016, "The Risk of Automation for Jobs in OECD Countries," table 5.

56. Council of Economic Advisers, 2016, "2016 Economic Report of the President," chapter 5, https://obamawhitehouse.archives.gov/sites/default/files/docs/ERP_2016_Chapter_5.pdf.

57. J. Furman, forthcoming, "Should We Be Reassured If Automation in the Future Looks Like Automation in the Past?," in *Economics of Artificial Intelligence*, ed. Ajay K. Agrawal, Joshua Gans, and Avi Goldfarb (Chicago: University of Chicago Press), 8.

58. M. Ford, 2015. *Rise of the Robots: Technology and the Threat of a Jobless Future* (New York: Basic Books), introduction, Kindle.

59. D. Remus and F. Levy, 2017, "Can Robots Be Lawyers? Computers, Lawyers, and the Practice of Law," *Georgetown Journal Legal Ethics* 30 (3): 526.

60. As we made clear, "we focus on estimating the share of employment that can potentially be substituted by computer capital, from a technological capabilities point of view, over some unspecified number of years. We make no attempt to estimate how many jobs will actually be automated. The actual extent and pace of computerisation will depend on several additional factors which were left unaccounted for" (C. B. Frey and Osborne, 2017, "The Future of Employment," 268).

61. See also D. H. Autor, 2014, "Skills, Education, and the Rise of Earnings Inequality among the 'Other 99 Percent,'" *Science* 344 (6186): 843–51.

62. W. K. Blodgett, 1918, "Doing Farm Work by Motor Tractor," *New York Times*, January 6.

63. D. P. Gross, 2018, "Scale Versus Scope in the Diffusion of New Technology: Evidence from the Farm Tractor," *RAND Journal of Economics* 49 (2): 449.

64. "17,000,000 Horses on Farms," 1921, *New York Times*, December 30.

65. T. Sorensen, P. Fishback, S. Kantor, and P. Rhode, 2008, "The New Deal and the Diffusion of Tractors in the 1930s" (Working paper, University of Arizona, Tucson).

66. R. Solow, 1987, "We'd Better Watch Out," *New York Times* Book Review, July 12; H. Gilman, 1987, "The Age of Caution: Companies Slow the Move to Automation," *Wall Street Journal*, June 12.

67. Quoted in ibid.

68. See, for example, T. F. Bresnahan, E. Brynjolfsson, and L. M. Hitt, 2002, "Information Technology, Workplace Organization, and the Demand for Skilled Labor: Firm-Level Evidence," *Quarterly Journal of Economics* 117 (1): 339–76; E. Brynjolfsson, L. M. Hitt, and S. Yang, 2002, "Intangible Assets: Computers and Organizational Capital," *Brookings Papers on Economic Activity* 2002 (1): 137–81; E. Brynjolfsson and L. M. Hitt, 2000, "Beyond Computation: Information Technology, Organizational Transformation and Business Performance," *Journal of Economic Perspectives* 14 (4): 23–48.

69. M. Hammer, 1990, "Reengineering Work: Don't Automate, Obliterate," *Harvard Business Review* 68 (4): 104–12.

70. On companies reengineering plans, see J. Rifkin, 1995, *The End of Work: The Decline of the Global Labor Force and the Dawn of the Post-market Era* (New York: G. P. Putnam's Sons).

71. P. A. David, 1990, "The Dynamo and the Computer: An Historical Perspective on the Modern Productivity Paradox," *American Economic Review* 80 (2): 355–61.

72. For a detailed discussion, see R. J. Gordon, 2005, "The 1920s and the 1990s in Mutual Reflection" (Working Paper 11778, National Bureau of Economic Research, Cambridge, MA).

73. S. D. Oliner and D. E. Sichel, 2000, "The Resurgence of Growth in the Late 1990s: Is Information Technology the Story?," *Journal of Economic Perspectives* 14 (4): 3–22.

74. W. D. Nordhaus, 2005, "The Sources of the Productivity Rebound and the Manufacturing Employment Puzzle" (Working Paper 11354, National Bureau of Economic Research, Cambridge, MA).

75. In the period 1993–2007, robots are estimated to have accounted for bit more than one-tenth of overall growth in gross domestic product (GDP) across seventeen countries. See G. Graetz and G. Michaels, forthcoming, "Robots at Work," *Review of Economics and Statistics*.

76. J. Bughin et al., 2017, "How Artificial Intelligence Can Deliver Real Value to Companies," McKinsey Global Institute, https://www.mckinsey.com/business-functions/mckinsey-analytics/our-insights/how-artificial-intelligence-can-deliver-real-value-to-companies.

77. It is true that many of the benefits brought by technology are unmeasured, which could in principle account for some of the productivity slowdown. In a recent study, the economists Austan Goolsbee and Peter Klenow used a novel approach to measure the value of internet-based technologies, examining the time people spend on the internet. Building on the intuition that consumption involves expenditure of both income and time, they estimated that the internet-related consumer surplus could be up to 3 percent (or $3,000 annually for the median person). See A. Goolsbee and P. Klenow, 2006, "Valuing Consumer Products by the Time Spent Using Them: An Application to the Internet," *American Economic Review* 96 (2): 108–13. Chad Syverson recently extended their value-of-time analysis, using the American Time Use Survey and data on personal disposable income. Applying the 3 percent estimate of Goolsbee and Klenow, he calculated an internet-related consumer surplus of around $3,900 per capita for 2105 (2017, "Challenges to Mismeasurement Explanations for the US Productivity Slowdown," *Journal of Economic Perspectives* 31 [2]: 165–86). All the same, it is not clear that mismeasurement has become greater in the computer era. Indeed, the Boskin Commission, appointed by the U.S. Senate in 1995, also found evidence of substantial unmeasured quality improvements. The question of whether the recent productivity slowdown is an artifact of mismeasurement is thus not a question of whether mismeasurement exists but one of whether it has gotten larger in recent years. Economists have shown that the answer is no. While there is surely mismeasurement, it seems to have gotten smaller, not larger. Mismeasurement associated with prices of computer hardware and related services as well as intangible assets (such as patents, trademarks, and advertising expenditures) only make the productivity slowdown worse. The decline in domestic production of computer-related goods and services since the period 1995–2004 means that despite mismeasurement's having worsened for some digital technologies, the mismeasurement problem was greater then than it is now. Together, these adjustments add 0.5 percentage point to the labor productivity numbers published for 1995–2004, but only 0.2 percentage point for 2004–14 (see D. M. Byrne, J. G. Fernald, and M. B. Reinsdorf, 2016, "Does the United States Have a Productivity Slowdown or a Measurement Problem?," *Brookings Papers on Economic Activity*, 2016 [1]: 109–82). Even if we factor in high-end estimates of consumer benefits from free services like those of Wikipedia, Google, Facebook, and so on, this can account for only about a third of the slowdown. Syverson has calculated that if the productivity deceleration had not happened, measured GDP would have been 16 percent higher in 2015, adding $2.9 trillion to the U.S. economy. This amounts to $9,100 for every citizen or $23,400 for every household (2017, "Challenges to Mismeasurement Explanations for the US Productivity Slowdown"). The bottom line is that mismeasurement might be large, but it is not sufficiently large to account for the productivity slowdown. The productivity slowdown appears structural and real.

78. Brynjolfsson, Rock, and Syverson, forthcoming, "Artificial Intelligence and the Modern Productivity Paradox," 25.

79. C. F. Kerry and J. Karsten, 2017, "Gauging Investment in Self-Driving Cars," Brookings Institution, October 16. https://www.brookings.edu/research/gauging-investment-in-self-driving-cars/.

80. Brynjolfsson, Rock, and Syverson, forthcoming, "Artificial Intelligence and the Modern Productivity Paradox," 25.

81. N. F. Crafts and T. C. Mills, 2017, "Trend TFP Growth in the United States: Forecasts versus Outcomes" (Discussion Paper 12029, Centre for Economic Policy Research, London). Their findings are consistent with the observation of Eric Bartelsman that productivity forecasts perform "horribly, with forecast standard errors being larger than ranges that would be useful for policy purposes" (2013, "ICT, Reallocation and Productivity" [Brussels: European Commission, Directorate-General for Economic and Financial Affairs]).

82. H. Jerome, 1934, "Mechanization in Industry" (Working Paper 27, National Bureau of Economic Research, New York), 19.

83. H. R. Varian, forthcoming, "Artificial Intelligence, Economics, and Industrial Organization," in *The Economics of Artificial Intelligence: An Agenda*, ed. Ajay K. Agrawal, Joshua Gans, and Avi Goldfarb (Chicago: University of Chicago Press), 1.

84. Ibid., 15.

85. Brynjolfsson, Rock, and Syverson, forthcoming, "Artificial Intelligence and the Modern Productivity Paradox."

86. N. F. Crafts, 2004, "Steam as a General Purpose Technology: A Growth Accounting Perspective," *Economic Journal* 114 (495): 338–51.

87. Quoted in J. L. Simon, 2000, *The Great Breakthrough and Its Cause* (Ann Arbor: University of Michigan Press), 108.

88. P. Colquhoun, 1815, *A Treatise on the Wealth, Power, and Resources of the British Empire*, Johnson Reprint Corporation), 68–69. See also J. Mokyr, 2011, *The Enlightened Economy; Britain and the Industrial Revolution, 1700–1850* (London: Penguin), chapter 5, Kindle. I am indebted to Joel Mokyr for pointing me to this reference.

89. Malthus, [1798] 2013, *An Essay on the Principle of Population*, 179.

90. R. Henderson, 2017, comment on "Artificial Intelligence and the Modern Productivity Paradox: A Clash of Expectations and Statistics, by E. Brynjolfsson, D. Rock and C. Syverson," National Bureau of Economic Research, http://www.nber.org/chapters/c14020.pdf.

91. J. M. Keynes, [1930] 2010, "Economic Possibilities for Our Grandchildren," in *Essays in Persuasion* (London: Palgrave Macmillan), 321–32.

92. V. A. Ramey and N. Francis, 2009, "A Century of Work and Leisure," *American Economic Journal: Macroeconomics* 1 (2): 189–224.

93. W. A. Sundstrom, 2006, "Hours and Working Conditions," in *Historical Statistics of the United States, Earliest Times to the Present: Millennial Edition Online*, ed. S. B. Carter et al. (New York: Cambridge University Press).

94. Ramey and Francis, 2009, "A Century of Work and Leisure."

95. These estimates are based on the age-year specific leisure measures and survival probabilities. See ibid.

96. These results depart somewhat from the estimates of Mark Aguiar and Erik Hurst, who find a larger increase in leisure for the period after 1965. The main reason is that they classify

child care as leisure rather than home production. See M. Aguiar and E. Hurst, 2007, "Measuring Trends in Leisure: The Allocation of Time Over Five Decades," *Quarterly Journal of Economics* 122 (3): 969–1006. Ramey and Francis also classify activities like talking to and playing with children as leisure, but they classify other child care tasks as home production. Given that people report much lower levels of enjoyment associated with such activities, this seems reasonable. See J. Robinson and G. Godbey, 2010, *Time for Life: The Surprising Ways Americans Use Their Time* (Philadelphia: Penn State University Press.)

97. Keynes, [1930] 2010, "Economic Possibilities for Our Grandchildren," 322.

98. R. L. Heilbroner, 1966, "Where Do We Go from Here?," *New York Review of Books*, March 17, https://www.nybooks.com/articles/1966/03/17/where-do-we-go-from-here/.

99. D. H. Autor, 2015, "Why Are There Still So Many Jobs? The History and Future of Workplace Automation," *Journal of Economic Perspectives* 29 (3): 8.

100. Heilbroner, 1966, "Where Do We Go from Here?"

101. B. Stevenson and J. Wolfers, 2013, "Subjective Well-Being and Income: Is There Any Evidence of Satiation?," *American Economic Review* 103 (3): 598–604.

102. H. Simon, 1966, "Automation," *New York Review of Books*, March 26, https://www.nybooks.com/articles/1966/05/26/automation-3/.

103. C. Stewart, 1960, "Social Implications of Technological Progress," in *Impact of Automation: A Collection of 20 Articles about Technological Change, from the* Monthly Labor Review (Washington, DC: Bureau of Labor Statistics), 12.

104. H. Voth, 2000, *Time and Work in England 1750–1830* (Oxford: Clarendon Press of Oxford University Press).

105. Aguiar and Hurst, 2007, "Measuring Trends in Leisure Measuring Trends in Leisure."

106. Quoted in C. Curtis, 1983, "Machines vs. Workers." *New York Times*, February 8.

107. F. Bastiat, 1850, "That Which Is Seen, and That Which Is Not Seen," https://mises.org/library/which-seen-and-which-not-seen.

108. D. Acemoglu and P. Restrepo, 2018b, "The Race between Man and Machine: Implications of Technology for Growth, Factor Shares, and Employment," *American Economic Review* 108 (6): 1488–542.

109. T. Berger and C. B. Frey, 2017a, "Industrial Renewal in the 21st Century: Evidence from US Cities," *Regional Studies* 51 (3): 404–13.

110. Brynjolfsson and McAfee, 2014, *The Second Machine Age*, 11.

111. A. Goolsbee, 2018, "Public Policy in an AI Economy" (Working Paper 24653, National Bureau of Economic Research, Cambridge, MA).

Chapter 13

1. See D. S. Landes, 1969, *The Unbound Prometheus: Technological Change and Development in Western Europe from 1750 to the Present* (Cambridge: Cambridge University Press), 4.

2. A. H. Hansen, 1939, "Economic Progress and Declining Population Growth," *American Economic Review* 29 (1): 10–11.

3. R. J. Gordon, 2016, *The Rise and Fall of American Growth: The U.S. Standard of Living Since the Civil War* (Princeton, NJ: Princeton University Press).

4. Landes, 1969, *The Unbound Prometheus*, 4.

5. F. Fukuyama, 2014, *Political Order and Political Decay: From the Industrial Revolution to the Globalization of Democracy* (New York: Farrar, Straus and Giroux), 450.

6. On workers rationally opposing replacing technologies, see A. Korinek and J. E. Stiglitz, 2017, "Artificial Intelligence and Its Implications for Income Distribution and Unemployment" (Working Paper 24174, National Bureau of Economic Research, Cambridge, MA).

7. A. Greif and M. Iyigun, 2013, "Social Organizations, Violence, and Modern Growth," *American Economic Review* 103 (3): 534–38.

8. Quoted in A. Greif and M.Iyigun, 2012, "Social Institutions, Violence and Innovations: Did the Old Poor Law Matter?" (Working paper, Stanford University, Stanford, CA), 4.

9. Malthus wrote: "To remedy the frequent distresses of the common people, the poor laws of England have been instituted; but it is to be feared, that though they may have alleviated a little the intensity of individual misfortune, they have spread the general evil over a much larger surface. . . . The poor laws of England tend to depress the general condition of the poor in these two ways. Their first obvious tendency is to increase [the] population without increasing the food for its support. . . . Secondly, the quantity of provisions consumed in workhouses upon a part of the society that cannot in general be considered as the most valuable part diminishes the shares that would otherwise belong to more industrious and more worthy members, and thus in the same manner forces more to become dependent" ([1798] 2013, *An Essay on the Principle of Population*, 55 and 62–63, Digireads.com, Kindle). In similar fashion, Ricardo argued that "the clear and direct tendency of the poor laws, is in direct opposition to these obvious principles: it is not, as the legislature benevolently intended, to amend the condition of the poor, but to deteriorate the condition of both poor and rich. . . . This pernicious tendency of these laws is no longer a mystery, since it has been fully developed by the able hand of Mr. Malthus; and every friend to the poor must ardently wish for their abolition" ([1817] 1911, *The Principles of Political Economy and Taxation*. Reprint. London: Dent, 33).

10. On brilliant versus mediocre technologies, see D. Acemoglu and P. Restrepo, 2018a, "Artificial Intelligence, Automation and Work" (Working Paper 24196, National Bureau of Economic Research, Cambridge, MA).

11. Daron Acemoglu and Pascual Restrepo decompose the sources underpinning the demand for labor, showing that the replacement of workers in manufacturing can explain a large part of the decoupling between wages and productivity. This process began in the 1980s and has intensified since the turn of the twenty-first century. At the same time, it is important to remember that we have seen similar episodes before. Much like the situation today, in the mid-nineteenth century America saw machines take over existing work more rapidly than new technologies were able to reinstate labor in new activities. See D. Acemoglu and P. Restrepo, forthcoming, "Automation and New Tasks: The Implications of the Task Content of Production for Labor Demand," *Journal of Economic Perspectives*. The authors' data do not allow them to go farther back than 1850, but (as noted in chapter 5) Britain saw a similar pattern in the early part of the nineteenth century, when textile machinery replaced artisan craftsmen in large numbers.

12. N. Eberstadt, 2016, *Men without Work: America's Invisible Crisis* (Conshohocken, PA: Templeton Press).

13. C. B. Frey and M. A. Osborne, 2017, "The Future of Employment: How Susceptible Are Jobs to Computerisation?," *Technological Forecasting and Social Change* 114:254–80.

14. R. Bowley, 2017, "The Fastest-Growing Jobs in the U.S. Based on LinkedIn Data," *LinkedIn Official Blog*, December 7, https://blog.linkedin.com/2017/december/7/the-fastest -growing-jobs-in-the-u-s-based-on-linkedin-data.

15. S. Murthy, 2014, "Top 10 Job Titles That Didn't Exist 5 Years Ago (Infographic)," *LinkedIn Talent Blog*, January 6, https://business.linkedin.com/talent-solutions/blog/2014/01/top-10 -job-titles-that-didnt-exist-5-years-ago-infographic.

16. M. Berg, 1976, "The Machinery Question," PhD diss., University of Oxford, 2.

17. L. Summers, 2017, "Robots Are Wealth Creators and Taxing Them Is Illogical," *Financial Times*, March 5.

18. C. Goldin and L. Katz, 2008, *The Race between Technology and Education* (Cambridge, MA: Harvard University Press), 1–2.

19. G. J. Duncan and R. J. Murnane, eds., 2011. *Whither Opportunity? Rising Inequality, Schools, and Children's Life Chances* (New York: Russell Sage Foundation).

20. J. D. Sachs, S. G. Benzell, and G. LaGarda, 2015, "Robots: Curse or Blessing? A Basic Framework" (Working Paper 21091, National Bureau of Economic Research, Cambridge, MA).

21. J. J. Heckman et al., 2010, "The Rate of Return to the HighScope Perry Preschool Program," *Journal of Public Economics* 94 (1–2), 114–28.

22. A. J. Reynolds et al., 2011, "School-Based Early Childhood Education and Age-28 Well-Being: Effects by Timing, Dosage, and Subgroups," *Science* 333 (6040): 360–64.

23. H. J. Holzer, D. Whitmore Schanzenbach, G. J. Duncan, and J. Ludwig, 2008, "The Economic Costs of Childhood Poverty in the United States," *Journal of Children and Poverty* 14 (1): 41–61.

24. A. M. Bell et al., 2017, "Who Becomes an Inventor in America? The Importance of Exposure to Innovation (Working Paper 24062, National Bureau of Economic Research, Cambridge, MA).

25. A. M. Bell et al., 2018, "Lost Einsteins: Who Becomes an Inventor in America?," *CentrePiece*, Spring, http://cep.lse.ac.uk/pubs/download/cp522.pdf, 11.

26. R. D. Putnam, 2016, *Our Kids: The American Dream in Crisis* (New York: Simon & Schuster), chapter 6.

27. K. L. Schlozman, S. Verba, and H. E. Brady, 2012, *The Unheavenly Chorus: Unequal Political Voice and the Broken Promise of American Democracy* (Princeton, NJ: Princeton University Press).

28. R. A. Dahl, 1998, *On Democracy* (New Haven, CT: Yale University Press), 76.

29. Quoted in G. R. Kremen, 1974, "MDTA: The Origins of the Manpower Development and Training Act of 1962," U.S. Department of Labor, https://www.dol.gov/general/aboutdol /history/mono-mdtatext.

30. O. Ashenfelter, 1978, "Estimating the Effect of Training Programs on Earnings," *Review of Economics and Statistics* 60 (1): 47–57.

31. Further complicating the matter is the fact that programs that target very different groups in the labor market cannot be compared. Workers from disadvantaged backgrounds and with less formal education naturally require more training and resources. What's more, the

effectiveness of training measures greatly depends on training content, local labor market characteristics, and the overall health of the economy.

32. B. S. Barnow and J. Smith, 2015, "Employment and Training Programs" (Working Paper 21659, National Bureau of Economic Research, Cambridge, MA).

33. R. J. LaLonde, 2007, *The Case for Wage Insurance* (New York: Council on Foreign Relations Press), 19.

34. On UBI versus the welfare state, see A. Goolsbee, 2018, "Public Policy in an AI Economy" (Working Paper 24653, National Bureau of Economic Research, Cambridge, MA).

35. On television and well-being, see B. S. Frey, 2008, *Happiness: A Revolution in Economics* (Cambridge, MA: MIT Press), chapter 9.

36. D. Graeber, 2018, *Bullshit Jobs: A Theory* (New York: Simon & Schuster). For survey evidence showing that people find meaning in their jobs, see R. Dur and M. van Lent, 2018, "Socially Useless Jobs" (Discussion Paper 18-034/VII, Amsterdam: Tinbergen Institute).

37. On happiness and unemployment, see B. S. Frey, 2008, *Happiness*, chapter 4.

38. I. Goldin, 2018, "Five Reasons Why Universal Basic Income Is a Bad Idea," *Financial Times*, February 11.

39. G. Hubbard, 2014, "Tax Reform Is the Best Way to Tackle Income Inequality," *Washington Post*, January 10.

40. For an overview of the effects of the EITC, see A. Nichols and J. Rothstein, 2015, "The Earned Income Tax Credit (EITC)" (Working Paper 21211, National Bureau of Economic Research, Cambridge, MA).

41. R. Chetty, N. Hendren, P, Kline, and E. Saez, 2014, "Where Is the Land of Opportunity? The Geography of Intergenerational Mobility in the United States," *Quarterly Journal of Economics* 129 (4): 1553–623.

42. L. Kenworthy, 2012, "It's Hard to Make It in America: How the United States Stopped Being the Land of Opportunity," *Foreign Affairs* 91(November/December): 97.

43. M. M. Kleiner, 2011, "Occupational Licensing: Protecting the Public Interest or Protectionism?" (Policy Paper 2011-009, Upjohn Institute, Kalamazoo, MI).

44. On occupational licensing and nonemployment among men in their prime, see B. Austin, E. L. Glaeser, and L. Summers, forthcoming, "Saving the Heartland: Place-Based Policies in 21st Century America," *Brookings Papers on Economic Activity*.

45. B. Fallick, C. A. Fleischman, and J. B. Rebitzer, 2006, "Job-Hopping in Silicon Valley: Some Evidence Concerning the Microfoundations of a High-Technology Cluster," *Review of Economics and Statistics* 88 (3): 472–81.

46. R. J. Gilson, 1999, "The Legal Infrastructure of High Technology Industrial Districts: Silicon Valley, Route 128, and Covenants Not to Compete," *New York University Law Review* 74 (August): 575.

47. S. Klepper, 2010, "The Origin and Growth of Industry Clusters: The Making of Silicon Valley and Detroit," *Journal of Urban Economics* 67 (1): 15–32.

48. T. Berger and C. B. Frey, 2017b, "Regional Technological Dynamism and Noncompete Clauses: Evidence from a Natural Experiment," *Journal of Regional Science* 57 (4): 655–68.

49. E. Moretti, 2012, *The New Geography of Jobs* (Boston: Houghton Mifflin Harcourt), 158–65.

50. C. T. Hsieh and E. Moretti, forthcoming, "Housing Constraints and Spatial Misallocation," *American Economic Journal: Macroeconomics*.

51. M. Rognlie, 2014, "A Note on Piketty and Diminishing Returns to Capital," unpublished manuscript, http://mattrognlie.com/piketty_diminishing_returns.pdf.

52. See, for example, E. L. Glaeser and J. Gyourko, 2002, "The Impact of Zoning on Housing Affordability (Working Paper 8835, National Bureau of Economic Research, Cambridge, MA); E. L. Glaeser, 2017, "Reforming Land Use Regulations" (Report in the Series on Market and Government Failures, Brookings Center on Regulation and Markets, Washington).

53. R. Chetty, N. Hendren, and L. F. Katz, 2016, "The Effects of Exposure to Better Neighborhoods on Children: New Evidence from the Moving to Opportunity Experiment," *American Economic Review* 106 (4): 855–902.

54. On place and the likelihood of becoming an inventor, see Bell et al., 2017, "Who Becomes an Inventor in America?," and 2018, "Lost Einsteins."

55. C. T. Hsieh and E. Moretti, 2017, "How Local Housing Regulations Smother the U.S. Economy, *New York Times*, September 6.

56. D. Etherington, 2018, "Hyperloop Transportation Technologies Signs First Cross-State Deal in the U.S.," TechCruch, https://techcrunch.com/2018/02/15/hyperloop-transportation -technologies-signs-first-cross-state-deal-in-the-u-s/?guccounter=1.

57. M. Busso, J. Gregory, and P. Kline, 2013, "Assessing the Incidence and Efficiency of a Prominent Place-Based Policy," *American Economic Review* 103 (2): 897–947.

58. For more on the TVA, see P. Kline and E. Moretti, 2013, "Local Economic Development, Agglomeration Economies, and the Big Push: 100 Years of Evidence from the Tennessee Valley Authority," *Quarterly Journal of Economics* 129 (1): 275–331.

59. E. Moretti, 2004, "Estimating the Social Return to Higher Education: Evidence from Longitudinal and Repeated Cross-Sectional Data," *Journal of Econometrics* 121 (1–2): 175–212.

60. S. Liu, 2015, "Spillovers from Universities: Evidence from the Land-Grant Program," *Journal of Urban Economics* 87 (May): 25–41.

61. K. K. Charles, E. Hurst, and M. J. Notowidigdo, 2016, "The Masking of the Decline in Manufacturing Employment by the Housing Bubble," *Journal of Economic Perspectives* 30 (2): 179–200.

BIBLIOGRAPHY

Abowd, J. M., P. Lengermann, and K. L. McKinney. 2003. "The Measurement of Human Capital in the US Economy." LEHD Program technical paper TP-2002-09, Census Bureau, Washington.

Abraham, K. G., and M. S. Kearney. 2018. "Explaining the Decline in the US Employment-to-Population Ratio: A Review of the Evidence." Working Paper 24333, National Bureau of Economic Research, Cambridge, MA.

Acemoglu, D., and D. H. Autor. 2011. "Skills, Tasks and Technologies: Implications for Employment and Earnings." In *Handbook of Labor Economics*, edited by David Card and Orley Ashenfelter, 4:1043–171. Amsterdam: Elsevier.

Acemoglu, D., S. Johnson, and J. Robinson. 2005. "The Rise of Europe: Atlantic Trade, Institutional Change, and Economic Growth." *American Economic Review* 95 (3): 546–79.

Acemoglu, D., and P. Restrepo. 2018a. "Artificial Intelligence, Automation and Work." Working Paper 24196, National Bureau of Economic Research, Cambridge, MA.

Acemoglu, D., and P. Restrepo. 2018b. "The Race between Man and Machine: Implications of Technology for Growth, Factor Shares, and Employment." *American Economic Review* 108 (6): 1488–542.

Acemoglu, D., and P. Restrepo. 2018c. "Robots and Jobs: Evidence from US Labor Markets." Working paper, Massachusetts Institute of Technology, Cambridge, MA.

Acemoglu, D., and P. Restrepo. Forthcoming. "Automation and New Tasks: The Implications of the Task Content of Production for Labor Demand." *Journal of Economic Perspectives*.

Acemoglu, D., and J. A. Robinson. 2006. "Economic Backwardness in Political Perspective." *American Political Science Review* 100 (1): 115–31.

Acemoglu, D., and J. A. Robinson. 2012. *Why Nations Fail: The Origins of Power, Prosperity and Poverty*. New York: Crown Business.

Agrawal, A., J. Gans, and A. Goldfarb. 2016. "The Simple Economics of Machine Intelligence." *Harvard Business Review*, November 17. https://hbr.org/2016/11/the-simple-economics-of-machine-intelligence.

Aguiar, M., and E. Hurst. 2007. "Measuring Trends in Leisure: The Allocation of Time over Five Decades." *Quarterly Journal of Economics* 122 (3): 969–1006.

Aidt, T., G. Leon, and M. Satchell. 2017. "The Social Dynamics of Riots: Evidence from the Captain Swing Riots, 1830–31." Working paper, Cambridge University.

Aidt, T., and R. Franck. 2015. "Democratization under the Threat of Revolution: Evidence from the Great Reform Act of 1832." *Econometrica* 83 (2): 505–47.

Akst, D. 2013. "What Can We Learn from Past Anxiety over Automation?" *Wilson Quarterly,* Summer. https://wilsonquarterly.com/quarterly/summer-2014-where-have-all-the-jobs-gone /theres-much-learn-from-past-anxiety-over-automation/.

Aldcroft, D. H., and Oliver, M. J. 2000. *Trade Unions and the Economy: 1870–2000.* Aldershot, UK: Ashgate.

Alexopoulos, M., and J. Cohen. 2011. "Volumes of Evidence: Examining Technical Change in the Last Century through a New Lens." *Canadian Journal of Economics/Revue Canadienne d'économique* 44 (2): 413–50.

Alexopoulos, M., and J. Cohen. 2016. "The Medium Is the Measure: Technical Change and Employment, 1909–1949." *Review of Economics and Statistics* 98 (4): 792–810.

Allen, R. C. 2001. "The Great Divergence in European Wages and Prices from the Middle Ages to the First World War." *Explorations in Economic History* 38 (4): 411–47.

Allen, R. C. 2007. "Pessimism Preserved: Real Wages in the British Industrial Revolution." Working Paper 314, Department of Economics, Oxford University.

Allen, R. C. 2009a. *The British Industrial Revolution in Global Perspective.* Cambridge: Cambridge University Press. Kindle.

Allen, R. C. 2009b. "Engels' Pause: Technical Change, Capital Accumulation, and Inequality in the British Industrial Revolution." *Explorations in Economic History* 46 (4): 418–35.

Allen, R. C. 2009c. "How Prosperous Were the Romans? Evidence from Diocletian's Price Edict (AD 301)." In *Quantifying the Roman Economic: Methods and Problems,* edited by Alan Bowman and Andrew Wilson, 327–45. Oxford: Oxford University Press.

Allen, R. C. 2009d. "The Industrial Revolution in Miniature: The Spinning Jenny in Britain, France, and India." *Journal of Economic History* 69 (4): 901–27.

Allen, R. C. 2017. "Lessons from History for the Future of Work." *Nature News* 550 (7676): 321–24.

Allen, R. C. Forthcoming. "The Hand-Loom Weaver and the Power Loom: A Schumpeterian Perspective." *European Review of Economic History.*

Allen, R. C., J. P. Bassino, D. Ma, C. Moll-Murata, and J. L. Van Zanden. 2011. "Wages, Prices, and Living Standards in China, 1738–1925: In Comparison with Europe, Japan, and India." *Economic History Review* 64 (January): 8–38.

Allison, G. 2017. *Destined for War: Can America and China Escape Thucydides's Trap?* Boston: Houghton Mifflin Harcourt. Kindle.

Alston, L. J., and T. J. Hatton. 1991. "The Earnings Gap between Agricultural and Manufacturing Laborers, 1925–1941." *Journal of Economic History* 51 (1): 83–99.

Anderson, M. 1990. "The Social Implications of Demographic Change." In *The Cambridge Social History of Britain, 1750–1950,* vol. 2: *People and Their Environment,* edited by F.M.L. Thompson, 1–70. Cambridge: Cambridge University Press.

Anelli, M., I. Colantone, and P.Stanig. 2018. "We Were the Robots: Automation in Manufacturing and Voting Behavior in Western Europe." Working paper, Bocconi University, Milan.

Annual Registrar or a View of the History, Politics, and Literature for the Year 1811. 1811. London: printed for Baldwin, Cradock, and Joy.

Armelagos, G. J., and M. N. Cohen. 1984. *Paleopathology at the Origins of Agriculture*, edited by G. J. Armelagos and M. N. Cohen, 235–69. Orlando, FL: Academic Press.

Arntz, M., T. Gregory, and U. Zierahn. 2016. "The Risk of Automation for Jobs in OECD Countries." OECD Social, Employment and Migration Working Paper 189, Organisation of Economic Co-operation and Development, Paris.

Ashenfelter, O. 1978. "Estimating the Effect of Training Programs on Earnings." *Review of Economics and Statistics* 60 (1): 47–57.

Ashraf, Q., and O. Galor. 2011. "Dynamics and Stagnation in the Malthusian Epoch." *American Economic Review* 101 (5): 2003–41.

Ashton, T. S. 1948. *An Economic History of England: The Eighteenth Century.* London: Routledge.

Austin, B., E. L. Glaeser, and L. Summers. Forthcoming. "Saving the Heartland: Place-Based Policies in 21st Century America." *Brookings Papers on Economic Activity.*

Autor, D. H. 2014. "Skills, Education, and the Rise of Earnings Inequality among the 'Other 99 Percent.'" *Science* 344 (6186): 843–51.

Autor, D. H. 2015. "Polanyi's Paradox and the Shape of Employment Growth." In *Re-evaluating Labor Market Dynamics*, 129–77. Kansas City: Federal Reserve Bank of Kansas City.

Autor, D. H. 2015. "Why Are There Still So Many Jobs? The History and Future of Workplace Automation." *Journal of Economic Perspectives* 29 (3): 3–30.

Autor, D. H., and A. Salomons. Forthcoming. "Is Automation Labor-Displacing? Productivity Growth, Employment, and the Labor Share." *Brookings Papers on Economic Activity.*

Autor, D. H., and D. Dorn. 2013. "The Growth of Low-Skill Service Jobs and the Polarization of the US Labor Market." *American Economic Review* 103 (5): 1553–97.

Autor, D. H., D. Dorn, and G. Hanson. Forthcoming. "When Work Disappears: Manufacturing Decline and the Falling Marriage-Market Value of Men." *American Economic Review: Insights.*

Autor, D. H., D. Dorn, G., Hanson, and K. Majlesi. 2016a. "Importing Political Polarization? The Electoral Consequences of Rising Trade Exposure." Working Paper 22637, National Bureau of Economic Research, Cambridge, MA.

Autor, D. H., D. Dorn, G. Hanson, and K. Majlesi. 2016b. "A Note on the Effect of Rising Trade Exposure on the 2016 Presidential Election." Appendix to "Importing Political Polarization? The Electoral Consequences of Rising Trade Exposure." Working Paper 22637, National Bureau of Economic Research, Cambridge, MA.

Autor, D. H., F. Levy, and R. J. Murnane. 2003. "The Skill Content of Recent Technological Change: An Empirical Exploration." *Quarterly Journal of Economics* 118 (4): 1279–333.

Babbage, C. 1832. *On the Economy of Machinery and Manufactures.* London: Charles Knight.

Bacci, M. L. 2017. *A Concise History of World Population.* Oxford: John Wiley and Sons.

Baines, E. 1835. *History of the Cotton Manufacture in Great Britain.* London: H. Fisher, R. Fisher, and P. Jackson.

Bairoch, P. 1991. *Cities and Economic Development: From the Dawn of History to the Present*. Chicago: University of Chicago Press.

Baldwin, G. B., and G. P. Schultz. 1960. "The Effects of Automation on Industrial Relations." In *Impact of Automation: A Collection of 20 Articles about Technological Change, from the* Monthly Labor Review. Washington, DC: Bureau of Labor Statistics, 47–49.

Balke, N. S., and R. J. Gordon. 1989. "The Estimation of Prewar Gross National Product: Methodology and New Evidence." *Journal of Political Economy* 97 (1): 38–92.

Barnow, B. S., and J. Smith. 2015. "Employment and Training Programs." Working Paper 21659, National Bureau of Economic Research, Cambridge, MA.

Barro, R. J., and X. Sala-i-Martin. 1992. "Convergence." *Journal of Political Economy* 100 (2): 223–51.

Bartels, L. M. 2016. *Unequal Democracy: The Political Economy of the New Gilded Age*. Princeton, NJ: Princeton University Press.

Bartelsman, E. J. 2013. "ICT, Reallocation and Productivity." Brussels: European Commission, Directorate-General for Economic and Financial Affairs.

Bastiat, F. 1850. "That Which Is Seen, and That Which Is Not Seen." Mises Institute. https://mises.org/library/which-seen-and-which-not-seen.

Becker, G. 1968 "Crime and Punishment: An Economic Approach." *Journal of Political Economy* 76 (2): 169–217.

Becker, S. O., E. Hornung, and L. Woessmann. 2011. "Education and Catch-Up in the Industrial Revolution." *American Economic Journal: Macroeconomics* 3 (3): 92–126.

Bell, A. M., R. Chetty, X. Jaravel, N. Petkova, and J. Van Reenen. 2017. "Who Becomes an Inventor in America? The Importance of Exposure to Innovation." Working Paper 24062, National Bureau of Economic Research, Cambridge, MA.

Bell, A. M., R. Chetty, X. Jaravel, N. Petkova, and J. Van Reenen. 2018. "Lost Einsteins: Who Becomes an Inventor in America?" *CentrePiece*, Spring, http://cep.lse.ac.uk/pubs/download/cp522.pdf.

Berg, M. 1976. "The Machinery Question." PhD diss., University of Oxford.

Berg, M. 2005. *The Age of Manufactures, 1700–1820: Industry, Innovation and Work in Britain*. London: Routledge.

Berger, T., and C. B. Frey. 2016. "Did the Computer Revolution Shift the Fortunes of U.S. Cities? Technology Shocks and the Geography of New Jobs." *Regional Science and Urban Economics* 57 (March): 38–45.

Berger, T., and C. B. Frey. 2017a. "Industrial Renewal in the 21st Century: Evidence from US Cities." *Regional Studies* 51 (3): 404–13.

Berger, T., and C. B. Frey. 2017b. "Regional Technological Dynamism and Noncompete Clauses: Evidence from a Natural Experiment." *Journal of Regional Science* 57 (4): 655–68.

Bernal, J. D. 1971. *Science in History*. Vol. 1: *The Emergence of Science*. Cambridge, MA: MIT Press.

Bernhofen, D. M., Z. El-Sahli, and R. Kneller. 2016. "Estimating the Effects of the Container Revolution on World Trade." *Journal of International Economics* 98 (January): 36–50.

Bessen, J. 2015. *Learning by Doing: The Real Connection between Innovation, Wages, and Wealth*. New Haven, CT: Yale University Press.

Bessen, J. 2018. "Automation and Jobs: When Technology Boosts Employment." Law and Economics Paper 17-09, Boston University School of Law.

Bivens, J., E. Gould, E. Mishel, and H. Shierholz. 2014. "Raising America's Pay." Briefing Paper 378, Economic Policy Institute, New York.

Blake, W. 1810. "Jerusalem." https://www.poetryfoundation.org/poems/54684/jerusalem-and -did-those-feet-in-ancient-time.

Boerner, L., and B. Severgnini. 2015. "Time for Growth." Economic History Working Paper 222/2015, London School of Economics and Political Science.

Boerner, L., and B. Severgnini. 2016. "The Impact of Public Mechanical Clocks on Economic Growth." Vox, October 10. https://voxeu.org/article/time-growth.

Bogart, D. 2005. "Turnpike Trusts and the Transportation Revolution in 18th Century England." *Explorations in Economic History* 42 (4): 479–508.

Boix, C., and F. Rosenbluth. 2014. "Bones of Contention: The Political Economy of Height Inequality." *American Political Science Review* 108 (1): 1–22.

Bolt, J., R. Inklaar, H. de Jong, and J. L. Van Zanden. 2018. "Rebasing 'Maddison': New Income Comparisons and the Shape of Long-Run Economic Development." Maddison Project Working Paper 10, Maddison Project Database, version 2018.

Bolt, J., and J. L. Van Zanden. 2014. "The Maddison Project: Collaborative Research on Historical National Accounts." *Economic History Review* 67 (3): 627–51.

Boserup, E. 1965. *The Condition of Agricultural Growth: The Economics of Agrarian Change under Population Pressure*. London: Allen and Unwin.

Bowen, H. R. 1966. *Report of the National Commission on Technology, Automation, and Economic Progress*. Vol. 1. Washington, DC: Government Printing Office.

Braverman, H. 1998. *Labor and Monopoly Capital: The Degradation of Work in the Twentieth Century*. 25th anniversary ed. New York: New York University Press.

Bresnahan, T. F., E. Brynjolfsson, and L. M. Hitt. 2002. "Information Technology, Workplace Organization, and the Demand for Skilled Labor: Firm-Level Evidence." *Quarterly Journal of Economics* 117 (1): 339–76.

Brown, J. 1832. *A Memoir of Robert Blincoe: An Orphan Boy; Sent From the Workhouse of St. Pancras, London at Seven Years of Age, to Endure the Horrors of a Cotton-Mill*. London: J. Doherty.

Brynjolfsson, E., and L. M. Hitt. 2000. "Beyond Computation: Information Technology, Organizational Transformation and Business Performance." *Journal of Economic Perspectives* 14 (4): 23–48.

Brynjolfsson, E., L. M. Hitt, and S. Yang. 2002. "Intangible Assets: Computers and Organizational Capital." *Brookings Papers on Economic Activity* 2002 (1): 137–81.

Brynjolfsson, E., and A. McAfee. 2014. *The Second Machine Age: Work, Progress, and Prosperity in a Time of Brilliant Technologies*. New York: W. W. Norton.

Brynjolfsson, E., and A. McAfee. 2017. *Machine, Platform, Crowd: Harnessing Our Digital Future*. New York: W. W. Norton.

Brynjolfsson, E., D. Rock, and C. Syverson. Forthcoming. "Artificial Intelligence and the Modern Productivity Paradox: A Clash of Expectations and Statistics." In *The Economics of Artificial*

Intelligence: An Agenda, edited by A. K. Agrawal, J. Gans, and A. Goldfarb, Chicago: University of Chicago Press.

Bryson, B. 2010. *At Home: A Short History of Private Life*. Toronto: Doubleday Canada.

Bughin, J., E. Hazan, S. Ramaswamy, M. Chui, T. Allas, P. Dahlström, N. Henke, et al. 2017. "How Artificial Intelligence Can Deliver Real Value to Companies." McKinsey Global Institute. https://www.mckinsey.com/business-functions/mckinsey-analytics/our-insights/how -artificial-intelligence-can-deliver-real-value-to-companies.

Busso, M., J. Gregory, and P. Kline. 2013. "Assessing the Incidence and Efficiency of a Prominent Place-Based Policy." *American Economic Review* 103 (2): 897–947.

Byrn, E. W. 1900. *The Progress of Invention in the Nineteenth Century*. New York: Munn and Company.

Byrne, D. M., J. G. Fernald, and M. B. Reinsdorf. 2016. "Does the United States Have a Productivity Slowdown or a Measurement Problem?" *Brookings Papers on Economic Activity* 2016 (1): 109–82.

Bythell, D. 1969. *The Handloom Weavers: A Study in the English Cotton Industry during the Industrial Revolution*. Cambridge: Cambridge University Press.

Cairncross, F. 2001. *The Death of Distance: 2.0: How the Communications Revolution Will Change Our Lives*. New York: Texere Publishing.

Cameron, R. 1993. *A Concise Economic History of the World from Paleolithic Times to the Present*. 2nd ed. New York: Oxford University Press.

Cannadine, D. 1977. "The Landowner as Millionaire: The Finances of the Dukes of Devonshire, c. 1800–c. 1926." *Agricultural History Review* 25 (2): 77–97.

Caprettini, B., and H. J. Voth. 2017. "Rage against the Machines: Labour-Saving Technology and Unrest in England, 1830–32." Working paper, University of Zurich.

Cardwell, D. 1972. *Turning Points in Western Technology: A Study of Technology, Science and History*. New York: Science History Publications.

Cardwell, D. 2001. *Wheels, Clocks, and Rockets: A History of Technology*. New York: W. W. Norton.

Case, A., and A. Deaton. 2015. "Rising Morbidity and Mortality in Midlife among White Non-Hispanic Americans in the 21st Century." *Proceedings of the National Academy of Sciences* 112 (49): 15078–83.

Case, A., and A. Deaton. 2017. "Mortality and Morbidity in the 21st Century." *Brookings Papers on Economic Activity* 1: 397–476.

Chapman, S. D. 1967. *The Early Factory Masters: The Transition to the Factory System in the Midlands Textile Industry*. Exeter: David and Charles.

Charles, K. K., E. Hurst, and M. J. Notowidigdo. 2016. "The Masking of the Decline in Manufacturing Employment by the Housing Bubble." *Journal of Economic Perspectives* 30 (2): 179–200.

Cherlin, A. J. 2013. *Labor's Love Lost: The Rise and Fall of the Working-Class Family in America*. New York: Russell Sage Foundation.

Chetty, R., D. Grusky, M. Hell, N, Hendren, R. Manduca, and J. Narang. 2017. "The Fading American Dream: Trends in Absolute Income Mobility Since 1940." *Science* 356 (6336): 398–406.

Chetty, R., and N. Hendren. 2018. "The Impacts of Neighborhoods on Intergenerational Mobility II: County-Level Estimates." *Quarterly Journal of Economics* 133 (3): 1163–228.

Chetty, R., N. Hendren, and L. F. Katz. 2016. "The Effects of Exposure to Better Neighborhoods on Children: New Evidence from the Moving to Opportunity Experiment." *American Economic Review* 106 (4): 855–902.

Chetty, R., N. Hendren, P. Kline, and E. Saez. 2014. "Where Is the Land of Opportunity? The Geography of Intergenerational Mobility in the United States." *Quarterly Journal of Economics* 129 (4): 1553–623.

Cipolla, C. M. 1972. Introduction to *The Fontana Economic History of Europe*, edited by C. M. Cipolla. 1:7–21. Collins.

Cisco. 2018. "Cisco Visual Networking Index: Forecast and Trends, 2017–2022." San Jose, CA: Cisco. https://www.cisco.com/c/en/us/solutions/collateral/service-provider/visual-net working-index-vni/complete-white-paper-c11-481360.html.

Clague, E. 1960. "Adjustments to the Introduction of Office Automation." *Bureau of Labor Statistics Bulletin*, no. 1276.

Clague, E., and W. J. Couper, 1931. "The Readjustment of Workers Displaced by Plant Shutdowns." *Quarterly Journal of Economics* 45 (2): 309–46.

Clark, A. E., E. Diener, Y. Georgellis, and R. E. Lucas. 2008. "Lags and Leads in Life Satisfaction: A Test of the Baseline Hypothesis." *Economic Journal* 118 (529): 222–43.

Clark, A. E., and A. J. Oswald. 1994. "Unhappiness and Unemployment." *Economic Journal* 104 (424): 648–59.

Clark, A. E., and A. J. Oswald. 1996. "Satisfaction and Comparison Income." *Journal of Public Economics* 61 (3): 359–81.

Clark, G. 2001. "The Secret History of the Industrial Revolution." Working paper, University of California, Davis.

Clark, G. 2005. "The Condition of the Working Class in England, 1209–2004." *Journal of Political Economy* 113 (6): 1307–40.

Clark, G. 2008. *A Farewell to Alms: A Brief Economic History of the World*. Princeton, NJ: Princeton University Press.

Clark, G., and G. Hamilton. 2006. "Survival of the Richest: The Malthusian Mechanism in Pre-Industrial England." *Journal of Economic History* 66 (3): 707–36.

Clark, G., M. Huberman, and P. H. Lindert. 1995. "A British Food Puzzle, 1770–1850." *Economic History Review* 48 (2): 215–37.

Cockburn, I. M., R. Henderson, and S. Stern. 2018. "The Impact of Artificial Intelligence on Innovation." Working Paper 24449, National Bureau of Economic Research, Cambridge, MA.

Collins, W. J., and W. H. Wanamaker. 2015. "The Great Migration in Black and White: New Evidence on the Selection and Sorting of Southern Migrants." *Journal of Economic History* 75 (4): 947–92.

Colquhoun, P. [1814] 1815. *A Treatise on the Wealth, Power, and Resources of the British Empire*. Reprint, London: Johnson Reprint Corporation.

Comin, D., and M. Mestieri, 2018. "If Technology Has Arrived Everywhere, Why Has Income Diverged?" *American Economic Journal: Macroeconomics* 10 (3): 137–78.

Cooper, M. R., G. T. Barton, and A. P. Brodell. 1947. "Progress of Farm Mechanization." USDA Miscellaneous Publication 630 (October).

Cortes, G. M., N. Jaimovich, C. J. Nekarda, and H. E. Siu. 2014. "The Micro and Macro of Disappearing Routine Jobs: A Flows Approach." Working Paper 20307 National Bureau of Economic Research, Cambridge, MA.

Cortes, G. M., N. Jaimovich, and H. E. Siu. 2017. "Disappearing Routine Jobs: Who, How, and Why?" *Journal of Monetary Economics.* 91 (September): 69–87.

Cortes, G. M., N. Jaimovich, and H. E. Siu. 2018. "The 'End of Men' and Rise of Women in the High-Skilled Labor Market." Working Paper 24274, National Bureau of Economic Research, Cambridge, MA.

Coudray, N., P. S. Ocampo, T. Sakellaropoulos, N. Narula, M. Snuderl, D. Fenyö, A. L. Moreira, et al. 2018. "Classification and Mutation Prediction from Non–Small Cell Lung Cancer Histopathology Images Using Deep Learning." *Nature Medicine* 24 (10): 1559–67.

Council of Economic Advisers. 2016. "2016 Economic Report of the President," chapter 5. https://obamawhitehouse.archives.gov/sites/default/files/docs/ERP_2016_Chapter_5.pdf.

Cowan, R. S. 1983. *More Work for Mother: The Ironies of Household Technology from the Open Hearth to the Microwave.* New York: Basic.

Cowie, J. 2016. *The Great Exception: The New Deal and the Limits of American Politics.* Princeton, NJ: Princeton University Press.

Cox, G. W. 2012. "Was the Glorious Revolution a Constitutional Watershed?" *Journal of Economic History* 72 (3): 567–600.

Crafts, N. F. 1985. *British Economic Growth during the Industrial Revolution.* Oxford: Oxford University Press.

Crafts, N. F. 1987. "British Economic Growth, 1700–1850: Some Difficulties of Interpretation." *Explorations in Economic History* 20 (4): 245–68.

Crafts, N. F. 2004. "Steam as a General Purpose Technology: A Growth Accounting Perspective." *Economic Journal* 114 (495): 338–51.

Crafts, N. F., and C. K. Harley. 1992. "Output Growth and the British Industrial Revolution: A Restatement of the Crafts-Harley View." *Economic History Review* 45 (4): 703–30.

Crafts, N. F., and T. C. Mills. 2017. "Trend TFP Growth in the United States: Forecasts versus Outcomes." Centre for Economic Policy Research Discussion Paper 12029, London.

Crouzet, F. 1985. *The First Industrialists: The Problems of Origins.* Cambridge: Cambridge University Press.

Dahl, R. A. 1961. *Who Governs? Democracy and Power in an American City.* New Haven, CT: Yale University Press.

Dahl, R. A. 1998. *On Democracy.* New Haven, CT: Yale University Press.

Dao, M. C., M. M. Das, Z. Koczan, and W. Lian. 2017. "Why Is Labor Receiving a Smaller Share of Global Income? Theory and Empirical Evidence." Working Paper No. 17/169, International Monetary Fund, Washington, DC.

Dauth, W., S. Findeisen, J. Südekum, and N. Woessner. 2017. "German Robots: The Impact of Industrial Robots on Workers." Discussion Paper DP12306, Center for Economic and Policy Research, London.

David, P. A. 1990. "The Dynamo and the Computer: An Historical Perspective on the Modern Productivity Paradox." *American Economic Review* 80 (2): 355–61.

David, P. A., and G. Wright. 1999. *Early Twentieth Century Productivity Growth Dynamics: An Inquiry into the Economic History of Our Ignorance*. Oxford: Oxford University Press.

Davis, J. J. 1927. "The Problem of the Worker Displaced by Machinery." *Monthly Labor Review* 25 (3): 32–34.

Davis, R. 1973. *English Overseas Trade 1500–1700*. London: Macmillan.

Day, R. H. 1967. "The Economics of Technological Change and the Demise of the Sharecropper." *American Economic Review* 57 (3): 427–49.

Deaton, A. 2013. *The Great Escape: Health, Wealth, and the Origins of Inequality*. Princeton, NJ: Princeton University Press.

Defoe, D. [1724] 1971. *A Tour through the Whole Island of Great Britain*. London: Penguin.

DeLong, B. 1998. "Estimating World GDP: One Million BC–Present." Working paper, University of California, Berkeley.

Dent, C. 2006. "Patent Policy in Early Modern England: Jobs, Trade and Regulation." *Legal History* 10 (1): 71–95.

Desmet, K., A. Greif, and S. Parente. 2018. "Spatial Competition, Innovation and Institutions: The Industrial Revolution and the Great Divergence." Working Paper 24727, National Bureau of Economic Research, Cambridge, MA.

Devine, W. D., Jr. 1983. "From Shafts to Wires: Historical Perspective on Electrification." *Journal of Economic History* 43 (2): 347–72.

De Vries, J. 2008. *The Industrious Revolution: Consumer Behavior and the Household Economy, 1650 to the Present*. Cambridge: Cambridge University Press.

Diamond, J. 1987. "The Worst Mistake in the History of the Human Race." *Discover*, May 1, 64–66.

Diamond, J. 1993. "Ten Thousand Years of Solitude." *Discover*, March 1, 48–57.

Diamond, J. 1998. *Guns, Germs and Steel: A Short History of Everybody for the Last 13,000 Years*. New York: Random House.

Dickens, C. [1854] 2017. *Hard Times*. Amazon Classics. Kindle.

Disraeli, B. 1844. *Coningsby*. A Public Domain Book. Kindle Edition.

Dittmar, J. E. 2011. "Information Technology and Economic Change: The Impact of the Printing Press." *Quarterly Journal of Economics* 126 (3): 1133–72.

Doepke, M., and F. Zilibotti. 2008. "Occupational Choice and the Spirit of Capitalism." *Quarterly Journal of Economics* 123 (2): 747–93.

Duncan, G. J., and R. J. Murnane, eds. 2011. *Whither Opportunity? Rising Inequality, Schools, and Children's Life Chances*. New York: Russell Sage Foundation.

Dur, R., and M. van Lent. 2018. "Socially Useless Jobs." Discussion Paper 18-034/VII, Tinbergen Institute, Amsterdam.

Duranton, G., and D. Puga. 2001. "Nursery Cities: Urban Diversity, Process Innovation, and the Life Cycle of Products." *American Economic Review* 91 (5): 1454–77.

Eberstadt, N. 2016. *Men without Work: America's Invisible Crisis*. Conshohocken, PA: Templeton.

Eden, F. M. 1797. *The State of the Poor; or, An History of the Labouring Classes in England*. 3 vols. London: B. and J. White.

Ehrlich, I. 1973. "Participation in Illegitimate Activities: A Theoretical and Empirical Investigation." *Journal of Political Economy* 81 (3): 521–65.

Ehrlich, I. 1996. "Crime, Punishment, and the Market for Offenses." *Journal of Economic Perspectives* 10 (1): 43–67.

Elsby, M. W., B. Hobijn, and A. Şahin. 2013. "The Decline of the US Labor Share." *Brookings Papers on Economic Activity* 2013 (2): 1–63.

Endrei, W., and W. v. Stromer. 1974. "Textiltechnische und hydraulische Erfindungen und ihre Innovatoren in Mitteleuropa im 14. / 15. Jahrhundert." *Technikgeschichte* 41:89–117.

Engels, F., [1844] 1943. *The Condition of the Working-Class in England in 1844*. Reprint, London: Allen & Unwin.

Epstein, R. C. 1928. *The Automobile Industry*. Chicago: Shaw.

Epstein, S. R. 1998. "Craft Guilds, Apprenticeship and Technological Change in Preindustrial Europe." *Journal of Economic History* 58 (3): 684–713.

Esteva, A., B. Kuprel, R. A. Novoa, J. Ko, S. M. Swetter, H. M. Blau, and S. Thrun. 2017. "Dermatologist-Level Classification of Skin Cancer with Deep Neural Networks." *Nature* 542 (7639): 115–18.

Eveleth, P., and J. M. Tanner. 1976. *Worldwide Variation in Human Growth*. Cambridge: Cambridge University Press.

Fallick, B., C. A. Fleischman, and J. B. Rebitzer. 2006. "Job-Hopping in Silicon Valley: Some Evidence Concerning the Microfoundations of a High-Technology Cluster." *Review of Economics and Statistics* 88 (3): 472–81.

Farber, H. S., D. Herbst, I. Kuziemko, and S. Naidu. 2018. "Unions and Inequality over the Twentieth Century: New Evidence from Survey Data." Working Paper 24587, National Bureau of Economic Research, Cambridge, MA.

Faunce, W. A. 1958a. "Automation and the Automobile Worker." *Social Problems* 6 (1): 68–78.

Faunce, W. A. 1958b. "Automation in the Automobile Industry: Some Consequences for In-Plant Social Structure." *American Sociological Review* 23 (4): 401–7.

Faunce, W. A., E. Hardin, and E. H. Jacobson. 1962. "Automation and the Employee." *Annals of the American Academy of Political and Social Science* 340 (1): 60–68.

Feinstein, C. H. 1990. "New Estimates of Average Earnings in the United Kingdom." *Economic History Review* 43 (4): 592–633.

Feinstein, C. H. 1991. "A New Look at the Cost of Living." In *New Perspectives on the Late Victorian Economy*, edited by J. Foreman-Peck, 151–79. Cambridge: Cambridge University Press.

Feinstein, C. H. 1998. "Pessimism Perpetuated: Real Wages and the Standard of Living in Britain during and after the Industrial Revolution." *Journal of Economic History* 58 (3): 625–58.

Ferguson, N. 2012. *Civilization: The West and the Rest*. New York: Penguin.

Ferrer-i-Carbonell, A. 2005. "Income and Well-Being: An Empirical Analysis of the Comparison Income Effect." *Journal of Public Economics* 89 (5–6): 997–1019.

Field, A. J. 2007. "The Origins of US Total Factor Productivity Growth in the Golden Age." *Cliometrica* 1 (1): 63–90.

Field, A. J. 2011. *A Great Leap Forward: 1930s Depression and U.S. Economic Growth*. New Haven, CT: Yale University Press.

Fielden, J. 2013. *Curse of the Factory System*. London: Routledge.

Finley, M. I. 1965. "Technical Innovation and Economic Progress in the Ancient World." *Economic History Review* 18 (1): 29–45.

Finley, M. I. 1973. *The Ancient Economy*. Berkeley: University of California Press.

Fisher, I. 1919. "Economists in Public Service: Annual Address of the President." *American Economic Review* 9 (1): 5–21.

Flamm, K. 1988. "The Changing Pattern of Industrial Robot Use." In *The Impact of Technological Change on Employment and Economic Growth*, edited by R. M. Cyert and D. C. Mowery, 267–328. Cambridge, MA: Ballinger Publishing.

Flink, J. J. 1988. *The Automobile Age*. Cambridge, MA: MIT Press.

Flinn, M. W. 1962. *Men of Iron: The Crowleys in the Early Iron Industry*. Edinburgh: Edinburgh University Press.

Flinn, M. W. 1966. *The Origins of the Industrial Revolution*. London: Longmans.

Floud, R. C., K. Wachter, and A. Gregory. 1990. *Height, Health, and History: Nutritional Status in the United Kingdom, 1750–1980*. Cambridge: Cambridge University Press.

Fogel, R. W. 1983. "Scientific History and Traditional History." In *Which Road to the Past?*, edited by R. W. Fogel and G. R. Elton, 5–70. New Haven, CT: Yale University Press.

Forbes, R. J. 1958. *Man: The Maker*. New York: Abelard-Schuman.

Ford, M. 2015. *Rise of the Robots: Technology and the Threat of a Jobless Future*. New York: Basic Books. Kindle.

Fortunato, M., M. G. Azar, B. Piot, J. Menick, I. Osband, A. Graves, V. Mnih, et al. 2017. "Noisy Networks for Exploration." Preprint, submitted. https://arxiv.org/abs/1706.10295.

Frey, B. S. 2008. *Happiness: A Revolution in Economics*. Cambridge, MA: MIT Press.

Frey, C. B., T. Berger, and C. Chen. 2018. "Political Machinery: Did Robots Swing the 2016 U.S. Presidential Election?" *Oxford Review of Economic Policy* 34 (3): 418–42.

Frey, C. B., and M. Osborne. 2018. "Automation and the Future of Work—Understanding the Numbers." Oxford Martin School. https://www.oxfordmartin.ox.ac.uk/opinion/view/404.

Frey, C. B., and M. A. Osborne. 2017. "The Future of Employment: How Susceptible Are Jobs to Computerisation?" *Technological Forecasting and Social Change* 114 (January): 254–80.

Friedman, T. L. 2006. *The World Is Flat: The Globalized World in the Twenty-First Century*. London: Penguin.

Friedrich, O. 1983. "The Computer Moves In (Machine of the Year)." *Time*, January 3, 14–24.

Fukuyama, F. 2014. *Political Order and Political Decay: From the Industrial Revolution to the Globalization of Democracy*. New York: Farrar, Straus and Giroux.

Furman, J. Forthcoming. "Should We Be Reassured If Automation in the Future Looks Like Automation in the Past?" In *Economics of Artificial Intelligence*, edited by A. K. Agrawal, J. Gans, and A. Goldfarb, Chicago: University of Chicago Press.

Füssel, S. 2005. *Gutenberg and the Impact of Printing*. Aldershot, UK: Ashgate.

Gadd, I. A., and P. Wallis. 2002. *Guilds, Society, and Economy in London 1450–1800*. London: Centre for Metropolitan History.

Galor, O. 2011. "Inequality, Human Capital Formation, and the Process of Development." In *Handbook of the Economics of Education*, edited by E. A. Hanushek, S. J. Machin, and L. Woessmann, vol. 4, 441–93. Amsterdam: Elsevier.

Galor, O., and D. N. Weil. 2000. "Population, Technology, and Growth: From Malthusian Stagnation to the Demographic Transition and Beyond." *American Economic Review* 90 (4): 806–28.

Ganong, P., and D. Shoag. 2017. "Why Has Regional Income Convergence in the U.S. Declined?" *Journal of Urban Economics* 102 (November):76–90.

Gaskell, E. C. 1884. *Mary Barton.* London: Chapman and Hall.

Gaskell, P. 1833. *The Manufacturing Population of England: Its Moral, Social, and Physical Conditions.* London: Baldwin and Cradock.

Gerschenkron, A. 1962. *Economic Backwardness in Historical Perspective: A Book of Essays.* Cambridge, MA: Belknap Press of Harvard University Press.

Gille, B. 1969. "The Fifteenth and Sixteenth Centuries in the Western World." In *A History of Technology and Invention: Progress through the Ages,* edited by M. Daumas and translated by E. B. Hennessy, 2:16–148. New York: Crown.

Gille, B. 1986. *History of Techniques.* Vol. 2, *Techniques and Sciences.* New York: Gordon and Breach Science Publishers.

Gilson, R. J. 1999. "The Legal Infrastructure of High Technology Industrial Districts: Silicon Valley, Route 128, and Covenants Not to Compete." *New York University Law Review* 74 (August): 575.

Giuliano, V. E. 1982. "The Mechanization of Office Work." *Scientific American* 247 (3): 148–65.

Glaeser, E. L. 1998. "Are Cities Dying?" *Journal of Economic Perspectives* 12 (2): 139–60.

Glaeser, E. L. 2013. Review of *The New Geography of Jobs,* by Enrico Moretti. *Journal of Economic Literature* 51 (3): 825–37.

Glaeser, E. L. 2017. "Reforming Land Use Regulations." Report in the Series on Market and Government Failures, Brookings Center on Regulation and Markets, Washington.

Glaeser, E. L., and J. D. Gottlieb. 2009. "The Wealth of Cities: Agglomeration Economies and Spatial Equilibrium in the United States." *Journal of Economic Literature* 47 (4): 983–1028.

Glaeser, E. L., and J. Gyourko. 2002. "The Impact of Zoning on Housing Affordability." Working Paper 8835, National Bureau of Economic Research, Cambridge, MA.

Goldin, C., and L. Katz. 2008. *The Race between Technology and Education.* Cambridge, MA: Harvard University Press.

Goldin, C., and R. A. Margo. 1992. "The Great Compression: The Wage Structure in the United States at Mid-Century." *Quarterly Journal of Economics* 107 (1): 1–34.

Goldin, C., and K. Sokoloff. 1982. "Women, Children, and Industrialization in the Early Republic: Evidence from the Manufacturing Censuses." *Journal of Economic History* 42 (4): 741–74.

Goldsmith, S., G. Jaszi, H. Kaitz, and M. Liebenberg. 1954. "Size Distribution of Income Since the Mid-Thirties." *Review of Economics and Statistics* 36 (1): 1–32.

Goldstein, A. 2018. *Janesville: An American Story.* New York: Simon & Schuster.

Goolsbee, A. 2018. "Public Policy in an AI Economy." Working Paper 24653, National Bureau of Economic Research, Cambridge, MA.

Goolsbee, A., and P. Klenow. 2006. "Valuing Consumer Products by the Time Spent Using Them: An Application to the Internet." *American Economic Review* 96 (2): 108–13.

Goos, M. A., and A. Manning. 2007. "Lousy and Lovely Jobs: The Rising Polarization of Work in Britain." *Review of Economics and Statistics* 89 (1): 118–33.

Goos, M., A. Manning, and A. Salomons. 2009. "Job Polarization in Europe." *American Economic Review* 99 (2): 58–63.

Goos, M., A. Manning, and A. Salomons. 2014. "Explaining Job Polarization: Routine-Biased Technological Change and Offshoring." *American Economic Review* 104 (8): 2509–26.

Gordon, R. J. 2005. "The 1920s and the 1990s in Mutual Reflection." Working Paper 11778, National Bureau of Economic Research, Cambridge, MA.

Gordon, R. J. 2014. "The Demise of U.S. Economic Growth: Restatement, Rebuttal, and Reflections." Working Paper 19895, National Bureau of Economic Research, Cambridge, MA.

Gordon, R. J. 2016. *The Rise and Fall of American Growth: The U.S. Standard of Living Since the Civil War*. Princeton, NJ: Princeton University Press.

Gould, E. D., B. A. Weinberg, and D. B. Mustard. 2002. "Crime Rates and Local Labor Market Opportunities in the United States: 1979–1997." *Review of Economics and Statistics* 84 (1): 45–61.

Graeber, D. 2018. *Bullshit Jobs: A Theory*. New York: Simon & Schuster.

Graetz, G., and G. Michaels. Forthcoming. "Robots at Work." *Review of Economics and Statistics*.

Gramlich, J. 2017. "Most Americans Would Favor Policies to Limit Job and Wage Losses Caused by Automation." Pew Research Center. http://www.pewresearch.org/fact-tank/2017/10/09/most-americans-would-favor-policies-to-limit-job-and-wage-losses-caused-by-automation/.

Greenwood, J., A. Seshadri, and M. Yorukoglu. 2005. "Engines of Liberation." *Review of Economic Studies* 72 (1): 109–33.

Greif, A., and M. Iyigun. 2012. "Social Institutions, Violence and Innovations: Did the Old Poor Law Matter?" Working paper, Stanford University, Stanford, CA.

Greif, A., and M. Iyigun. 2013. "Social Organizations, Violence, and Modern Growth." *American Economic Review* 103 (3): 534–38.

Grier, D. A. 2005. *When Humans Were Computers*. Princeton, NJ: Princeton University Press.

Gross, D. P. 2018. "Scale Versus Scope in the Diffusion of New Technology: Evidence from the Farm Tractor." *RAND Journal of Economics* 49 (2): 427–52.

Habakkuk, H. J. 1962. *American and British Technology in the Nineteenth Century: The Search for Labour Saving Inventions*. Cambridge: Cambridge University Press.

Hacker, J. S., and P. Pierson. 2010. *Winner-Take-All Politics: How Washington Made the Rich Richer—and Turned Its Back on the Middle Class*. New York: Simon & Schuster.

Hammer, M. 1990. "Reengineering Work: Don't Automate, Obliterate." *Harvard Business Review* 68 (4): 104–12.

Hansard, T. C. 1834. *General Index to the First and Second Series of Hansard's Parliamentary Debates: Forming a Digest of the Recorded Proceedings of Parliament, from 1803 to 1820*. London: Kraus Reprint Co.

Hansen, A. H. 1939. "Economic Progress and Declining Population Growth." *American Economic Review* 29 (1): 1–15.

Harper, K. 2017. *The Fate of Rome: Climate, Disease, and the End of an Empire*. Princeton, NJ: Princeton University Press.

Hartz, L. 1955. *The Liberal Tradition in America: An Interpretation of American Political Thought Since the Revolution*. Boston: Houghton Mifflin Harcourt.

Hawke, G. R. 1970. *Railways and Economic Growth in England and Wales, 1840–1870*. Oxford: Clarendon Press of Oxford University Press.

Headrick, D. R. 2009. *Technology: A World History*. New York: Oxford University Press.

Heaton, H. 1936. *Economic History of Europe*. New York: Harper and Brothers.

Heckman, J. J., S. H. Moon, R. Pinto, P. A. Savelyev, and A. Yavitz. 2010. "The Rate of Return to the HighScope Perry Preschool Program." *Journal of Public Economics* 94 (1–2): 114–28.

Heilbroner, R. L. 1966. "Where Do We Go From Here?" *New York Review of Books*, March 17. https://www.nybooks.com/articles/1966/03/17/where-do-we-go-from-here/.

Henderson, R. 2017. Comment on "Artificial Intelligence and the Modern Productivity Paradox: A Clash of Expectations and Statistics," by E. Brynjolfsson, D. Rock and C. Syverson," National Bureau of Economic Research. http://www.nber.org/chapters/c14020.pdf.

Hibbert, F. A. 1891. *The Influence and Development of English Guilds*. New York: Sentry.

Himmelfarb, G. 1968. *Victorian Minds*. New York: Knopf.

Hobbes, T. 1651. *Leviathan*, chapter 13, https://ebooks.adelaide.edu.au/h/hobbes/thomas/h68l/chapter13.html.

Hobsbawm, E. 1962. *The Age of Revolution: Europe 1789–1848*. London: Weidenfeld and Nicolson. Kindle.

Hobsbawm, E. 1968. *Industry and Empire: From 1750 to the Present Day*. New York: New Press. Kindle.

Hobsbawm, E., and G. Rudé. 2014. *Captain Swing*. New York: Verso.

Hodgen, M. T. 1939. "Domesday Water Mills." *Antiquity* 13 (51): 261–79.

Hodges, H. 1970. *Technology in the Ancient World*. New York: Barnes & Noble.

Holzer, H. J., D. Whitmore Schanzenbach, G. J. Duncan, and J. Ludwig. 2008. "The Economic Costs of Childhood Poverty in the United States." *Journal of Children and Poverty* 14 (1): 41–61.

Hoppit, J. 2008. "Political Power and British Economic Life, 1650–1870." In *The Cambridge Economic History of Modern Britain*, vol. 1, *Industrialisation, 1700–1870*, edited by R. Floud, J. Humphries, and P. Johnson, 370–71. Cambridge: Cambridge University Press.

Horn, J. 2008. *The Path Not Taken: French Industrialization in the Age of Revolution, 1750–1830*. Cambridge, MA: MIT Press. Kindle.

Hornbeck, R. 2012. "The Enduring Impact of the American Dust Bowl: Short- and Long-Run Adjustments to Environmental Catastrophe." *American Economic Review* 102 (4): 1477–507.

Hornbeck, R., and S. Naidu. 2014. "When the Levee Breaks: Black Migration and Economic Development in the American South." *American Economic Review* 104 (3): 963–90.

Horrell, S. 1996. "Home Demand and British Industrialisation." *Journal of Economic History* 56 (September): 561–604.

Hounshell, D. 1985. *From the American System to Mass Production, 1800–1932: The Development of Manufacturing Technology in the United States*. Baltimore, MD: Johns Hopkins University Press.

Hsieh, C. T., and E. Moretti. Forthcoming. "Housing Constraints and Spatial Misallocation." *American Economic Journal: Macroeconomics*.

Humphries, J. 2010. *Childhood and Child Labour in the British Industrial Revolution*. Cambridge: Cambridge University Press.

Humphries, J. 2013. "The Lure of Aggregates and the Pitfalls of the Patriarchal Perspective: A Critique of the High Wage Economy Interpretation of the British Industrial Revolution." *Economic History Review* 66 (3): 693–714.

Humphries, J., and T. Leunig. 2009. "Was Dick Whittington Taller Than Those He Left Behind? Anthropometric Measures, Migration and the Quality of Life in Early Nineteenth Century London." *Explorations in Economic History* 46 (1): 120–31.

Humphries, J., and B. Schneider. Forthcoming. "Spinning the Industrial Revolution." *Economic History Review*.

International Chamber of Commerce. 1925. "Report of the American Committee on Highway Transport, June, 1925." Washington, DC: American Section, International Chamber of Commerce.

International Federation of Robotics. 2016. "World Robotics: Industrial Robots [dataset]." https://ifr.org/worldrobotics/.

Jackson, R. 1806. *The Speech of R. Jackson Addressed to the Committee of the House of Commons Appointed to Consider of the State of the Woollen Manufacture of England, on Behalf of the Cloth-Workers and Sheermen of Yorkshire, Lancashire, Wiltshire, Somersetshire and Gloucestershire*. London: C. Stower.

Jacobson, L. S., R. J. LaLonde, and D. G. Sullivan. 1993. "Earnings Losses of Displaced Workers." *American Economic Review* 83 (4): 685–709.

Jaimovich, N., and H. E. Siu. 2012. "Job Polarization and Jobless Recoveries." Working Paper 18334, National Bureau of Economic Research, Cambridge, MA.

Jakubauskas, E. B. 1960. "Adjustment to an Automatic Airline Reservation System." In *Impact of Automation: A Collection of 20 Articles about Technological Change, from the* Monthly Labor Review, 93–96. Washington, DC: Bureau of Labor Statistics.

Jerome, H. 1934. "Mechanization in Industry." Working Paper 27. National Bureau of Economic Research, Cambridge, MA.

Johnson, G. E. 1975. "Economic Analysis of Trade Unionism." *American Economic Review* 65 (2): 23–28.

Johnson, L. B. 1964. "Remarks upon Signing Bill Creating the National Commission on Technology, Automation, and Economic Progress." August 19. http://archive.li/F9iX8.

Johnston, L., and S. H. Williamson. 2018. "What Was the U.S. GDP Then?" MeasuringWorth .com. http://www.measuringworth.org/usgdp/.

Kaitz, K. 1998. "American Roads, Roadside America." *Geographical Review* 88 (3): 363–87.

Kaldor, N. 1957. "A Model of Economic Growth." *Economic Journal* 67 (268): 591–624.

Kanefsky, J., and J. Robey. 1980. "Steam Engines in 18th-Century Britain: A Quantitative Assessment." *Technology and Culture* 21 (2): 161–86.

Karabarbounis, L., and B. Neiman. 2013. "The Global Decline of the Labor Share." *Quarterly Journal of Economics* 129 (1): 61–103.

Katz, L. F., and R. A. Margo. 2013. "Technical Change and the Relative Demand for Skilled Labor: The United States in Historical Perspective." Working Paper 18752, National Bureau of Economic Research, Cambridge, MA.

Kaufman, B. E. 1982. "The Determinants of Strikes in the United States, 1900–1977." *ILR Review* 35 (4): 473–90.

Kay-Shuttleworth, J.P.K. 1832. *The Moral and Physical Condition of the Working Classes Employed in the Cotton Manufacture in Manchester*. Manchester: Harrisons and Crosfield.

Kealey, E. J. 1987. *Harvesting the Air: Windmill Pioneers in Twelfth-Century England*. Berkeley: University of California Press.

Kelly, M., C. Ó Gráda. 2016. "Adam Smith, Watch Prices, and the Industrial Revolution." *Quarterly Journal of Economics* 131 (4): 1727–52.

Kendrick, J. W. 1961. *Productivity Trends in the United States*. Princeton, NJ: Princeton University Press.

Kendrick, J. W. 1973. *Postwar Productivity Trends in the United States, 1948–1969*. Cambridge, MA: National Bureau of Economic Research.

Kennedy, J. F. 1960. "Papers of John F. Kennedy. Pre-Presidential Papers. Presidential Campaign Files, 1960. Speeches and the Press. Speeches, Statements, and Sections, 1958–1960. Labor: Meeting the Problems of Automation." https://www.jfklibrary.org/asset-viewer/archives /JFKCAMP1960/1030/JFKCAMP1960-1030-036.

Kennedy, J. F. 1962. "News Conference 24." https://www.jfklibrary.org/archives/other -resources/john-f-kennedy-press-conferences/news-conference-24.

Kenworthy, L. 2012. "It's Hard to Make It in America: How the United States Stopped Being the Land of Opportunity." *Foreign Affairs* 91 (November/December): 97–109.

Kerry, C. F., and J. Karsten. 2017. "Gauging Investment in Self-Driving Cars." Brookings Institution, October 16. https://www.brookings.edu/research/gauging-investment-in-self-driving -cars/.

Keynes, J. M. [1930] 2010. "Economic Possibilities for Our Grandchildren." In *Essays in Persuasion*, 321–32. London: Palgrave Macmillan.

Klein, M. 2007. *The Genesis of Industrial America, 1870–1920*. Cambridge: Cambridge University Press.

Kleiner, M. M. 2011. "Occupational Licensing: Protecting the Public Interest or Protectionism?" Policy Paper 2011-009, Upjohn Institute, Kalamazoo, MI.

Klemm, F. 1964. *A History of Western Technology*. Cambridge, MA: MIT Press.

Klepper, S. 2010. "The Origin and Growth of Industry Clusters: The Making of Silicon Valley and Detroit." *Journal of Urban Economics* 67 (1): 15–32.

Kline, P., and E. Moretti. 2013. "Local Economic Development, Agglomeration Economies, and the Big Push: 100 Years of Evidence from the Tennessee Valley Authority." *Quarterly Journal of Economics* 129 (1): 275–331.

Koch, C. 2016. "How the Computer Beat the Go Master." *Scientific American* 27 (4): 20–23.

Komlos, J. 1998. "Shrinking in a Growing Economy? The Mystery of Physical Stature during the Industrial Revolution." *Journal of Economic History* 58 (3): 779–802.

Komlos, J., and B. A'Hearn. 2017. "Hidden Negative Aspects of Industrialization at the Onset of Modern Economic Growth in the US." *Structural Change and Economic Dynamics* 41 (June): 43–52.

Korinek, A., and J. E. Stiglitz. 2017. "Artificial Intelligence and Its Implications for Income Distribution and Unemployment." Working Paper 24174, National Bureau of Economic Research, Cambridge, MA.

Kremen, G. R. 1974. "MDTA: The Origins of the Manpower Development and Training Act of 1962." Washington, DC: Department of Labor. https://www.dol.gov/general/aboutdol /history/mono-mdtatext.

Kremer, M. 1993. "The O-Ring Theory of Economic Development." *Quarterly Journal of Economics* 108 (3): 551–75.

Krugman, P. R. 1995. *Peddling Prosperity: Economic Sense and Nonsense in the Age of Diminished Expectations*. New York: Norton.

Kuznets, S. 1955. "Economic Growth and Income Inequality." *American Economic Review* 45(1): 1–28.

LaLonde, R. J. 2007. *The Case for Wage Insurance*. New York: Council on Foreign Relations Press.

Lamont, M. 2009. *The Dignity of Working Men: Morality and the Boundaries of Race, Class, and Immigration*. Cambridge, MA: Harvard University Press.

Landels, J. G. 2000. *Engineering in the Ancient World*. Berkeley: University of California Press.

Landes, D. S. 1969. *The Unbound Prometheus: Technological Change and Development in Western Europe from 1750 to the Present*. Cambridge: Cambridge University Press.

Langdon, J. 1982. "The Economics of Horses and Oxen in Medieval England." *Agricultural History Review* 30 (1): 31–40.

Langton, J., and R. J. Morris. 2002. *Atlas of Industrializing Britain, 1780–1914*. London: Routledge.

Larsen, C. S. 1995. "Biological Changes in Human Populations with Agriculture." *Annual Review of Anthropology* 24 (1): 185–213.

Lebergott, S. 1993. *Pursuing Happiness: American Consumers in the Twentieth Century*. Princeton, NJ: Princeton University Press.

Lee, D. 1973. "Science, Philosophy, and Technology in the Greco-Roman World: I." *Greece and Rome* 20 (1): 65–78.

Lee, J. 2014. "Measuring Agglomeration: Products, People, and Ideas in U.S. Manufacturing, 1880–1990." Working paper, Harvard University.

Lee, R., and M. Anderson. 2002. "Malthus in State Space: Macroeconomic-Demographic Relations in English History, 1540 to 1870." *Journal of Population Economics* 15 (2): 195–220.

Lee, T. B. 2016. "This Expert Thinks Robots Aren't Going to Destroy Many Jobs. And That's a Problem." *Vox*. https://www.vox.com/a/new-economy-future/robert-gordon-interview.

Le Goff, J. 1982. *Time, Work, and Culture in the Middle Ages*. Chicago: University of Chicago Press.

Leighton, A. C. 1972. *Transport and Communication in Early Medieval Europe AD 500–1100*. London: David and Charles Publishers.

Lenoir, T. 1998. "Revolution from Above: The Role of the State in Creating the German Research System, 1810–1910." *American Economic Review* 88 (2): 22–27.

Leunig, T. 2006. "Time Is Money: A Re-Assessment of the Passenger Social Savings from Victorian British Railways." *Journal of Economic History* 66 (3): 635–73.

Levy, F. 2018. "Computers and Populism: Artificial Intelligence, Jobs, and Politics in the Near Term." *Oxford Review of Economic Policy* 34 (3): 393–417.

Levy, F., and R. J. Murnane. 2004. *The New Division of Labor: How Computers Are Creating the Next Job Market*. Princeton, NJ: Princeton University Press.

Lewis, D. L. 1986. "The Automobile in America: The Industry." *Wilson Quarterly*, 10 (5): 47–63.

Lewis, H. G. 1963. *Unionism and Relative Wages in the U.S.: An Empirical Inquiry*. Chicago: Chicago University Press.

Lilley, S. 1966. *Men, Machines and History: The Story of Tools and Machines in Relation to Social Progress*. Paris: International Publishers.

Lin, J. 2011. "Technological Adaptation, Cities, and New Work." *Review of Economics and Statistics* 93 (2): 554–74.

Lindert, P. H., 1986. Unequal English wealth since 1670. *Journal of Political Economy*, 94(6): 1127–62.

Lindert, P. H. 2000a. "Three Centuries of Inequality in Britain and America." In *Handbook of Income Distribution*, edited by A. B. Atkinson and F. Bourguignon, vol. 1, 167–216. Amsterdam: Elsevier.

Lindert, P. H. 2000b. "When Did Inequality Rise in Britain and America?" *Journal of Income Distribution* 9 (1): 11–25.

Lindert, P. H. 2004. *Growing Public*, vol. 1: *The Story: Social Spending and Economic Growth Since the Eighteenth Century*. Cambridge: Cambridge University Press.

Lindert, P. H., and J. G. Williamson. 1982. "Revising England's Social Tables 1688–1812." *Explorations in Economic History* 19 (4): 385–408.

Lindert, P. H., and J. G. Williamson. 1983. "Reinterpreting Britain's Social Tables, 1688–1913." *Explorations in Economic History* 20 (1): 94–109.

Lindert, P. H., and J. G. Williamson. 2012. "American Incomes 1774–1860." Working Paper 18396, National Bureau of Economic Research, Cambridge, MA.

Lindert, P. H., and J. G. Williamson. 2016. *Unequal Gains: American Growth and Inequality Since 1700*. Princeton, NJ: Princeton University Press.

Liu, S. 2015. "Spillovers from Universities: Evidence from the Land-Grant Program." *Journal of Urban Economics* 87 (May): 25–41.

Long, J. 2005. "Rural-Urban Migration and Socioeconomic Mobility in Victorian Britain." *Journal of Economic History* 65 (1): 1–35.

Lordan, G., and D. Neumark. 2018. "People versus Machines: The Impact of Minimum Wages on Automatable Jobs." *Labour Economics* 52 (June): 40–53.

Lubin, I. 1929. *The Absorption of the Unemployed by American Industry*. Washington, DC: Brookings Institution.

Luttmer, E. F. 2005. "Neighbors as Negatives: Relative Earnings and Well-Being." *Quarterly Journal of Economics* 120 (3): 963–1002.

Lyman, P., and H. R. Varian. 2003. "How Much Information?" berkeley. edu/research/projects/how-much-info-2003.

Machlup, F. 1962. *The Production and Distribution of Knowledge in the United States*. Princeton, NJ: Princeton University Press.

MacLeod, C. 1998. *Inventing the Industrial Revolution: The English Patent System, 1660–1800*. Cambridge: Cambridge University Press.

Maddison, A. 2002. *The World Economy: A Millennial Perspective*. Paris: Organisation for Economic Co-operation and Development.

Maddison, A. 2005. *Growth and Interaction in the World Economy: The Roots of Modernity*. Washington, DC: AEI Press.

Maehl, W. H. 1967. *The Reform Bill of 1832: Why Not Revolution?* New York: Holt, Rinehart and Winston.

Malthus, T. [1798] 2013. *An Essay on the Principle of Population.* Digireads.com. Kindle.

Mandel, M., and B. Swanson. 2017. "The Coming Productivity Boom—Transforming the Physical Economy with Information." Washington, DC: Technology CEO Council.

Mann, F. C., and L. K. Williams. 1960. "Observations on the Dynamics of a Change to Electronic Data-Processing Equipment." *Administrative Science Quarterly* 5 (2): 217–56.

Manson, S., Schroeder, J., Van Riper, D., and Ruggles, S. (2018). IPUMS National Historical Geographic Information System: Version 13.0 [Database]. Minneapolis: University of Minnesota. http://doi.org/10.18128/D050.V13.0

Mantoux, P. 1961. *The Industrial Revolution in the Eighteenth Century: An Outline of the Beginnings of the Modern Factory System in England.* Translated by M. Vernon. London: Routledge.

Manuelli, R. E., and A. Seshadri. 2014. "Frictionless Technology Diffusion: The Case of Tractors." *American Economic Review* 104 (4): 1368–91.

Martin, T. C. 1905. "Electrical Machinery, Apparatus, and Supplies." In *Census of Manufactures, 1905.* Washington, DC: United States Bureau of the Census.

Marx, K. [1867] 1999. *Das Kapital.* Translated by S. Moore and E. Aveling. New York: Gateway edition. Kindle.

Marx, K., and F. Engels. [1848] 1967. *The Communist Manifesto.* Translated by S. Moore. London: Penguin.

Massey, D. S. 2007. *Categorically Unequal: The American Stratification System.* New York: Russell Sage Foundation.

Massey, D. S., J. Rothwell, and T. Domina. 2009. "The Changing Bases of Segregation in the United States." *Annals of the American Academy of Political and Social Science* 626 (1): 74–90.

Mathibela, B., P. Newman, and I. Posner. 2015. "Reading the Road: Road Marking Classification and Interpretation." *IEEE Transactions on Intelligent Transportation Systems* 16 (4): 2072–81.

Mathibela, B., M. A. Osborne, I. Posner, and P. Newman. 2012. "Can Priors Be Trusted? Learning to Anticipate Roadworks." IEEE Conference on Intelligent Transportation Systems, 927–932. https://ori.ox.ac.uk/learning-to-anticipate-roadworks/.

McCarty, N., K. T. Poole, and H. Rosenthal. 2016. *Polarized America: The Dance of Ideology and Unequal Riches.* Cambridge, MA: MIT Press.

McCloskey, D. N. 2010. *The Bourgeois Virtues: Ethics for an Age of Commerce.* Chicago: University of Chicago Press.

Mendels, F. F. 1972. "Proto-industrialization: The First Phase of the Industrialization Process." *Journal of Economic History* 32 (1): 241–61.

Merriam, R. H. 1905. "Bicycles and Tricycles." In *Census of Manufactures, 1905,* 289–97. Washington, DC: United States Bureau of the Census.

Milanovic, B. 2016a. "All the Ginis (ALG) Dataset." https://datacatalog.worldbank.org/dataset /all-ginis-dataset, Version October 2016.

Milanovic, B. 2016b. *Global Inequality: A New Approach for the Age of Globalization.* Cambridge, MA: Harvard University Press.

Milanovic, B., P. H. Lindert, and J. G. Williamson. 2010. "Pre-Industrial Inequality." *Economic Journal* 121 (551): 255–72.

Millet, D. J. 1972. "Town Development in Southwest Louisiana, 1865–1900." *Louisiana History*, 13 (2): 139–68.

Mills, F. C. 1934. Introduction to "Mechanization in Industry," by H. Jerome. Cambridge, MA: National Bureau of Economic Research.

Mitch, D. F. 1992. *The Rise of Popular Literacy in Victorian England: The Influence of Private Choice and Public Policy*. Philadelphia: University of Pennsylvania Press.

Mitch, D. F. 1993. "The Role of Human Capital in the First Industrial Revolution." In *The British Industrial Revolution: An Economic Perspective*, edited by J. Mokyr, 241–80. Boulder, CO: Westview Press.

Mitchell, B. 1975. *European Historical Statistics, 1750–1970*. London: Macmillan.

Mitchell, B. 1988. *British Historical Statistics*. Cambridge: Cambridge University Press.

Mokyr, J. 1992a. *The Lever of Riches: Technological Creativity and Economic Progress*. New York: Oxford University Press.

Mokyr, J. 1992b. "Technological Inertia in Economic History." *Journal of Economic History* 52 (2): 325–38.

Mokyr, J. 1998. "The Political Economy of Technological Change." In *Technological Revolutions in Europe: Historical Perspectives*, edited by K. Bruland and M. Berg, 39–64. Cheltenham: Edward Elgar.

Mokyr, J. 2000. "Why 'More Work for Mother?" Knowledge and Household Behavior, 1870–1945." *Journal of Economic History* 60 (1): 1–41.

Mokyr, J. 2001. "The Rise and Fall of the Factory System: Technology, Firms, and Households Since the Industrial Revolution." *Carnegie-Rochester Conference Series on Public Policy* 55 (1): 1–45.

Mokyr, J. 2002. *The Gifts of Athena: Historical Origins of the Knowledge Economy*. Princeton, NJ: Princeton University Press.

Mokyr, J. 2011. *The Enlightened Economy: Britain and the Industrial Revolution, 1700–1850*. London: Penguin. Kindle.

Mokyr, J., and H. Voth. 2010. "Understanding Growth in Europe, 1700–1870: Theory and Evidence." In *The Cambridge Economic History of Modern Europe*, edited by S. Broadberry and K. O'Rourke, 1:7–42. Cambridge: Cambridge University Press.

Mom, G. P., and D. A. Kirsch. 2001. "Technologies in Tension: Horses, Electric Trucks, and the Motorization of American Cities, 1900–1925." *Technology and Culture* 42 (3): 489–518.

Moore, B., Jr. 1993. *Social Origins of Dictatorship and Democracy: Lord and Peasant in the Making of the Modern World*. Boston: Beacon Press.

Moravec, H. 1988. *Mind Children: The Future of Robot and Human Intelligence*. Cambridge, MA: Harvard University Press.

Moretti, E. 2004. "Estimating the Social Return to Higher Education: Evidence from Longitudinal and Repeated Cross-Sectional Data." *Journal of Econometrics* 121 (1–2): 175–212.

Moretti, E. 2010. "Local Multipliers." *American Economic Review* 100 (2): 373–77.

Moretti, E. 2012. *The New Geography of Jobs*. Boston: Houghton Mifflin Harcourt.

Morse, H. B. 1909. *The Guilds of China*. London: Longmans, Green and Co.

Mumford, L. 1934. *Technics and Civilization*. New York: Harcourt, Brace and World.

Mummert, A., E. Esche, J. Robinson, and G. J. Armelagos. 2011. "Stature and Robusticity During the Agricultural Transition: Evidence from the Bioarchaeological Record." *Economics and Human Biology* 9 (3): 284–301.

Murray, C. 2013. *Coming Apart: The State of White America, 1960–2010.* New York: Random House Digital.

Mutz, D. C. 2018. "Status Threat, Not Economic Hardship, Explains the 2016 Presidential Vote." *Proceedings of the National Academy of Sciences* 115 (19): 4330–39.

Myers, R. J. 1929. "Occupational Readjustment of Displaced Skilled Workmen." *Journal of Political Economy* 37 (4): 473–89.

Nadiri, M. I., and T. P. Mamuneas. 1994. "Infrastructure and Public R&D Investments, and the Growth of Factor Productivity in U.S. Manufacturing Industries." Working Paper 4845, National Bureau of Economic Research, Cambridge, MA.

Nardinelli, C. 1986. "Technology and Unemployment: The Case of the Handloom Weavers." *Southern Economic Journal* 53 (1): 87–94.

Neddermeyer, U. 1997. "Why Were There No Riots of the Scribes?" *Gazette du Livre Médiéval* 31 (1): 1–8.

Nedelkoska, L., and G. Quintini. 2018. "Automation, Skills Use and Training." OECD Social, Employment and Migration Working Paper 202, Organisation of Economic Co-operation and Development, Paris.

Nelson, D. 1995. *Farm and Factory: Workers in the Midwest, 1880–1990.* Bloomington: Indiana University Press.

Nicolini, E. A. 2007. "Was Malthus Right? A VAR Analysis of Economic and Demographic Interactions in Pre-Industrial England." *European Review of Economic History* 11 (1): 99–121.

Nichols, A., and J. Rothstein. 2015. "The Earned Income Tax Credit (EITC)." Working Paper 21211, National Bureau of Economic Research, Cambridge, MA.

Nordhaus, W. D. 1996. "Do Real-Output and Real-Wage Measures Capture Reality? The History of Lighting Suggests Not." In *The Economics of New Goods*, edited by T. F. Bresnahan and R. J. Gordon, 27–70. Chicago: University of Chicago Press.

Nordhaus, W. D. 2005. "The Sources of the Productivity Rebound and the Manufacturing Employment Puzzle." Working Paper 11354, National Bureau of Economic Research, Cambridge, MA.

Nordhaus, W. D. 2007. "Two Centuries of Productivity Growth in Computing." *Journal of Economic History* 67 (1): 128–59.

North, D. C. 1991. "Institutions." *Journal of Economic Perspectives* 5 (1): 97–112.

North, D. C., and B. R. Weingast. 1989. "Constitutions and Commitment: The Evolution of Institutions Governing Public Choice in Seventeenth-Century England." *Journal of Economic History* 49 (4): 803–32.

Nuvolari, A., and M. Ricci. 2013. "Economic Growth in England, 1250–1850: Some New Estimates Using a Demand Side Approach." *Rivista di Storia Economica* 29 (1): 31–54.

Nye, D. E. 1990. *Electrifying America: Social Meanings of a New Technology, 1880–1940.* Cambridge, MA: MIT Press.

Nye, D. E. 2013. *America's Assembly Line.* Cambridge, MA: MIT Press.

Oestreicher, R. 1988. "Urban Working-Class Political Behavior and Theories of American Electoral Politics, 1870–1940." *Journal of American History* 74 (4): 1257–86.

Officer, L. H., and S. H. Williamson. 2018. "Annual Wages in the United States, 1774–Present." MeasuringWorth. https://www.measuringworth.com/datasets/uswage/ http://www.measuringworth.com/uswages/.

Ogilvie, S. 2019. *The European Guilds: An Economic Analysis*. Princeton, NJ: Princeton University Press.

Oliner, S. D., and D. E. Sichel. 2000. "The Resurgence of Growth in the Late 1990s: Is Information Technology the Story?" *Journal of Economic Perspectives* 14 (4): 3–22.

Olmstead, A. L., and P. W. Rhode. 2001. "Reshaping the Landscape: The Impact and Diffusion of the Tractor in American Agriculture, 1910–1960." *Journal of Economic History* 61 (3): 663–98.

Owen, W. 1962. "Transportation and Technology." *American Economic Review* 52 (2): 405–13.

Parsley, C. J. 1980. "Labor Union Effects on Wage Gains: A Survey of Recent Literature." *Journal of Economic Literature* 18 (1): 1–31.

Patterson, R. 1957. "Spinning and Weaving." In *From the Renaissance to the Industrial Revolution, c. 1500–c. 1750*, edited by C. Singer, E. J. Holmyard, A. R. Hall, and T. I. Williams, 191–200. Vol. 3 of *A History of Technology*. New York: Oxford University Press.

Peri, G. 2012. "The Effect of Immigration on Productivity: Evidence from US States." *Review of Economics and Statistics* 94 (1), 348–58.

Peri, G. 2018. "Did Immigration Contribute to Wage Stagnation of Unskilled Workers?" *Research in Economics* 72 (2): 356–65.

Peterson, W., and Y. Kislev. 1986. "The Cotton Harvester in Retrospect: Labor Displacement or Replacement?" *Journal of Economic History* 46 (1): 199–216.

Phelps, E. S. 2015. *Mass Flourishing: How Grassroots Innovation Created Jobs, Challenge, and Change*. Princeton, NJ: Princeton University Press.

Phyllis, D., and W. A. Cole. 1962. *British Economic Growth, 1688–1959: Trends and Structure*. Cambridge: Cambridge University Press.

Piketty, T. 2014. *Capital in the Twenty-First Century*. Cambridge, MA: Harvard University Press.

Piketty, T. 2018. "Brahmin Left vs. Merchant Right: Rising Inequality and the Changing Structure of Political Conflict." Working paper, Paris School of Economics.

Piketty, T., and E. Saez. 2003. "Income Inequality in the United States, 1913–1998." *Quarterly Journal of Economics* 118 (1): 1–41.

Polanyi, M. 1966. *The Tacit Dimension*. New York: Doubleday.

Prashar, A. 2018. "Evaluating the Impact of Automation on Labour Markets in England and Wales." Working paper, Oxford University.

President's Advisory Committee on Labor-Management Policy. 1962. *The Benefits and Problems Incident to Automation and Other Technological Advances*. Washington, DC: Government Printing Office.

Price, D. de S. 1975. *Science Since Babylon*. New Haven, CT: Yale University Press.

Putnam, R. D., ed. 2004. *Democracies in Flux: The Evolution of Social Capital in Contemporary Society*. Oxford: Oxford University Press.

Putnam, R. D. 2016. *Our Kids: The American Dream in Crisis*. New York: Simon & Schuster.

Rajan, R. G. 2011. *Fault Lines: How Hidden Fractures Still Threaten the World Economy*. Princeton, NJ: Princeton University Press.

Ramey, V. A. 2009. "Time Spent in Home Production in the Twentieth-Century United States: New Estimates from Old Data." *Journal of Economic History* 69 (1): 1–47.

Ramey, V. A., and N. Francis. 2009. "A Century of Work and Leisure." *American Economic Journal: Macroeconomics* 1 (2): 189–224.

Randall, A. 1991. *Before the Luddites: Custom, Community and Machinery in the English Woollen Industry, 1776–1809*. Cambridge: Cambridge University Press.

Rasmussen, W. D. 1982. "The Mechanization of Agriculture." *Scientific American* 247 (3): 76–89.

Rector, R., and R. Sheffield. 2011. "Air Conditioning, Cable TV, and an Xbox: What Is Poverty in the United States Today?" Washington, DC: Heritage Foundation.

Reich, R. 1991. *The Work of Nations: Preparing Ourselves for Twenty-First Century Capitalism*. New York: Knopf.

Remus, D., and F. Levy. 2017. "Can Robots Be Lawyers: Computers, Lawyers, and the Practice of Law." *Georgetown Journal Legal Ethics* 30 (3): 501–45.

Reuleaux, F. 1876. *Kinematics of Machinery: Outlines of a Theory of Machines*. Translated by A.B.W. Kennedy. London: MacMillan.

Reynolds, A. J., J. A. Temple, S. R. Ou, I. A. Arteaga, and B. A. White. 2011. "School-Based Early Childhood Education and Age-28 Well-Being: Effects by Timing, Dosage, and Subgroups." *Science* 333 (6040): 360–64.

Ricardo, D. [1817] 1911. *The Principles of Political Economy and Taxation*. Reprint. London: Dent.

Rifkin, J. 1995. *The End of Work: The Decline of the Global Labor Force and the Dawn of the Post-market Era*. New York: G. P. Putnam's Sons.

Robinson, J., and G. Godbey. 2010. *Time for Life: The Surprising Ways Americans Use Their Time*. Philadelphia: Penn State University Press.

Rodrik, D. 2016. "Premature Deindustrialization." *Journal of Economic Growth* 21 (1): 1–33.

Rodrik, D. 2017a. "Populism and the Economics of Globalization." Working Paper 23559, National Bureau of Economic Research, Cambridge, MA.

Rodrik, D. 2017b. *Straight Talk on Trade: Ideas for a Sane World Economy*. Princeton, NJ: Princeton University Press.

Rognlie, M. 2014. "A Note on Piketty and Diminishing Returns to Capital," unpublished manuscript. http://mattrognlie.com/piketty_diminishing_returns.pdf.

Roosevelt, F. D. 1940. "Annual Message to the Congress," January 3. By G. Peters and J. T. Woolley. The American Presidency Project. https://www.presidency.ucsb.edu/documents/annual-message-the-congress.

Rosenberg, N. 1963. "Technological Change in the Machine Tool Industry, 1840–1910." *Journal of Economic History* 23 (4): 414–43.

Rosenberg, N., and L. E. Birdzell. 1986. *How the West Grew Rich: The Economic Transformation of the Western World*. London: Basic.

Rostow, W. W. 1960. *The Stages of Growth: A Non-Communist Manifesto*. Cambridge: Cambridge University Press.

Rothberg, H. J. 1960. "Adjustment to Automation in Two Firms." In *Impact of Automation: A Collection of 20 Articles about Technological Change, from the Monthly Labor Review*, 79–93. Washington, DC: Bureau of Labor Statistics.

Rousseau, J. J. [1755] 1999. *Discourse on the Origin of Inequality*. Oxford: Oxford University Press.

Ruggles, S., S. Flood, R. Goeken, J. Grover, E. Meyer, J. Pacas, and M. Sobek. 2018. IPUMS USA. Version 8.0 [dataset]. https://usa.ipums.org/usa/.

Russell, B. 1946. *History of Western Philosophy and Its Connection with Political and Social Circumstances: From the Earliest Times to the Present Day*. New York: Simon & Schuster.

Sachs, J. D., S. G. Benzell, and G. LaGarda. 2015. "Robots: Curse or Blessing? A Basic Framework." Working Paper 21091, National Bureau of Economic Research, Cambridge, MA.

Sanderson, M. 1995. *Education, Economic Change and Society in England 1780–1870*. Cambridge: Cambridge University Press.

Scheidel, W. 2018. *The Great Leveler: Violence and the History of Inequality from the Stone Age to the Twenty-First Century*. Princeton, NJ: Princeton University Press.

Scheidel, W., and S. J. Friesen. 2009. "The Size of the Economy and the Distribution of Income in the Roman Empire." *Journal of Roman Studies* 99 (March): 61–91.

Schlozman, K. L., S. Verba, and H. E. Brady. 2012. *The Unheavenly Chorus: Unequal Political Voice and the Broken Promise of American Democracy*. Princeton, NJ: Princeton University Press.

Schumpeter, J. A. 1939. *Business Cycles*. Vol. 1. New York: McGraw-Hill.

Schumpeter, J. A. [1942] 1976. *Capitalism, Socialism and Democracy*. 3rd ed. New York: Harper Torchbooks.

Scoville, W. C. 1960. *The Persecution of Huguenots and French Economic Development 1680–1720*. Berkeley: University of California Press.

Shannon, C. E. 1950. "Programming a Computer for Playing Chess." *Philosophical Magazine* 41 (314): 256–75.

Shaw-Taylor, L., and A. Jones. 2010. "The Male Occupational Structure of Northamptonshire 1777–1881: A Case of Partial De-Industrialization?" Working paper, Cambridge University.

Simon, H. 1966. "Automation." *New York Review of Books*, March 26. https://www.nybooks.com /articles/1966/05/26/automation-3/.

Simon, H. [1960] 1985. "The Corporation: Will It Be Managed by Machines?" In *Management and the Corporation*, edited by M. L. Anshen and G. L. Bach, 17–55. New York: McGraw-Hill.

Simon, J. L. 2000. *The Great Breakthrough and Its Cause*. Ann Arbor: University of Michigan Press.

Smil, V. 2005. *Creating the Twentieth Century: Technical Innovations of 1867–1914 and Their Lasting Impact*. New York: Oxford University Press.

Smiles, S. 1865. *Lives of Boulton and Watt*. Philadelphia: J. B. Lippincott.

Smith, A. [1776] 1976. *An Inquiry into the Nature and Causes of the Wealth of Nations*. Chicago: University of Chicago Press.

Smolensky, E., and R. Plotnick. 1993. "Inequality and Poverty in the United States: 1900 to 1990." Paper 998–93, University of Wisconsin Institute for Research on Poverty, Madison.

Snooks, G. D. 1994. "New Perspectives on the Industrial Revolution." In *Was the Industrial Revolution Necessary?*, edited by G. D. Snooks, 1–26. London: Routledge.

Sobek, M. 2006. "Detailed Occupations—All Persons: 1850–1990 (Part 2). Table Ba1396-1439." In *Historical Statistics of the United States, Earliest Times to the Present: Millennial Edition,* edited by S. B. Carter, S. S. Gartner, M. R. Haines, A. Olmstead, R. Sutch, and G. Wright. New York: Cambridge University Press.

Solow, R. M. 1956. "A Contribution to the Theory of Economic Growth." *Quarterly Journal of Economics* 70 (1): 65–94.

Solow, R. 1987. "We'd Better Watch Out." *New York Times* Book Review, July 12.

Solow, R. M. 1965. "Technology and Unemployment." *Public Interest* 1 (Fall): 17–27.

Sorensen, T., P. Fishback, S. Kantor, and P. Rhode. 2008. "The New Deal and the Diffusion of Tractors in the 1930s." Working paper, University of Arizona, Tucson.

Southall, H. R. 1991. "The Tramping Artisan Revisits: Labour Mobility and Economic Distress in Early Victorian England." *Economic History Review* 44 (2): 272–96.

Spence, M., and S. Hlatshwayo. 2012. "The Evolving Structure of the American Economy and the Employment Challenge." *Comparative Economic Studies* 54 (4): 703–38.

Stasavage, D. 2003. *Public Debt and the Birth of the Democratic State: France and Great Britain 1688–1789.* Cambridge: Cambridge University Press.

Steckel, R. H. 2008. "Biological Measures of the Standard of Living." *Journal of Economic Perspectives* 22 (1): 129–52.

Stephenson, J. Z. 2018. "'Real' Wages? Contractors, Workers, and Pay in London Building Trades, 1650–1800." *Economic History Review* 71 (1): 106–32.

Stevenson, B., and J. Wolfers. 2013. "Subjective Well-Being and Income: Is There Any Evidence of Satiation?" *American Economic Review* 103 (3): 598–604.

Stewart, C. 1960. "Social Implications of Technological Progress." In *Impact of Automation: A Collection of 20 Articles about Technological Change, from the* Monthly Labor Review, 11–15. Washington, DC: Bureau of Labor Statistics.

Stokes, Bruce. 2017. "Public Divided on Prospects for Next Generation." Pew Research Center, Spring 2017 Global Attitudes Survey, June 5. http://www.pewglobal.org/2017/06/05/2 -public-divided-on-prospects-for-the-next-generation/.

Strasser, S. 1982. *Never Done: A History of American Housework.* New York: Pantheon.

Sullivan, D., and T. von Wachter. 2009. "Job Displacement and Mortality: An Analysis Using Administrative Data." *Quarterly Journal of Economics* 124 (3): 1265–1306.

Sundstrom, W. A. 2006. "Hours and Working Conditions." In *Historical Statistics of the United States, Earliest Times to the Present: Millennial Edition Online,* edited by S. B. Carter, S. S. Gartner, M. R. Haines, A. L. Olmstead, R. Sutch, and G. Wright, 301–35. New York: Cambridge University Press.

Swetz, F. J. 1987. *Capitalism and Arithmetic: The New Math of the 15th Century.* La Salle, IL: Open Court.

Syverson, C. 2017. "Challenges to Mismeasurement Explanations for the US Productivity Slowdown." *Journal of Economic Perspectives* 31 (2): 165–86.

Szreter, S., and G. Mooney. 1998. "Urbanization, Mortality, and the Standard of Living Debate: New Estimates of the Expectation of Life at Birth in Nineteenth-Century British Cities." *Economic History Review* 51 (1): 84–112.

Taft, P., P. Ross. 1969. "American Labor Violence: Its Causes, Character, and Outcome." In *Violence in America: Historical and Comparative Perspectives*, edited by H. D. Graham, and T. R. Gurr, 1:221–301. London: Corgi.

Taine, H. A. 1958. *Notes on England, 1860–70*. Translated by E. Hyams. London: Strahan.

Tella, R. D., R. J. MacCulloch, and A. J. Oswald. 2003. "The Macroeconomics of Happiness." *Review of Economics and Statistics* 85 (4): 809–27.

Temin, P. 2006. "The Economy of the Early Roman Empire." *Journal of Economic Perspectives* 20 (1): 133–51.

Temin, P. 2012. *The Roman Market Economy*. Princeton, NJ: Princeton University Press.

Thernstrom, S. 1964. *Poverty and Progress: Social Mobility in a Nineteenth Century City*. Cambridge, MA: Harvard University Press.

Thomas, R., and N. Dimsdale. 2016. "Three Centuries of Data–Version 3.0." London: Bank of England. https://www.bankofengland.co.uk/statistics/research-datasets.

Thompson, E. P. 1963. *The Making of the English Working Class*. New York: Victor Gollancz, Vintage Books.

Tilly, C. 1975. *The Formation of National States in Western Europe*. Princeton, NJ: Princeton University Press.

Tinbergen, J. 1975. *Income Distribution: Analysis and Policies*. Amsterdam: North Holland.

Tocqueville, A. de. 1840. *Democracy in America*. Translated by H. Reeve. Vol. 2. New York: Alfred A. Knopf.

Toffler, A. 1980. *The Third Wave*. New York: Bantam Books.

Trajtenberg, M. 2018. "AI as the Next GPT: A Political-Economy Perspective." Working Paper 24245, National Bureau of Economic Research, Cambridge, MA.

Treat, J. R., N. J. Castellan, R. L. Stansifer, R. E. Mayer, R. D. Hume, D. Shinar, S. T. McDonald, et al. 1979. *Tri-Level Study of the Causes of Traffic Accidents: Final Report*, vol. 2: *Special Analyses*. Bloomington, IN: Institute for Research in Public Safety.

Tolley, H. R., and Church, L. M. 1921. "Corn-Belt Farmers' Experience with Motor Trucks." United States Department of Agriculture, Bulletin No. 931, February 25.

Tucker, G. 1837. *The Life of Thomas Jefferson, Third President of the United States: With Parts of His Correspondence Never Before Published, and Notices of His Opinions on Questions of Civil Government, National Policy, and Constitutional Law*. Vol. 2. Philadelphia: Carey, Lea and Blanchard.

Tuttle, C. 1999. *Hard at Work in Factories and Mines: The Economics of Child Labor during the British Industrial Revolution*. Boulder, CO: Westview Press.

Twain, M., and C. D. Warner. [1873] 2001. *The Gilded Age: A Tale of Today*. New York: Penguin.

Twain, M. 1835. "Taming the Bicycle." The University of Adelaide Library, last updated March 27, 2016. https://ebooks.adelaide.edu.au/t/twain/mark/what_is_man/chapter15.html.

Ure, A. 1835. *The Philosophy of Manufactures*. London: Charles Knight.

U.S. Bureau of the Census. 1960. D785, "Work-injury Frequency Rates in Manufacturing, 1926–1956," and D.786–790, "Work-injury Frequency Rates in Mining, 1924–1956." In *Historical Statistics of the United States, Colonial Times to 1957*. Washington, DC: Government Printing Office. https://www.census.gov/library/publications/1960/compendia/hist_stats_colonial-1957.html.

U.S. Congress. 1955. "Automation and Technological Change." Hearings before the Subcommittee on Economic Stabilization of the Congressional Joint Committee on the Economic Report (84th Cong., 1st sess.), pursuant to sec. 5(a) of Public Law 304, 79th Cong. Washington, DC: Government Printing Office.

U.S. Congress. 1984. "Computerized Manufacturing Automation: Employment, Education, and the Workplace." No. 235. Washington, DC: Office of Technology Assessment.

U.S. Department of Agriculture. 1963. *1962 Agricultural Statistics*. Washington, DC: Government Printing Office.

Usher, A. P. 1954. *A History of Mechanical Innovations*. Cambridge, MA: Harvard University Press.

Van Zanden, J. 2004. "Common Workmen, Philosophers and the Birth of the European Knowledge Economy." Paper for the Global Economic History Network Conference, Leiden, September 16–18.

Van Zanden, J. L., E. Buringh, and M. Bosker. 2012. "The Rise and Decline of European Parliaments, 1188–1789." *Economic History Review* 65 (3): 835–61.

Varian, H. R. Forthcoming. "Artificial Intelligence, Economics, and Industrial Organization." In *The Economics of Artificial Intelligence: An Agenda, edited by* A. K. Agrawal, J. Gans, and A. Goldfarb. Chicago: University of Chicago Press.

Vickers, C., and N. L. Ziebarth. 2016. "Economic Development and the Demographics of Criminals in Victorian England." *Journal of Law and Economics* 59 (1): 191–223.

Von Tunzelmann, G. N. 1978. *Steam Power and British Industrialization to 1860*. Oxford: Oxford University Press.

Voth, H. 2000. *Time and Work in England 1750–1830*. Oxford: Clarendon Press of Oxford University Press.

Wadhwa, V., and A. Salkever. 2017. *The Driver in the Driverless Car: How Our Technology Choices Will Create the Future*. San Francisco: Berrett-Koehler.

Walker, C. R. 1957. *Toward the Automatic Factory: A Case Study of Men and Machines*. New Haven, CT: Yale University Press.

Wallis, P. 2014. "Labour Markets and Training." In *The Cambridge Economic History of Modern Britain*, 1:178–210, *Industrialisation, 1700–1870*, edited by R. Floud, J. Humphries, and P. Johnson. Cambridge University Press.

Walmer, O. R. 1956. "Workers' Health in an Era of Automation." *Monthly Labor Review* 79 (7): 819–23.

Weber, M. 1927. *General Economic History*. New Brunswick, NJ: Transaction Books.

Weinberg, B. A. 2000. "Computer Use and the Demand for Female Workers." *ILR Review* 53 (2): 290–308.

Weinberg, E. 1960. "A Review of Automation Technology." In *Impact of Automation: A Collection of 20 Articles about Technological Change, from the* Monthly Labor Review, 3–10. Washington, DC: Bureau of Labor Statistics.

Weinberg, E. 1956. "An Inquiry into the Effects of Automation." *Monthly Labor Review* 79 (January): 7–14.

Weinberg, E. 1960. "Experiences with the Introduction of Office Automation." *Monthly Labor Review* 83 (4): 376–80.

Weingroff, R. F. 2005. "Designating the Urban Interstates." Federal Highway Administration Highway History. https://www.fhwa.dot.gov/infrastructure/fairbank.cfm.

White, K. D. 1984. *Greek and Roman Technology*. Ithaca, NY: Cornell University Press.

White, L. 1962. *Medieval Technology and Social Change*. Oxford: Oxford University Press.

White, L. 1967. "The Historical Roots of Our Ecologic Crisis." *Science* 155 (3767): 1203–7.

White, L. A. 2016. *Modern Capitalist Culture*. London: Routledge.

White, W. J. 2001. "An Unsung Hero: The Farm Tractor's Contribution to Twentieth-Century United States Economic Growth." PhD diss., Ohio State University.

Wiener, N. 1988. *The Human Use of Human Beings: Cybernetics and Society*. New York: Perseus Books Group.

Williamson, J. G. 1987. "Did English Factor Markets Fail during the Industrial Revolution?" *Oxford Economic Papers* 39 (4): 641–78.

Williamson, J. G. 2002. *Coping with City Growth during the British Industrial Revolution*. Cambridge: Cambridge University Press.

Wilson, W. J. 1996. "When Work Disappears." *Political Science Quarterly* 111 (4): 567–95.

Wilson, W. J. 2012. *The Truly Disadvantaged: The Inner City, the Underclass, and Public Policy*. Chicago: University of Chicago Press.

Woirol, G. R. 1980. "Economics as an Empirical Science: A Case Study." Working paper, University of California, Berkeley.

Woirol, G. R. 2006. "New Data, New Issues: The Origins of the Technological Unemployment Debates." *History of Political Economy* 38 (3): 473–96.

Woirol, G. R. 2012. "Plans to End the Great Depression from the American Public." *Labor History* 53 (4): 571–77.

Wolman, L. 1933. "Machinery and Unemployment." *Nation*, February 22, 202–4.

World Bank Group. 2016. *World Development Report 2016: Digital Dividends*. Washington, DC: World Bank Publications.

World Health Organization. 2015. "Road Traffic Deaths." http://www.who.int/gho/road_safety/mortality/en.

Wright, Q. 1942. *A Study of War*. Vol. 1. Chicago: University of Chicago Press.

Wrigley, E. A. 2010. *Energy and the English Industrial Revolution*. Cambridge: Cambridge University Press.

Wu, Y., M. Schuster, Z. Chen, Q. V. Le, M. Norouzi, W. Macherey, M. Krikun, et al. 2016. "Google's Neural Machine Translation System: Bridging the Gap between Human and Machine Translation." Preprint, submitted September 26. https://arxiv.org/abs/1609.08144.

Xiong, W., L. Wu, F. Alleva, J. Droppo, X. Huang, and A. Stolcke. 2017. "The Microsoft 2017 Conversational Speech Recognition System." Microsoft AI and Research Technical Report MSR-TR-2017-39, August 2017.

Young, A. 1772. *Political Essays Concerning the Present State of the British Empire*. London: printed for W. Strahan and T. Cadell.

Zhang, X., M. Li, J. H. Lim, Y. Weng, Y.W.D. Tay, H. Pham, and Q. C. Pham. 2018. "Large-Scale 3D Printing by a Team of Mobile Robots." *Automation in Construction* 95 (November): 98–106.

INDEX

453